Integrating Artificial and Human Intelligence through Agent Oriented Systems Design

This book departs from the assumption that Artificial Intelligence (AI) systems will provide a maximum advantage by replacing human cognitive processing. Instead, this book subscribes to the assumption that AI systems will provide a maximal advantage when the system is specifically designed to augment human intelligence. It provides methods for designing effective systems that include one or more humans and one or more AI entities and uses the approach that assumes automation does not replace human activity but fundamentally changes the structure of human work when AI is added to existing systems.

Integrating Artificial and Human Intelligence through Agent Oriented Systems Design discusses the potential impact of AI on human work and life and explores why teamwork is necessary today for complex work environments. The book explains the processes and methods humans employ to effectively team with one another and presents the elements of artificial agents that permit them to function as team members in joint human and artificial teams. It discusses design goals and illustrates how the methods that have been used to model the complex interactions among human and artificial agents can be expanded to enable the design of interaction between them to make possible the attainment of the shared goals. Model-Based Systems Engineering (MBSE) tools that provide logical designs of human–agent teams, the AI within these teams, training to be deployed for human and artificial agent team members, and the interfaces between human and artificial agent team members are all covered. MBSE files containing profiles and examples for building MBSE models used in the design approach are featured on the author's website (https://lodesterresci.com/hat).

This book is an ideal read for students, professors, engineers, and project managers associated with designing and developing AI systems or systems that seek to incorporate AI.

Systems Innovation Book Series

Series Editor: Adedeji Badiru

Systems Innovation refers to all aspects of developing and deploying new technology, methodology, techniques, and best practices in advancing industrial production and economic development. This entails such topics as product design and development, entrepreneurship, global trade, environmental consciousness, operations and logistics, introduction and management of technology, collaborative system design, and product commercialization. Industrial innovation suggests breaking away from the traditional approaches to industrial production. It encourages the marriage of systems science, management principles, and technology implementation. Particular focus will be the impact of modern technology on industrial development and industrialization approaches, particularly for developing economics. The series will also cover how emerging technologies and entrepreneurship are essential for economic development and society advancement.

Project Management for Scholarly Researchers
Systems, Innovation, and Technologies
Adedeji B. Badiru

Industrial Engineering in Systems Design
Guidelines, Practical Examples, Tools, and Techniques
Brian Peacock and Adedeji B. Badiru

Leadership Matters
An Industrial Engineering Framework for Developing and Sustaining Industry
Adedeji B. Badiru and Melinda Tourangeau

Systems Engineering
Influencing Our Planet and Reengineering Our Actions
Adedeji B. Badiru

Total Productive Maintenance, Second Edition
Strategies and Implementation Guide
Tina Agustiady and Elizabeth A. Cudney

Assessing Innovation
Metrics, Rubrics and Standards
Adedeji B. Badiru and Melinda Tourangeau

Integrating Artificial and Human Intelligence through Agent Oriented Systems Design
Michael E. Miller and Christina F. Rusnock

Integrating Artificial and Human Intelligence through Agent Oriented Systems Design

Michael E. Miller and Christina F. Rusnock

CRC Press is an imprint of the
Taylor & Francis Group, an **informa** business

Designed cover image: Shutterstock—Elnur

First edition published 2025
by CRC Press
2385 NW Executive Center Drive, Suite 320, Boca Raton FL 33431

and by CRC Press
4 Park Square, Milton Park, Abingdon, Oxon, OX14 4RN

CRC Press is an imprint of Taylor & Francis Group, LLC

ISBN: 978-1-032-54657-5 (hbk)
ISBN: 978-1-032-54936-1 (pbk)
ISBN: 978-1-003-42818-3 (ebk)

DOI: 10.1201/9781003428183

Typeset in Times LT Std
by Apex CoVantage, LLC

Access the Support Material: https://lodesterresci.com/hat

Contents

Preface

Automation, the replacement of human activity with activity by powered equipment, has transformed the human experience over the past few centuries. This replacement began with the simple substitution of repetitive human physical activities, often with more effective or efficient systems. As a classic example, walking or riding on walking animals was replaced by transportation on steam-powered trains during the 19th century. Not only did trains replace human locomotion as a primary means of long-distance travel, but they also increased the effectiveness of transportation, permitting the average speed of locomotion to increase by as much as an order of magnitude.

The application of aircraft in the following century increased the speed of locomotion by another order of magnitude. As a result of these and other similar innovations in automation, we, as humans, have come to expect that automation not only reduces human effort but also significantly improves our capabilities and the efficiency of our activities.

With the innovation of the thin-film transistor and the microprocessor in the latter half of the 20th century, automation began a significant transformation. Suddenly, manufactured equipment could leverage logic within microprocessors not only to replace human physical activity but also to support or replace human cognitive activity. The field of Artificial Intelligence (AI) was born and garnered substantial investment. While the initial foray into AI created significant interest, the field quickly faded as the promise of a general intelligence system was rapidly dispelled. However, the resulting systems revolutionized process equipment which operated in highly constrained manufacturing and production environments where the necessary system response was highly predictable.

A second awakening of AI occurred in the 1980s with the advent of expert systems. Again, the technology garnered significant investment as these systems promised the encapsulation of expert human knowledge into systems that could be employed in less predictable environments. The promise of AI once again faded as the field failed to deliver the dramatic increases in effectiveness and efficiency provided by systems that replaced human physical activity. However, these systems extended the application of AI beyond constrained process environments. These automations impacted systems in less predictable environments, including automating aspects of commercial aviation.

Currently, we are experiencing the third wave of investment in AI. Once again, a recent technology, specifically deep learning, has shown significant promise in enabling systems to respond to the variability in certain real-world applications. Deep learning applies significant advancements in specialized graphic processing units with the availability of large data sets to enable the training of multilayer neural networks. While smaller neural networks with smaller numbers of layers have existed for decades, recent advances have enabled the technology to complete certain tasks, such as object recognition, with a performance that rivals or surpasses the capability of humans. However, training these algorithms require thousands of labeled images, unlike humans who can learn a new object with a single example.

As with the previous waves of investment, the emphasis of these systems is on the automation of human cognitive processing. That is, they seek to replace not only the physical labor performed by humans but also the thought processes we employ when approaching a problem. Inherent in this approach is the assumption that these systems will not only significantly outperform the human in these cognitive tasks but also exhibit error-free performance across the same variety of contexts as human performance. However, several decades of experience has not fully supported this assumption. Instead, we see that when consistent, predictable responses are required, these systems can outperform humans. However, in less predictable environments, these systems can fail catastrophically, requiring the human operator to step in and "save the day" when the automation fails to adapt to unforeseen conditions.

As a result, AI proponents often point to other potential advantages of artificial intelligence, such as the reduction of workforce costs as AI reduces the numbers of humans required to operate a system or the increases in system predictability as the most unpredictable part of most system, the human, is replaced. These proponents often neglect the additional labor required to operate and maintain the AI, along with the additional tasks created by the AI interfacing with other parts of the system or system-of-systems. As for increasing system predictability by removing the human, this relies upon the common simplistic perspective of accident investigations to point to human error as the root cause for accidents. Accident investigation boards often attribute accidents to the fact that the human either fails to adapt procedures or inappropriately adapts procedures during an automation, hardware, or software failure, leading to the accident. Lost in this analysis is the fact that the original failure sequence often begins with an automation, hardware, or software failure. Further, humans underreport automation failures and simply "deal with it" when the automation fails. Among the accidents are incidents where the human over-relies on the automation, for example, human is distracted by their smart phone while operating a semiautonomous car on "autopilot", or under-relies on automation, that is, assumes control during a time where the automation is fully capable of managing the existing circumstances. Thus, when a designer simply adds AI to a human system without considering its impact on human work, the modification can result in undesirable outcomes. Designers are required to design AI into human systems, rather than replacing humans to allow the human to vet both the inputs and outputs. In the case of generative AI technology, such as ChatGPT, the AI is able to ingest and mimic large amounts of data to apparently "create" novel behavior but does not inherently understand the information, is not able to reason across the information, and is not able to anticipate the outcome when creating new responses. Thus, these generative AI systems still require a human to "sanity check" the results.

This text departs from the underlying assumption that systems employing AI will provide maximum advantage by replacing human cognitive processing. Instead, this text subscribes to the underlying assumption that achieving maximal advantage requires designing and employing AI to augment human intelligence. This text provides methods for designing effective systems that include one or more humans and one or more AI entities, working collaboratively as a team to accomplish system goals. The proposed approach assumes that AI should not simply replace human activity but fundamentally change the structure of human work to augment human's performance. To provide a design method, this text borrows from publications in systems engineering, software engineering, human factors engineering, cognitive systems engineering, and human–robot interaction to discuss the important aspects of designing AI systems and the interaction between humans and these intelligent systems to achieve a symbiotic human–AI team. To aid the design of future systems, this text additionally provides a meta-model and a modeling approach to aid the design of systems that support effective interactive human–AI teams.

Many popular works of art have shown ominous futures where AI controls the world after freeing itself from its human masters. Alternatively works of art such as Pixar Studio's WALL-E show futures where automation takes over all the need for human activity, leaving humans with nothing to do but lie poolside while robots address their every need. While we might question the credibility of these futures, works of art such as Miranda Lambert's country song "Automatic" provide prescient warnings of the impact of automation on society. This song provides insight into at least three automation influences including:

- The loss of personal accomplishment when automation simplifies work removing difficulty as a source of motivation and enjoyment, where previous processes provided humans the opportunity to challenge themselves and excel in difficult activities;
- The loss of skill and knowledge acquisition as automation replaces or simplifies human activity; and
- The loss of pride in craftmanship, which occurs as society emphasizes automation's speed and productivity over quality.

This text emphasizes the importance of designing systems with a focus on effective, collaborative, interactions between humans, automation, and AI with the goal to enable a future where we not only learn to live with but also embrace automation and AI to significantly enhance our way of life. However, various futures are possible, including those where AI replaces human cognitive task performance, such that humans lose valuable skills and involvement, diminishing life experiences. It is our hope that by examining the uniqueness of humans and of AI that we can begin to envision a richer future where AI augments our abilities and enhances our lives, providing more time for memorable experiences that make life worth living. We cannot forget that facing and overcoming challenges are one of the things that make human life worth living. Each one of us is unique and responds to different challenges to emphasize this uniqueness. Therefore, AI should seek not only to replace work but also to challenge us, leading to faster and greater personal and organizational learning and growth.

How to Use This Book

This book describes a method and considerations for designing human–AI agent teams. The text is intended to be used in advanced undergraduate or graduate courses in systems engineering, multidiscipline, computer science, or human factors engineering courses.

The book's topics and chronology assume multidisciplinary or systems engineering courses. We recommend courses in these areas use the book as designed, following the current chapter order. For multidisciplinary courses, we recommend organizing the students into multidisciplinary teams, where each team focuses on the design of a particular human–AI agent team. Ideally, these teams will include individuals who have an exposure to previous courses in cognitive task analysis, and these students might be expected to lead definition of the "As-Is" system description. Students with experience in the Unified Markup Language or the Systems Modeling Language would lead the modeling portions of the project. Computer science and other engineering students would lead concept prototyping for agents to participate in the human–AI agent team.

When applying this text in computer science courses, faculty may consider exposing the students to the Unified Markup Language and possibly Systems Modeling Language prior to a course that applies this text or deliberately incorporate external sources providing an introduction to these languages as part of this course. Instructors in these programs may find the description of agents and Artificial Intelligence methods in Chapter 6 rudimentary. Faculty may decide to augment or replace this material with external reading or to use courses in AI or agent design as prerequisites to the course that applies the current text.

When applying this text in human factors engineering programs, the instructor may decide not to emphasize the model-based systems engineering portions of this text and rather emphasize the human–AI-agent teaming portions of the book. Instructors in these curricula may find it useful to assign the portions of Chapter 4 and perhaps Chapter 11 early in the course, perhaps even prior to Chapter 2 or to augment this book with additional exercises and education in cognitive task analysis. It is expected that human factors students may find the material in Chapters 4 and 11 more relatable, making it easier for these students to engage with the remaining content. Instructors in these courses may find it useful to emphasize the perceptual loop and the utility of either delegating goals to individual agents which complete the perceptual loop independent of other agents or delegating shared goals to multiple agents which synchronously participate in the perceptual loop to complete these goals in an interdependent fashion.

About the Authors

Michael E. Miller is a professor of systems engineering at the Air Force Institute of Technology (AFIT), Dayton, Ohio, the United States. Dr. Miller has over 30 years of experience designing human-friendly devices and teaching human systems integration. He holds a PhD in Industrial and Systems Engineering with a human factors engineering specialization from Virginia Tech and an MS and BS in the same discipline from Ohio University. Early in his career, he contributed to the design of first-generation IBM Thinkpads, handwriting tablets, and systems employing gaze tracking as a mouse replacement. At Eastman Kodak Company, he contributed to the user interface design and image processing for digital cameras, image processing for digital photofinishing systems, interface design for immersive imaging systems, and the design of and algorithm development for OLED televisions, developing the RGBW technology used in LG's OLED Televisions. At AFIT, he has conducted research on various human interface technologies, much of which has involved algorithm development or human–AI agent interaction, and has taught a course on this topic for over a decade. Mike has contributed to over 100 issued U.S. patents, 50 peer-reviewed journal articles, and numerous conference papers. He is a member of the Human Factors and Ergonomics Society, the International Council on Systems Engineering, and the Society for Information Display.

Christina F. Rusnock is an adjunct associate professor of systems engineering at the Air Force Institute of Technology (AFIT). Dr. Rusnock commissioned into the United States Air Force as an acquisitions officer in 2004. Her career has included a wide variety of acquisition, staff, command, and academic assignments involving space, aircraft, and software systems. She earned a PhD in Industrial Engineering in 2013 from the University of Central Florida, Orlando, Florida, the United States; an MS degree in Industrial Engineering in 2011 from UCF; an MS degree in Research and Development Management in 2008 from AFIT in Dayton, Ohio; and a BA in Economics-Government in 2004 from Claremont McKenna College, Claremont, California, the United States. Her research interests include human performance modeling, mental workload, trust in automation, and situation awareness, with a focus on applications in human–machine teaming for autonomous systems. She has contributed to more than 50 books, peer-reviewed journal articles, and conference proceedings.

The views expressed are those of the authors and do not reflect the official guidance or position of the United States Government, the Department of Defense, the United States Air Force, or the United States Space Force.

1 Introducing Human–AI Teaming

1.1 WHY ARE WE CONCERNED ABOUT HUMAN–AI TEAMING?

From natural language processing systems, such as Siri or Alexa, to self-driving automobiles, and GPT4, *Artificial Intelligence* (AI) has recently received significant attention as a technology. Innovations in artificial intelligence have influenced our society for more than six decades. Artificial intelligence was first proposed as a replacement to human cognition in a 1955 research proposal for using neural networks on computers to use natural language, abstract data to concepts, and engage in self-improvement and creativity (Mccarthy et al., 2006). Marvin Minsky, who contributed to this proposal, later expanded on this concept as a superior replacement to human cognition (Minsky, 1961), and this viewpoint has received significant attention since. Nonetheless, research during the intervening decades has illustrated that the vision of creating generalized AI, where the AI can adapt to unknown circumstances, was more difficult to achieve than originally believed.

It is important to note that Minsky's original vision was to create generalizable AI that would outthink and therefore replace a human. As a result, AI technologists often do not consider the interaction between a human and AI and sometimes ignore the fact that AI must be integrated into larger workflows and system processes that include humans. As a result, AI technologists have spent significant resources to create AI that they failed to successfully implement in the target environment, even though other AI technology has transformed our everyday lives in meaningful ways. Although the media often suggests that current-day AI can outthink a human, in fact, many AI systems must be carefully programmed or trained for every situation to be encountered (United States Government Accountability Office, 2022). In a specific example of the limitations of AI, a recent article suggested that after investments of over $100 billion, self-driving cars have hit a performance ceiling that will prevent the broad adoption of this technology in the foreseeable future (Chafkin, 2022).

J.C.R. Licklider proposed a competitive vision near the time that Marvin Minsky stated his vision to have AI surpass human cognitive ability. Licklider suggested that AI acting alone will not achieve optimal cognitive performance. Instead, Licklider proposed that improved cognitive performance would be achieved by enabling robust "cooperative interaction between men and electronic computers" (Licklider & Clark, 1962). Licklider's vision led to the development of technologies that dramatically influenced the interaction between humans and computers, including the graphical user interface and the internet. Impressively, Licklider not only envisioned the symbiotic interaction between a single computer embodied with AI and a single human. Instead, Licklider understood the power of connecting multiple cognitive entities, both human and artificial, together to solve large, complex, or time-constrained

DOI: 10.1201/9781003428183-1

problems. It is this vision, originally termed human–machine *symbiosis*, that we now refer to as *human–agent teaming or human–AI teaming.*

In human–AI teaming, at least one human and one AI agent employ diversified knowledge, skills, and abilities to accomplish work that is complex or critical enough that it requires greater knowledge, skill, ability, capacity, or redundancy than a single agent can provide in the available time frame (Cooke et al., 2013; Salas et al., 1992). These agents perform this work in an interdependent fashion where the actions that the human takes depend on the actions of the AI agent and the AI agent's actions depend on the actions of the human.

In addition to human–machine symbiosis, human–agent teaming, and human–AI teaming, other terms are used to describe these same or similar concepts. These include human–machine teaming, augmented cognition, augmented intelligence, intelligence augmentation, amplified intelligence, human–agent collaboration, hybrid intelligence, human-in-the-loop AI, interactive machine learning, guardian angel systems, mixed-initiative systems, human/AI centaurs, human-centered AI, and human-centered machine learning.

Although we do not have broad adoption of completely autonomous vehicles, we do have broad inclusion of AI in modern vehicles, including AI controls of our windshield wipers, lights, and speed. The presence of these agents in our vehicle changes the tasks we perform while driving, and our interaction with these agents and the automobile changes the way the AI agents perform. AI in these vehicles warns us when we leave our lane and even automatically applies brakes in emergency situations. Therefore, human–AI teaming makes driving easier and safer for millions of automobile drivers, demonstrating that AI can support and team with us in our everyday lives. While today's semiautonomous vehicles have made great strides in automating the complete task of driving, the recognition of the brittleness of this automation in edge case situations necessitates a human supervisory teammate. Although the popular media does not broadly discuss the concept of human–AI teaming, the academic literature discusses the use of AI to improve human capabilities.

Academic literature has shown that human–AI teams can be "superior" to the best humans or AI agents in gaming. Gaming is a natural arena for AI to excel, since games have defined boundaries, constraints, rules, and operating parameters. Further, developers design games to challenge human mental capacities. One of the first demonstrations of AI superiority was IBM's "Deep Blue" chess win against the grand master Gary Kasparov. However, after this match, Kasparov sponsored "freestyle chess" championships in which chess players could partner with AI. Grand masters using supercomputers were featured in and expected to win one such tournament in 2005. However, the tournament also included lesser-known players. In the end, "a pair of amateur American chess players using three computers" won the tournament (DeCremer & Kasparov, 2021). The lesson from the tournament was that the interaction process between humans and AI determines the effectiveness of the team, such that teams constructed with better processes can win even when they have less capable humans or AI than other teams. Similar observations have been made by bestselling author Alex Bates in his book, which discusses the design of human–AI teams to avoid disasters that stem from system failures in industries and perhaps even more importantly to enhance the human creativity (Bates, 2018). Thus, even in areas that are ideal for AI superior performance, human–AI teaming provides an opportunity for increased performance over areas that have AI or humans alone.

Since Licklider's time, a few approaches have been developed to understand the aspects of the cooperative interaction required to form an effective ***human–AI team***. Despite these advances, examples abound where improper integration of AI to support human–AI teaming has resulted in user frustrations, accidents, or life-ending circumstances. A contributing factor is that design tools and methods for these human–AI teams have remained elusive. This text synthesizes several bodies of literature regarding human–AI teaming to formulate a philosophy, model, and approach to the design of human–AI teams. The proposed philosophy assumes that humans and AI can, and should, interact to jointly fulfill a set of desired goals. This philosophy leads to the development of a language and a meta-model that describes the interaction between two classes of cognitive entities: humans and AI agents, as goal-seeking "peers" within a larger complex system. We introduce a design method that leverages this language and meta-model to develop system level specifications for the AI agents as well as the user interface, personnel selection, and training requirements for both classes of agents. The development and application of this meta-model provide a formalized, efficient, and repeatable method to support the conceptual design of systems that support human–AI teaming as originally envisioned by Licklider. At the very least, the discussion of this philosophy, model, and design approach provides the reader with insight into important aspects that they should consider when integrating AI into existing systems to automate the aspects of human work.

The integration of work across several fields inspired the development of the meta-model and design approach proposed in this text. We drew specific inspiration from the fields of cognitive systems engineering (Crandall et al., 2006; Johnson et al., 2014; Woods & Hollnagel, 2006), human factors engineering (Norman & Draper, 1986), agent-oriented software design (Deloach, 2004; Garcia-Ojeda & DeLoach, 2010), information systems (Sterling & Taveter, 2009), model-based systems engineering, (Delligatti, 2013) and human–robot interaction (Johnson et al., 2011, 2018). Each of these bodies of literature provided insight into the function of the human–AI team or the design and modeling approach discussed in this text. The resulting model and design method are an amalgamation of these approaches into a more cohesive framework.

1.2 A HUMAN–AI TEAM EXAMPLE

Before proceeding further, it is useful to ask: "Do human–AI teams already exist?" Consider a typical automobile with traditional cruise control. The purpose or goal of this system is to transport one or more people from a location to a destination. The automation, that is, the cruise control, cannot achieve this goal by itself. Instead, humans must control most aspects of the automobile to achieve this purpose.

However, when the driver wants to maintain constant speed, he or she can enable the automation which then monitors the speed of the automobile and controls the flow of fuel to the engine to maintain this constant speed. In automobiles with traditional cruise control, the cruise control automates the task of controlling the speed of the vehicle. However, it does not monitor anything about the environment but only monitors a system attribute, that is, speed, of the automobile. Therefore, traditional cruise control does not fulfill the definition of an agent, which we will expand on later in this chapter.

Modern automobiles often include a feature called adaptive cruise control. In these systems, the automation applies sensors to determine the speed of and distance to the automobile with respect to any automobile on the road ahead. It then adjusts the speed of the automobile based on this relative speed and distance in addition to the desired speed indicated by the driver. In this case, adaptive cruise control meets the definition of an AI agent as it actively senses the environment (i.e., the car in front), reasons with respect to the environment, and takes actions indicated by its reasoning engine. Thus, as the human drives the automobile and employs this feature, they are working in a human–AI team. There is interdependence between the driver and the adaptive cruise control as the speed selected by the AI agent affects the driver's steering, and the driver's steering affects the speed selected by the AI agent, particularly when the driver changes lanes to pass a slower vehicle as the agent will increase the speed from the speed of the slow vehicle it is following to the speed set by the driver as the driver changes lanes. In this case, the driver and the adaptive cruise control agent are working cooperatively to control the automobile in a way that helps the team achieve the goal of moving toward a destination.

Most modern automobiles contain multiple agents that can interact with distinct aspects of driving, and human drivers can interact with multiple agents simultaneously. Agents in modern automobiles are narrow in scope, capable of helping the driver perform very particular tasks, many of which help the driver perform uncommon activities. For instance, automatic windshield wipers and high-low beam adjustment at night are so effective in one of the author's newest automobiles that he finds it disorienting to perform these actions manually in his older automobile. However, some functions do not consider enough contextual information and become a nuisance. For example, when driving on the rural, winding, two-lane roads near his home, he commonly encroaches on the centerline when the road is empty. This helps to reduce the amount of steering and lateral movement of the automobile, which reduces the likelihood that passengers will become motion sick. However, the automated lane-keeping agent in his automobile does not understand this context and constantly sounds alerts and tries to guide him into the center of the lane, producing an annoyance that makes this agent a hindrance rather than an aid. Therefore, he typically disables this "feature" in his automobile. This is an example of AI not being context-aware, and designers not always considering all of factors that users will consider important.

Reviewing this discussion on agents in an automobile, each agent in this example is controlling a subsystem within the car, for example, the accelerator, the windshield wipers, or the headlights. This concept of control in human–AI teams is important, and we will discuss this in more depth in a subsequent chapter. Also, important here is that the automobile described implements a simplistic human–AI team. It involves a single human who must interact with multiple AI agents to receive the benefits these agents provide. This changes the work the human must perform to drive, ideally simplifying the work by having AI agents assuming control for subsystems in the automobile, and relieving the human from the work associated with controlling these subsystems. Relieving the human of these duties reserves more time and mental capacity for the human to devote to higher-level activities, such as controlling the path of the automobile and navigating in complex environments. This idea of hierarchical arrangements of functions and goals, where different agents assume responsibility for different subgoals, is also important. We will discuss this hierarchical arrangement in depth when discussing patterns, particularly the subsumption pattern, in Chapter 8. The agents provide narrow functions within the example automotive system and interact little, if at all, with one another. This lack of interaction between artificial agents not only simplifies the design of the system but also leaves the operator to perform all integrative functions.

Although many AI leaders view augmenting humans through human–AI teams as a short-term solution on the path to a future powered by AI, this vision ignores the fact that this technology should support societal goals. It is important that developers design technology to support human goals and consider both short- and long-term implications of the design philosophies they employ.

As we discuss human–AI teams, we are interested in understanding not only relatively simple human–AI teams, as exemplified by this automobile, but also more complex teams. These more complex teams will often include multiple humans who must interact with one another to accomplish joint work, and these humans can then interact with a complex array of AI agents which perform a mix of work that involves low-level and higher-level functions. Further, these systems may operate in environments that vary in complexity and the level of risk that the environment imposes on the operator or other humans in the environment. As these human–AI teams become more complex, designers must place more effort on structuring these teams to ensure that the team will reliably and successfully fulfill the team's goal or purpose.

1.3 DEVELOPING A FRAME OF REFERENCE THROUGH DEFINITION

As the current text draws from multiple bodies of literature, it is important to agree upon certain important definitions to help us frame the problem. Therefore, much of the remainder of this chapter provides definitions and points of view that are important to understand before we proceed.

1.3.1 System Structure

Within the proposed framework, we assume a conceptual system structure comprising a technological hardware/software or machine subsystem as well as a human–AI team. The technological subsystem provides physical capabilities. The cognitive capabilities of one or more humans and one or more AI, goal-seeking agents harness these physical capabilities. The left panel of Figure 1.1 illustrates the system composition of this viewpoint, while the right panel of Figure 1.1 illustrates the assumed communication structure.

In the current text, we will use the word ***system*** to represent the holistic sum of interdependent parts that interact to achieve a specific purpose or goal (Dawson, 2020). The interaction of the parts and the fact that they depend on one another are inherent properties of a system. As shown in Figure 1.1, the system under design exists within and interacts with an environment. The system comprises a human–AI team and optionally includes a machine which interacts with the environment. Within the human–AI team, one or more humans and AI agents interact with one another through a user interface in an interdependent fashion. The user interface will likely include displays or other output devices, input devices, and computing resources to support the interface and potentially the AI. We will use the term ***interdependent*** or ***interdependence*** to reflect those parts of the system that depend upon one another to achieve the system purpose or goal. Therefore, the system cannot achieve its primary purpose when one of the interdependent parts of the system is not present. It is also important to note that many properties of the system emerge as the components of the system interact. ***Emergence*** is a phenomenon where interactions among objects generate new phenomena. For instance, individual neurons in the human brain do not create thought. Thoughts are, instead, an emergent property of the human brain that the interaction of neurons creates.

It is important when considering Figure 1.1 that ***machine*** is a generic manufactured physical object that interacts with the environment. It might be, for example, a robot, an automobile, or an aircraft. The form of this machine is not important, but what is important is that AI agents (embedded or separate) or humans can provide commands and control to the machine. Further, the human can provide commands and receive status from the AI agent, and the AI agent can interpret these commands

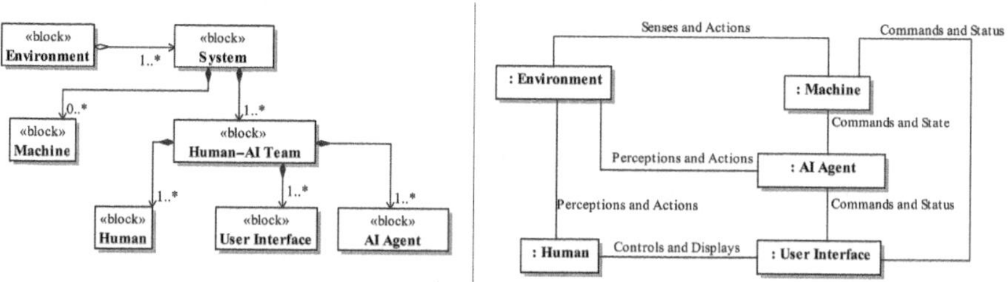

FIGURE 1.1 The left panel shows the structure of a system employing a human–AI team in a block definition diagram. The right panel shows the interaction between system components in an internal block definition diagram.

to control the machine. As shown, the system may contain zero or more machines, indicating that the machine is an optional component. The logical view of the AI agent that we propose in this diagram, where the AI is a near peer to the human and is viewed as existing outside the machine, is important to our framework. However, we acknowledge that this logical view often differs from the physical view of the system where the AI agent might be physically embedded in the hardware and software of the system.

As shown in the right panel of Figure 1.1, one or more humans interact with an interface to control the machine subsystem and can interact with the AI agents through either an integrated interface or separate, specialized interfaces. This viewpoint of the system is important, as it permits the flexibility for the humans or the AI agents to command the physical capabilities of the machine subsystem to accomplish a specific activity. However, humans and AI agents can cooperate and perform actions directly in the environment, independent of the machine, to achieve desired goals. Of course, the physical implementation of the system does not require the AI agents to be physically separate from the machine subsystem. This ***logical view*** of the system is important to create maximum flexibility during system design. It is also important that humans and, in certain instances, AI agents may directly perceive and act on the environment without relying on sensors in the machine.

1.3.2 Automation and Design Philosophy

Automation is the process of substituting an activity originally performed by a human with an activity performed by a man-made artifact or system (Parsons, 1985). This definition does not include the construction of systems that perform activities a person is incapable of performing. Parsons refers to machine activities that perform activities a person cannot as non-automation mechanization. While this definition of mechanization differs from common use, Parson's distinction between automation and mechanization is useful when describing the design of human–AI agent teams.

The differentiation of automation and non-automation mechanization can be subtle. For example, we all agree that a human cannot fly. Based upon this point of view and the definition of automation, it was not possible for the original aircraft designers to automate flight. People are, however, capable of locomotion. We might consider flight an advanced mode of locomotion, making flight an automation of locomotion. However, this distinction is critical to the method discussed in this text as the method requires a designer to be able to exploit human knowledge of the domain of interest to understand basic system requirements for systems that include AI.

As a trivial example of exploiting the human experience to understand system requirements, it was not possible for the original aircraft designers to leverage human flight experience to inform system design, as no human had experienced the process of flight. It was possible, however, to utilize human knowledge of locomotion to inform cognitive needs of flight systems. Specifically, it was possible to utilize human experience to understand the importance of determining geolocation and the ability to project the steps necessary to arrive at a desired location. An insightful designer may have understood the importance of generalizing self-location to include not only geolocation but also altitude and attitude during flight. Such a generalization may have permitted the insightful designer to project the need for altitude and attitude indicators before humans took to the skies. Although trivial, this example illustrates the general design philosophy promoted in this text. That is, a designer can apply AI or human–AI teams to improve human cognitive activity. Philosophically, we can apply the knowledge of existing human cognitive activity to understand the requirements for the human–AI team and the AI that we will need to create to achieve our design goals. This implies a process that includes the steps of:

1. examining the existing system in which humans operate to understand the cognitive work they perform and to derive a description of the current or "As-Is" system;
2. generalizing observations from the "As-Is" system in the context of AI technologies to select a portion of the existing cognitive work plus future cognitive work to automate;
3. identifying design implications for human–AI teaming, given the automation of portions of the cognitive work; and
4. applying the learnings from this generalization to develop design requirements for the future or "To-Be" system.

This text will typically assume a design process that includes these steps.

1.3.3 Automation, AI Agents, and Autonomy

Automation is any artificial system that applies a process that replaces human activity with activity performed by the synthetic artifact or system. Therefore, we can apply this term to a broad range of technology. The process in this text applies to a narrow category of automation systems, specifically those employing AI agents. As such, it is important to clearly define an AI agent.

An ***agent*** is typically defined as an entity capable of perceiving its environment, processing the perceived environmental information to determine a desired action, and performing the desired action in response to the environment (Weiss, 2013). Although not strictly required, agents will typically perceive multiple attributes of the environment and react to these changes to achieve a goal. Therefore, within this text, we will define an agent as an entity that can (1) perceive the environment in which it participates, (2) apply this information to engage in reasoning relative to a set of goals, and (3) apply this reasoning to drive actions in the environment.

Importantly, humans are agents according to this definition. ***AI agents*** are synthetic artifacts that fulfill these three basic requirements. In these AI agents, AI provides nondeterministic reasoning. In the current text, an AI agent must additionally be persistent. That is, the AI agent must be capable of taking action in response to changes within its environment without being activated by any other entity.

The literature often uses the term "human–machine teaming". This text focuses on AI agents rather than any general machine. This distinction is important as the ability to sense and react to changes in the agent's environment provides unique capabilities to the AI agent. Further, these AI agents are becoming ubiquitous due to the recent advent of low-cost digital sensors and actuators that accompany the ever-decreasing cost of computing power. Another trend, which has made AI agents particularly desirable, is the ability to leverage social media and tools such as Amazon Mechanical Turk to obtain exceptionally large data sets to train systems that employ some specific types of AI, specifically AI that employs "*Deep Learning*".

According to the definition provided earlier, agents do not have to be autonomous or exhibit agency. *Autonomy* is typically defined as the *authority* to select or perform an action free from external control or influence (Bradshaw, 2013). Importantly, this definition assumes that agents exhibiting autonomy decide or perform the necessary action. Therefore, *autonomous agents* can be defined as agents that create and pursue their own goals, as opposed to being controlled by another agent (Sterling & Taveter, 2009). In practice, autonomy requires the four preconditions shown in Figure 1.2 (Stevenson et al., 2023). As shown in this figure, besides having authority to select and take appropriate actions to achieve the goal, an agent must additionally be available to perform the activities and *capable* of selecting and performing the activities. Finally, environments are not stable, and an agent may be capable of selecting and performing activities under some, but not all, circumstances. For instance, if a human or AI agent must divide their cognitive resources across multiple tasks, they may temporarily not have resources available to select and perform certain tasks from this array of tasks. Therefore, they temporarily lack the *capacity* to select and perform the tasks. In another example, an agent may be capable of performing a task, such as lane keeping under most circumstances, but it does not have the capacity to perform lane keeping in construction zones without lane markings or during heavy fog or rain. Therefore, it is incapable of acting autonomously under these conditions and does not have the capacity to behave autonomously.

AI agents in the current text are designed to exhibit what Sterling and Taveter define as *controlled autonomy*, that is, the agent should act autonomously only to the extent that it is beneficial to the physical entity (e.g., human) who is responsible for controlling it (Sterling & Taveter, 2009). This definition is important as it is recognized that AI agents cannot be held responsible for their actions and, therefore, should not be granted authority for actions which could reasonably result in harm (Inagaki, 2006). This is evident in consumer products, such as automobiles with automatic driving features, as police and other authorities will hold the individual

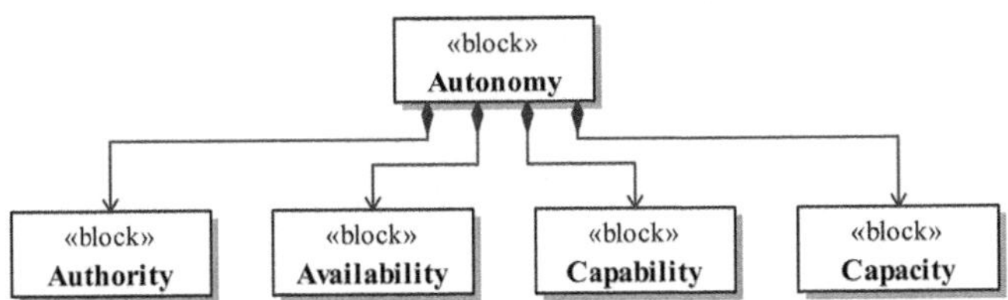

FIGURE 1.2 A block definition diagram illustrating the four essential components of autonomy.

Source: adapted from Stevenson et al. (2023)

in the driver's seat responsible if their vehicle violates a law. In these systems, the driver is legally responsible even if automated driving features were engaged and controlling the automobile at the time of the violation.

1.4 MODEL-BASED SYSTEMS ENGINEERING

The discipline of *systems engineering* was born from the complex systems that came into being during the Second World War. The discipline sought to develop methods and tools to facilitate the design of complex physical systems. With the development of software-centric systems, the discipline adopted and expanded upon model-based software engineering tools. Specifically, the discipline has adopted and expanded on the *Unified Markup Language (UML)*, which was developed to aid the design and specification of object-based software (Larman, 2005).

The dominant systems engineering extension to the UML specification is referred to as *Systems Markup Language (SysML)* (Delligatti, 2013). SysML provides specific diagrams for describing the structure, behavior, and requirements of any system (Friedenthal et al., 2014). The object management group did not design SysML Version 1 to replace UML but to augment UML. Specifically, SysML permits the Systems Engineer to describe the entire system, including the structure, behavior, and requirements of the system, as well as relationships among these representations. Therefore, SysML models provide a description of the basic behavior and requirements for the entire system, including software and non-software components, within the system. However, software engineers typically apply UML to design the structure and behavior of software systems.

Recently, software engineering has developed tools to model software systems employing software agents. Therefore, agent-based software engineering languages, such as Multi-Agent Software Engineering (MASE) or Organization-Based Multi-Agent Software Engineering (O-MaSE), have been discussed within the software engineering literature (Deloach, 2004; Deloach & García-Ojeda, 2010; Garcia-Ojeda & DeLoach, 2010).

We do not intend the modeling language and methodology in this text to replace the SysML, UML, or agent-based software engineering tools. Instead, we intend the *Agent-Oriented Systems Modeling (AOSM)* to extend SysML for the purpose of designing systems that support the teaming interaction between humans and AI agents. The AOSM meta-model discussed in this text is a system engineering tool for engineering systems that employ multiple cognitive entities (i.e., agents), whether they are human or AI. Further, we intend the AOSM language to be a tool to understand the requirements and user stories for the underlying physical and software systems necessary to support the cognitive activity that the model captures and represents. The AOSM method permits the partitioning of work and the design of interaction between humans and AI agents. Specifically, the resulting model provides insight into requirements for the human interface, training, and personnel selection, as well as requirements for the AI agents, including their knowledge, sensors, processing, and actuators. We expect that designers will need to apply SysML, UML, and agent-based software engineering tools to describe and design the lower-level functionality of the hardware and software components of the system. Thus, we view the AOSM language as an extension to SysML for aiding the development and modeling of the human–AI team (Miller et al., 2020).

1.5 NEED FOR A COMMON FRAMEWORK

As we design human–AI teams (HATs), it quickly becomes apparent that to create the *interdependence* necessary for teamwork, we need to design tightly coupled systems, which increases the design *complexity*. As such, it becomes useful to describe the structure and behavior of these systems at a high level before beginning a detailed design. In both software and systems engineering, it is common to facilitate this design using systems modeling languages, processes, and methods (Delligatti, 2013; Friedenthal et al., 2014; Larman, 2005). Many traditional system languages and methods permit one to begin a design process by iterating system definition, requirement, and conceptual design prior to detail design with the goal of reducing the amount of iteration and redesign later in the design process, when design changes can be costly (Walden et al., 2015). Therefore, it is useful to consider a modeling language that permits the examination of human–AI teams and specifically a language that supports the conceptual design of these teams.

This text develops and discusses the AOSM language and method. The AOSM language provides a set of design artifacts or terms which can apply to all agents, including both human and AI agents. This terminology allows us to consider the system independently of who performs specific functions. Without terminology that enables us to decouple the "who" from the "what", we would need to allocate functions between humans and AI agents early in the design process. However, early *allocation* often precludes us from considering new and interesting alternatives, and, therefore, it is our goal to delay allocation until later in the design process. In fact, the Systems Engineering process promotes the creation of multiple system configurations, which vary by allocation, to permit desirable configurations to be evaluated and selected during the design process (Blanchard & Fabrycky, 2006).

Therefore, the terminology and design artifacts must accurately depict the use of the system regardless of whether it is composed only of humans interacting with tools in an environment or whether it includes humans interacting with AI agents within the environment. The language described within this book extends the Unified Markup Language (UML) and the Systems Modeling Language (SysML) (Delligatti, 2013) to support the design of human–AI teams, incorporating concepts from multi-agent architectures (Weiss, 2013) and multi-agent software engineering (Deloach, 2004; Garcia-Ojeda & DeLoach, 2010; Sterling & Taveter, 2009). If you are unfamiliar with these languages, other resources are available, which describe the intricacies of these languages. However, reviewing this material is not necessary as this text will introduce you to the aspects of these languages that are important to employ the AOSM language and method.

1.6 ORGANIZATION OF THIS BOOK

This text seeks to provide the reader, that is, a human–AI agent team designer or design team, with foundational material to engage in human–AI team design prior to the detailed system design. Additionally, the text introduces the AOSM modeling language to permit the reader to engage in design as they apply the process discussed in this text. We have organized this text around a proposed design process. Figure 1.3 illustrates this process. As shown, this design process includes six steps that begin with defining the scope of the project by defining the system boundary and ends with generating specifications for components within the desired human–AI team. These specifications include the definition of software agent capabilities, training and procedures for humans and AI agents, and interaction requirements for the humans and AI agents within the human–AI team.

As shown in Figure 1.3, the design process begins with defining the scope of the "system" for design by defining the system boundary. The second step is to develop

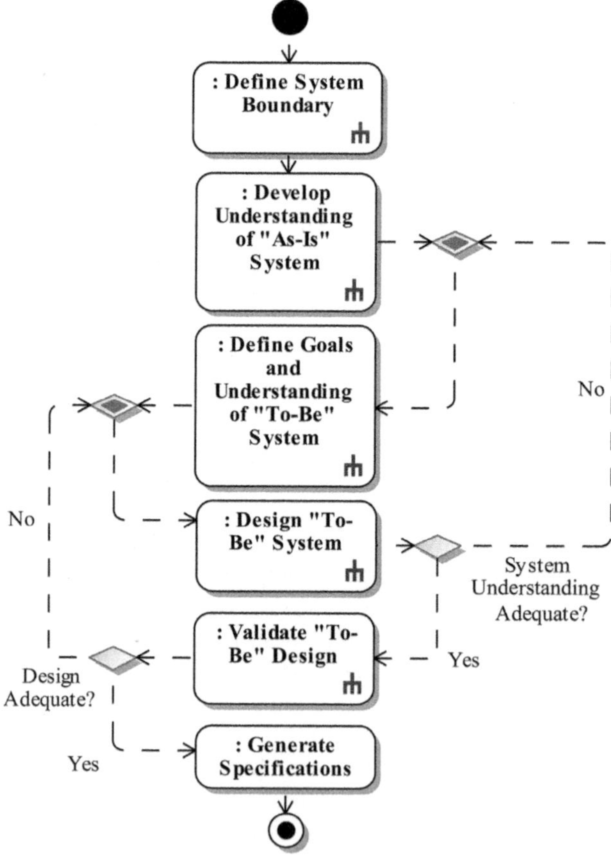

FIGURE 1.3 A SysML activity diagram illustrating the six high-level process steps of the Agent-Oriented Systems Modeling (AOSM) Conceptual Design Method.

an understanding of the existing or "As-Is" system. During this step, we recognize and document the cognitive work system and process we wish to augment with AI. Further, we analyze the way that humans in the target environment currently conduct cognitive work. Chapter 2 discusses the initial design artifacts and methods that are useful in defining the system boundary and begins to analyze the work performed by individuals in the "As-Is" system. Chapter 3 provides a deeper dive into understanding human goals and control structures before providing methods to create an understanding of the current cognitive work that individuals perform in the "As-Is" system. Chapter 4 explores the "As-Is" system. We delve further into teaming, communication, and shared goals within Chapter 4.

The next step in the AOSM method is to reimagine the work that humans in the "As-Is" system conduct. As we reimagine this work, we describe the system we wish to produce. Analyzing the existing system can produce several insights. For example, identifying times when humans in the existing system struggle to perform the work, perhaps where they may frequently become task saturated and unable to truly complete all necessary work within time limitations or identifying circumstances in which their skills might not permit them to fully analyze a situation. Shortcomings of the current work processes then become areas of work that must be reimagined. Additionally, stakeholders who are funding the design of the human–AI team may have other motivations for creating a new system, such as improving safety or reducing workforce that may also motivate necessary changes. These goals drive the need for changes in our system, which we must understand and communicate to others as our design goals before we continue the design. We discuss applying each of these goals to design in Chapter 5.

The fourth step in the AOSM method is to create an initial system design, including allocating the work among agents, both humans and AI. We discuss teaming with AI agents, architectural patterns for human–AI teams, and decision support systems to aid decision-making and allocation during design in Chapters 6 through 9. The fifth step is to assess the design, adjusting the design to overcome any identified shortcomings. We discuss the topics necessary to support this step in Chapters 10 through 12. These chapters discuss designing for exceptions, the coactive design process, and humans as team members. We examine our design to provide further specifications in the last step of the process. This includes specifying the design of components, including artificial agents, human training, and user interface elements. Finally, we conclude the book with a chapter in which we look at the future of human–agent teaming. As indicated by the arrows in Figure 1.3, this process is not a linear process but often requires that we revisit earlier activities as we discover the need for more information or the need to revisit earlier design decisions.

This book will exemplify this method using a high-level design example. Specifically, the text illustrates this design process by examining the work processes for detecting cancer. This example provides a useful illustration, because the system includes the need for multiple human and AI agents working together as a team. The design description is maintained at a very high level and is not meant to imply that the final design is recommended for implementation. However, it provides a rich example of a complex human–AI team environment that deserves further study. We intend this illustration to provide insight into the application of the AOSM design approach. This illustration should enable readers to employ the design language and method provided in this text to consistently perform useful conceptual design of human–AI teams. We acknowledge that this conceptual design will make various assumptions about both AI and human capabilities. It is important for the design team to evaluate the robustness of these assumptions either in parallel with or subsequently to the conceptual design process to arrive at a robust human–AI team.

We intend this book to be used as a textbook. Therefore, each chapter ends with both a section labeled "Questions and Discussion Topics" and a section labeled "Project Questions and Assignments". We provide questions to test the students' understanding of the material as well as topics that may be more philosophical in nature in the first of these two sections. For a student to understand the modeling language and method, we believe it is important for the student to practice using these tools. Therefore, we assume that a class employing this text should include a class project in which the students form design teams where each team works with a coach to select a system to redesign and practice using the tools discussed in this text. Therefore, the "Project Questions and Assignments" section of each chapter prescribes activities that the design team should perform with respect to their redesign project as they complete each chapter. It is our experience that the coach should structure the design team in such a way, if possible, to include a mixture of students who are interested in AI development and students who are interested in system design.

1.7 CHAPTER SUMMARY

In addition to introducing the need for and defining human–AI teaming, this chapter introduced the need for tools to aid the design of systems employing humans and AI agents as team members in a human–AI team. The tools introduced in this text provide a structured language useful for discovering and describing the design of systems employing teams of humans and AI agents. As discussed, we will also introduce one method, AOSM, to employ this language to arrive at a useful design for a human–AI team. We believe the language and the associated method support the design of human and AI agent teams with improved human–machine symbiosis, leading to a team performance that exceeds the performance of a team with either entity alone.

This chapter also provided a thorough discussion of agents and the rationale for focusing on human–AI teaming. This discussion provides an understanding of different methods and terms applied to describe agents and some of the structures and limitations of AI agents. An automotive example sought to provide an understanding of potential advantages and concerns around human–AI teaming, at least in a simple human–AI team. The reader should note, however, that the modeling techniques this book introduces are more attuned to designing the interactions between teams that include multiple humans and AI agents and are therefore more complex to understand and design than the human–AI team in the automobile example.

While tools to aid the design of human–AI teams and the associated infrastructure are important to realize the full promise of J.C.R. Licklider's vision, they are not sufficient. In addition, one must understand the intricacies of teaming and important considerations in the design and human–AI teams. Therefore, this text will address these foundational issues while simultaneously introducing the use of a standard language and modeling approach, AOSM, for describing human–AI team designs. The language and method in this text are particularly useful when developing HATs that are large enough to require large development teams. Additionally, the standard language documents the rationale for specific design decisions, permitting the engineers and scientists who implement the detailed design to understand the context of components within the larger system. SysML was designed to provide designers with a way to decompose and communicate the system components and their behavior within a cohesive system. The AOSM SysML language extension is intended to provide this same benefit to designers of human-AI teams.

1.8 QUESTIONS AND DISCUSSION TOPICS

1. Provide a list of three or more systems you interact with, which includes AI or AI agents, describing the AI or AI agents in each system.
2. For one of these systems, describe the automation for the system or subsystem that the AI or AI agent controls. Indicate whether the system also permits you to control this system or subsystem, and how this aspect of the design affects your ability to achieve your goals with the system.
3. Discuss a system that includes general automation and a system that includes one or more AI agents. Describe why you believe one of these systems includes an agent and the other does not.
4. Describe a situation in which you behave autonomously and a comparable situation in which you do not behave autonomously, indicating why you hold each of these beliefs. Be sure to indicate how your ability to behave autonomously relates to the four stated requirements for autonomous behavior.

1.9 PROJECT QUESTIONS AND ASSIGNMENTS

1. Using the human–AI agent architecture presented in Figure 1.1, as applicable, provide a brief description of the applicable human, machine, AI agents, user interface, and world/environment components of your current ("As-Is") system.
2. Provide and justify an example of the interdependence between the agents of your current ("As-Is") system.
3. Provide and justify an example of an emergent behavior of your current ("As-Is") system.

REFERENCES

Bates, A. (2018). *Augmented mind: AI, humans and the superhuman revolution*. Neocortex Ventures LLC.

Blanchard, B. S., & Fabrycky, W. (2006). *Systems engineering and analysis* (4th ed.). Pearson Education, Inc.

Bradshaw, J. M. (2013). From knowledge science to symbiosis science. *International Journal of Human Computer Studies*, *71*(2), 171–176. https://doi.org/10.1016/j.ijhcs.2012.10.003

Chafkin, M. (2022). Even after $100 billion, self-driving cars are going nowhere. *Bloomberg News*. https://www.bloomberg.com/news/features/2022-10-06/even-after-100-billion-self-driving-cars-are-going-nowhere?leadSource=uverify wall

Cooke, N. J., Gorman, J. C., Myers, C. W., & Duran, J. L. (2013). Interactive team cognition. *Cognitive Science*, *37*(2), 255–285. https://doi.org/10.1111/cogs.12009

Crandall, B., Klein, G., & Hoffman, R. R. (2006). *Working minds: A practitioner's guide to cognitive task analysis*. MIT Press.

Dawson, M. P. (2020). *Thinking in systems and mental models*. Independently Published.

DeCremer, D., & Kasparov, G. (2021). AI should augment human intelligence, not replace it. *Harvard Business Review*. https://hbr.org/2021/03/ai-should-augment-human-intelligence-not-replace-it

Delligatti, L. (2013). *SysML distilled: A brief guide to the systems modeling language*. Addison-Wesley.

Deloach, S. A. (2004). The MaSE methodology. In F. Bergenti, M. P. Gleizes, & F. Zambonelli (Eds.), *Methodologies and software engineering for agent systems* (pp. 107–125). Springer.

Deloach, S. A., & García-Ojeda, J. C. (2010). O-MaSE: A customisable approach to designing and building complex, adaptive multi-agent systems. *International Journal of Agent-Oriented Software Engineering*, *4*(3), 244–280. http://citeseerx.ist.psu.edu/viewdoc/download?doi=10.1.1.460.7287&rep=rep1&type=pdf

Friedenthal, S., Moore, A., & Steiner, R. (2014). *A practical guide to SysML*. Elsevier Science and Technology.

Garcia-Ojeda, J. C., & DeLoach, S. A. (2010). The O-MaSE process: A standard view. *CEUR Workshop Proceedings*, *627*(39), 55–66.

Inagaki, T. (2006). Design of human-machine interactions in light of domain-dependence of human-centered automation. *Cognition, Technology and Work*, *8*(3), 161–167. https://doi.org/10.1007/s10111-006-0034-z

Johnson, M., Bradshaw, J. M., & Feltovich, P. J. (2018). Tomorrow's human–machine design tools: From levels of automation to interdependencies. *Journal of Cognitive Engineering and Decision Making*, *12*(1), 77–82. https://doi.org/10.1177/1555343417736462

Johnson, M., Bradshaw, J. M., Feltovich, P. J., Hoffman, R. R., Jonker, C., Riemsdijk, B. Van, & Sierhuis, M. (2011). Beyond cooperative robotics: The central role of interdependence in coactive design. *IEEE Intelligent Systems*, *26*(3), 81–88. https://doi.org/10.1109/MIS.2011.47

Johnson, M., Bradshaw, J. M., Feltovich, P. J., Jonker, C. M., Van Riemsdijk, M. B., & Sierhuis, M. (2014). Coactive design: Designing support for interdependence in joint activity. *Journal of Human-Robot Interaction*, *3*(1), 43–69. https://doi.org/10.5898/JHRI.3.1.Johnson

Larman, C. (2005). *Applying UML and patterns: An introduction to object-oriented analysis and design and iterative development* (3rd ed.). Prentice Hall.

Licklider, J. C. R., & Clark, W. E. (1962). Online man-computer communication. *AFIPS Proceedings—1962 Spring Joint Computer Conference, AFIPS 1962*, 113–128. https://doi.org/10.1145/1460833.1460847

Mccarthy, J., Minsky, M. L., Rochester, N., & Shannon, C. E. (2006). Dartmouth summer. *AI Magazine*, *27*(4), 12–14.

Miller, M. E., McGuirl, J. M., Schneider, M. F., & Ford, T. C. (2020). Systems modeling language extension to support modeling of human-agent teams. *Systems Engineering*, *23*(5), 519–533. https://doi.org/10.1002/sys.21546

Minsky, M. (1961). Steps toward artificial intelligence. *Proceedings of the IRE*, *49*(1), 8–30. https://doi.org/10.1109/JRPROC.1961.287775

Norman, D., & Draper, S. (1986). *User-centered system design: A new perspective on human-computer interaction* (D. Norman & S. Draper, Eds.). Lawrence Erlbaum Associates.

Parsons, M. H. (1985). Automation and the individual: Comprehensive and comparative views. *Human Factors*, 27(1), 99–112. https://doi.org/10.1177/001872088502700109

Salas, E., Dickinson, T. L., Converse, S. A., & Tannenbaum, S. I. (1992). Toward an understanding on team performance and training. In R. W. Swezey & E. Salas (Eds.), *teams: Their training and performance* (pp. 3–29). Ablex.

Sterling, L. S., & Taveter, K. (2009). *The art of agent-oriented modeling*. MIT Press.

Stevenson, J. A., McGuirl, J., & Miller, M. E. (2023). Interface design for collaboration with semi-autonomous agents from an airborne aircraft. *22nd International Symposium on Aviation Psychology*, 1–6.

United States Government Accountability Office. (2022). *Artificial intelligence: Status of developing and acquiring capabilities for weapon systems* (February 2022 United States Government Accountability Office Report to the Committee on Armed Services, U.S. Senate ARTIFICIAL INTELLIGENCE Status of Developing and Acquiring Capabilities for Weapon Systems GAO-22–104765; Issue February). https://www.gao.gov/assets/gao-22-104765.pdf

Walden, D. D., Roedler, G. J., Forsbert, K. J., Hamelin, R. D., & Shortell, T. M. (2015). *Systems engineering handbook: A guide for system lifecycle processes and activities*. International Council on Systems Engineering (INCOSE).

Weiss, G. (2013). *Multiagent systems*. MIT Press.

Woods, D. D., & Hollnagel, E. (2006). *Joint cognitive systems: Patterns in cognitive systems engineering—David D. Woods, Erik Hollnagel—Google Books*. CRC, Taylor & Francis Group.

2 Defining Your System

2.1 META-MODEL OF AGENT-ORIENTED SYSTEMS MODELING

To explore the application of the AOSM method to model human–AI teams, it is useful to understand the structure of the model used to describe these teams. Figure 2.1 shows a meta-model of the elements in the modeling language. This figure shows each of the elements in the meta-model with distinct types of connectors.

The connectors in the meta-model follow the SysML convention as shown in the legend of Figure 2.1. As shown, lines having a filled diamond at one end indicate that the item connected to the filled diamond comprises the elements at the other end of the connector. For example, *human–AI teams* comprise *Agents*. Connectors with a hollow diamond indicate aggregation; that is, the item connected to the hollow diamond includes the items at the other end of the connector. Case in point, *Agent* definitions require *Training* and *Procedures*, but the *Training* and *Procedures* are not actually part of the *Agents*. Connectors with a hollow arrow at one end indicate generalization. For example, an *Agent* is a generalization of a *Human,* or a *Human* is a type of *Agent*. Connectors with the open-ended arrows indicate associations. For example, the *System* operates in an *Environment*; therefore, there is an association between the *System* and the *Environment*. Finally, the dashed line indicates association with the interaction between the individual elements. *OPD Elements* which appear near the bottom right of the figure are associated with the interaction between *Humans* and *AI agents*.

Understanding the connectors, we can now review the items in Figure 2.1. In this figure, the *System* appears near the top of the figure. Many human–AI teams work to fulfill goals but utilize a machine to fulfill these goals. As shown in Figure 2.1, the *System* is composed of the *human–AI team* and the *Machine*. As shown, this system might include one or more *human–AI teams*, and it may optionally include zero or more machines. The machine represents a physical entity that can act, altering the physical environment. Remembering our automotive example in Chapter 1, the automobile represents the machine, but a knowledge agent, like Amazon's Alexa, might not have a machine that can act in a physical environment. Thus, the *System* may exist without including a *Machine*, but the system contains user interface elements with embedded digital processing. Further, the *System* may also comprise other *Systems*. The system operates in an *Environment* and interacts with other items in a *Domain*. Also as shown, the *Domain* is a part of the *Environment* in which the *System* operates. We will discuss the domain in more detail later in this chapter.

The *human–AI team* comprises two or more *Agents*. The *human–AI team* may also include other smaller *human–AI teams,* as necessary. The diagram also shows that each *Agent* can, optionally, interact with a *Machine*, which is part of the *System*. Some *Agents* will be necessary to control any *Machines* within the *System*.

DOI: 10.1201/9781003428183-2

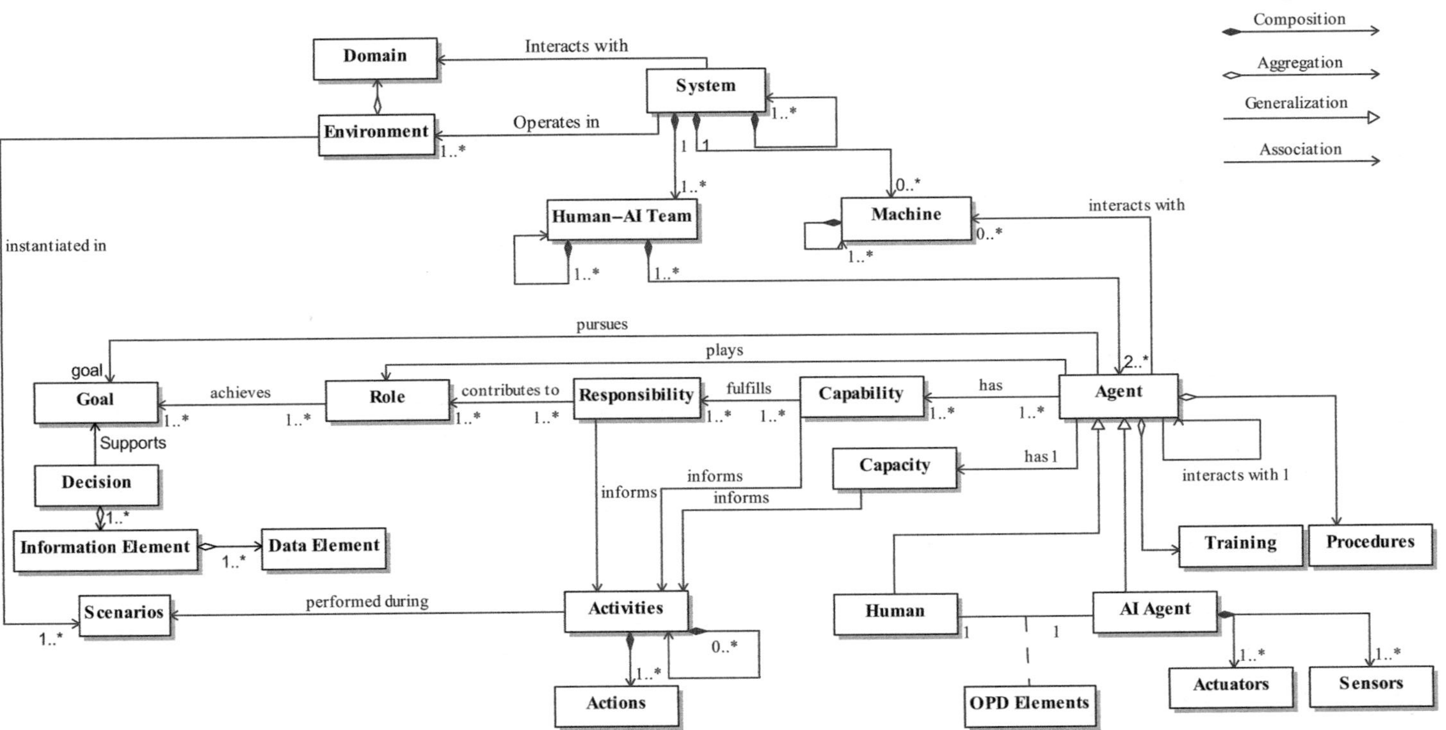

FIGURE 2.1 A depiction of the human–AI team meta-model.

Source: adapted from Miller et al. (2020)

As shown, *Agents* are a generalization of *Humans* and *AI agents*, that is, each *Human* or *AI agent* is a type of *Agent*. During design, we will need to think about the *AI agent*'s ability to sense or act within any environment. Therefore, we see that each *AI agent* comprises at least one *Actuator* and one *Sensor*. When a machine is not present in the system, the AI agents sense information; in Alexa's case, it receives verbal requests and provides information to other agents. Thus, the actuator can involve only information exchange or physical actions.

Figure 2.1 also shows that the *Humans* and *AI agents* interact with one another, requiring *OPD Elements*, which indicate the ability of the agents to observe, predict, and direct each other's actions (Johnson et al., 2018; Johnson & Bradshaw, 2021). *Observability* refers to the ability to obtain information to understand the state of your teammate. *Predictability* refers to the ability to understand what your teammate will do in the future. *Directability* refers to the ability to require or request your teammate to perform a desired activity. Additionally, each agent, whether human or AI, requires *Training* to develop their knowledge, skills, and abilities to perform tasks, and each agent follows *Procedures*. These *Procedures* are processes which encode the organization's protocols. The organization will typically train *Humans* to follow these procedures and provide *Humans* these procedures for reference during task execution. Similarly, these procedures are encoded into the logic of *AI agents*.

Returning to Figure 2.1, the *Agent* has one or more *Capabilities*, where a *Capability* is a set of knowledge, skills, and abilities necessary to fulfill a requirement. *Capabilities* are important as they permit agents to fulfill their *Responsibilities*. Agents playing a *Role* execute abstract items that must be accomplished, referred to as *Responsibilities* (e.g., a pilot must be able to perform the process necessary for take-off). We intend *Responsibilities* to be functionally oriented, without any presumed temporal sequence. Further, *Responsibilities* should not imply a particular implementation. A *Role* is a set of *Responsibilities* necessary to fulfill a *Goal* or a set of *goals* wherein the *Responsibilities* require knowledge, skills, and abilities which are unique from those necessary to perform other *Roles*. In human teams, *Roles* provide a mechanism to compartmentalize work to reduce training and the communication necessary between team members. Importantly, an *Agent* may play multiple *Roles*, and a *Role* requires the acting agent(s) to be *Capable* of fulfilling the associated *Responsibilities* (i.e., it must have capabilities that permit it to fulfill the associated responsibilities). As discussed previously, as an *Agent* plays a *Role*, they pursue the completion of the *Goals*. *Goals* are a desirable state that one or more *Agents* desire to achieve (e.g., arrive on schedule). Importantly, *Goals* describe desired end states but do not attempt to describe how the agent achieves the *Goals*. Further, these goals can be pursued by all agents, including human and AI agents (Sterling & Taveter, 2009).

To achieve goals, agents make decisions. As shown in Figure 2.1, Decisions support the completion of goals. To make proper decisions, the agents must integrate information elements; thus, decisions require the aggregation of information elements. Information elements are formed from individual elements of data from within the environment or from the system. Thus, information elements aggregate data elements.

In this modeling language, the link between goals, roles, responsibilities, capabilities, and agents, which flow through the center of this diagram, serves as the focus for design. Goals, roles, and responsibilities exist regardless of how the designers implement the human–AI team. Therefore, one can start the design process by understanding the goals, roles, and responsibilities that exist in any current system implementation. One can also seek to understand the agents within this environment and the capabilities that they have which permit them to achieve their goals through the role–responsibility linkages. As we begin to design or redesign the human–AI team, the presence of these links within the chain provides the flexibility in the way we describe the function of the team as we can consider the potential for many-to-one linkages between elements in this chain, which provides flexibility for teaming relationships. For example, for a single responsibility, one may design two or more agents to have the capability to fulfill the responsibility, in which case they must also be able to interact to dynamically fulfill the responsibility. Alternatively, we might only have one agent that has the capability to fulfill a responsibility in which case we do not have to design interaction among the agents to fulfill the responsibility. However, if there are circumstances where the agent may not have the capacity to perform the responsibility when required, then this becomes a source of potential brittleness in our system, as the agent may not fulfill the responsibility in a timely manner. Decisions, information elements, and data elements enter this discussion as well, as agents must have appropriate sensors to sense the data elements necessary to support decision-making.

Capacity is the extent to which an agent possesses the knowledge, skills, and ability to perform a specified duty or task at a specific moment in time and within a specific context. It is important to recognize that even though an agent may have the capability to fulfill a requirement (i.e., the knowledge, skills, and ability), they will have limited capacity, and therefore, they may not be able to always fulfill that requirement. For example, they might need to fulfill multiple requirements at the same time, and therefore the amount of work the agent needs to perform is greater than it can perform at one time. In another example, an agent might not be able to perceive data which is necessary to fulfill the requirement (e.g., a radar sensor on a car cannot detect the presence of a car in front of it due to the presence of heavy rain which clutters the radar feed). These and other limitations provide capacity limitations which prevent an agent from fulfilling a responsibility even though they have the capability to do so under alternate conditions. These capacity limitations occur when the agent is situated in a real-world environment and circumstance. Therefore, the designer cannot consider these capacity limitations in the abstract but must evaluate them within the context of use.

To understand capacity limitations, the AOSM language permits the definitions of scenarios within the environment where a scenario is a particular sequence of tasks which, when executed in a particular circumstance, fulfills a goal. The language then permits us to describe activities that the agents might perform within the context of the scenarios to help understand where points of brittleness exist and the desired interactions among the agents. Therefore, the language includes actions, which are discrete steps taken by an agent to execute a higher-level action or activity. An activity is a special type of action that defines a sequence of actions that the agent achieves to fulfill the responsibilities. We model these activities and actions within the context of a scenario so that we can consider the behavior of the agents as they are situated in an environment, permitting us to design the team to adapt to circumstances that present capacity limitations.

This section described each of the model constructs shown in Figure 2.1. In the end, our desire is to define the human–AI team, which includes the agents in this team with their capabilities, training, procedures, interaction elements (i.e., OPDs), and primary actuators and sensors. The model constructs to the left of these elements within Figure 2.1, including goals, roles, responsibilities, activities, and scenarios, provide the context that influences and permits us to evaluate our design.

2.2 SYSTEM EXAMPLE

Throughout this textbook, we will use a medical example to illustrate the implementation of this meta-model. Specifically, the example human–AI team will seek to perform cancer diagnosis using mammography. The use of AI for this application has several decades of history, with a significant emphasis on applying AI to replace the need for radiologists, rather than to form human–AI teams. Like most AI applications, this desire to replace the radiologist has been unsuccessful, even though technologists have successfully applied AI in this field. In fact, Rodney Brooks, a leading robotics researcher, recently criticized the public prognostications that AI would replace radiologists. These prognostications suppressed the development of medical professionals in this field to the point that there is now a shortage of radiologists (Zorpette, 2023). Thus, AI, which technologists originally discussed as a tool to enhance the effectiveness of radiology in applications, such as mammography, has had both negative as well as positive effects on the health care system and patient health.

2.3 SYSTEM BOUNDARY AND DOMAIN

Having defined the meta-model, the designer can apply this model in the first step of our process. Figure 2.2 shows a further decomposition of the first step in our process, which we labeled "Define System Boundary". As shown, in the right panel of Figure 2.2, this step comprises six sub-steps, including Developing Stakeholder Narratives, Depicting the System in a Block Definition Diagram (BDD), Identifying Domain and System Use Cases, and Identifying Difficult Decisions. Once again, completing the diagrams in each of these steps creates a better understanding of the others, and so it may be useful to revisit and revise each of the diagrams to be sure they are consistent with one another.

In systems engineering, it is important that we establish a system boundary before we begin to design the items inside our boundary. Failure to define and communicate the boundary around our system often leads to significant confusion between members of the design team and other stakeholders, including project sponsors, future users, and complimentary design teams. We have asked you to take an initial step toward defining this boundary for a project that you have in mind. Specifically, we asked you to think about the human–AI team architecture of your current or "As-Is system" at the end of Chapter 1. We will address this question as we define our boundary.

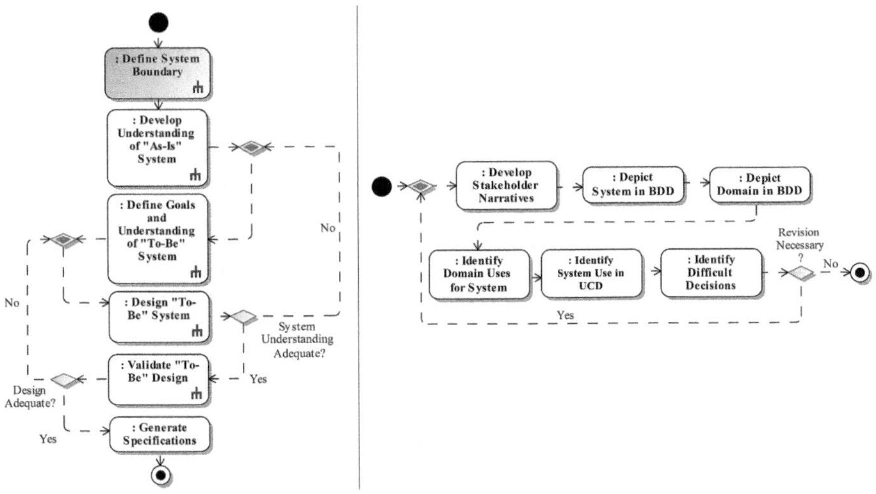

FIGURE 2.2 A depiction of the AOSM process on the left with the sub-steps associated with defining the system boundary shown on the right.

It is important to recognize that not all systems and boundaries are equal. Systems are often described as open or closed and as hard or soft (Hitchins, 1992). Traditional systems engineering viewed systems as closed and hard, while most human-oriented system designers view systems as open and soft. In a *closed system*, the system is unaffected by external influences. An example might be a traditional battery-operated clock which, once we insert batteries, operates free of influence from items beyond its boundary, assuming no catastrophic damage occurs. Therefore, the system is a closed system as nothing passes through the boundary to affect the performance of the clock. In *open systems*, external influences affect the performance of the system. As we assume the human–AI teams of interest respond to the environment, they are inherently open systems, and we must acknowledge that their performance will depend on the behavior of other people, systems, or environmental influences external to our system boundary.

Hard systems have predictable behavior, and we can optimize this behavior, while *soft systems* often include items with variable behavior, making traditional optimization of the system less relevant. Another attribute that differentiates hard from soft systems is once one removes a part or replaces it with a slightly different part, the replacement significantly alters the performance of a hard system, sometimes to the point that it no longer supports its original purpose. For example, if we simply replace the tires on our car with tires having a different diameter, the car might work, but some features, such as the speedometer, may fail to meet their intended function as the system is not designed to adapt to this change. In soft systems, one of the design criteria is resilience to change, so we might remove a component, for example, one person, and insert a different, perhaps less experienced, person. In this case, the designer should design the system to fulfill its basic functions in the face of this change. Thus, as we describe our boundary, we will recognize that it is permeable and that the components, structure, and behavior of components inside our system boundary may vary. Therefore, the system design should be robust to at least predictable sources of external and internal variation.

As we are defining our ***system boundary***, it is useful to consider key actors and systems that lie on both sides of the boundary—internal and external. Our ***system*** is a collection of interdependent parts that interact to accomplish specific goal(s). From a design perspective, we have control over the attributes, composition, and interactions of the entities that are within the system. Entities external to the system, which provide input, purpose, constraints, directions, and/or receive outputs, comprise the *domain* of our system. The system domain is the subset of the environment that the system operates in. The domain includes our system, its stakeholders, customers, primary users, and other systems in the environment.

2.4 STAKEHOLDER NARRATIVE

For illustrative purposes, we utilize a specific example, that of a radiology diagnostic center, which we apply throughout the remainder of this text to illustrate the design approach. As stated earlier, it is important that the illustrative example includes multiple individuals who are operating as a team that could benefit from the insertion of artificial agents.

To facilitate this example, let us begin by developing a short "As-Is" stakeholder narrative around the system we are considering. Most systems will have multiple stakeholders, and we could write these narratives from the point of view of many different stakeholders. Further, we could identify people who approach our system with multiple goals or motivations as is often typical in user design (Levy, 2015). In our case, let us examine the stakeholder narrative around mammography from a patient's point of view.

> Note the following example is for illustrative purposes, and it is not intended to be complete or does not recommend a specific design. However, it is recognized that these systems have life or death implications as mammary gland cancer diagnosis before malignancies spread beyond the mammary gland have a 99% survival rate, while this rate drops to 85% once the cancer spreads to the lymph nodes and to less than 30% when malignancies spread to other parts of the body (Killingsworth & Bohil, 2021).

Jane is a late middle-aged woman with a pair of nearly grown children who works a demanding job while running a busy household. She is concerned about her health and maintains an annual physical with her *Primary Care Physician*. During this visit, the physician reminds Jane of the need for her annual mammogram and submits a request for this specialized examination through the *Electronic Patients Record* platform. After her appointment, Jane runs off to work.

A few days later, Jane remembers to schedule the mammogram. While she is hesitant as she knows this will disrupt her schedule further, she calls the radiological *Diagnostic Center* and schedules an appointment. A few weeks later, she reports to the *Diagnostic Center* and is greeted by an assistant who has her complete a paper questionnaire and then escorts her to the mammography area. She stores her items, removes her clothes above her waist, and puts on a gown while waiting. The *Radiology Technician* eventually greets her and escorts her into the room with the mammography unit. The *Radiology Technician* positions Jane in the mammography unit, records mammograms using a *2D Radiology Machine*, and views the images to be sure they are clear and useful for diagnosis. If not, the *Radiology Technician* adjusts the machine, or the position of the patient and captures another mammogram. The *Radiology Technician* repeats the process until he or she records an adequate set of mammograms. The *Radiology Technician* then tells the patient that the *Radiologist* will view the mammograms and notify her with the results. The *Radiology Technician* then escorts Jane back to the waiting area, where she dresses and heads to work.

Unobserved by Jane, the *Radiologist* later views a traditional two-dimensional mammography image. While doing so, the radiologist activates a digital algorithm that serves as a *Diagnostic Aid* by scanning the image to identify any pattern typical of mammary gland cancer. The aid identifies clusters of calcifications in the mammogram, indicated by areas of high reflectance. Observing that the *Diagnostic Aid* identified a cluster of calcifications, but knowing that calcifications are common, the *Radiologist* might compare this year's mammogram image to previous year's mammogram to see if any changes have occurred, if a prior mammogram image is available. The *Radiologist* may also compare it to the mammogram from the other side of Jane's chest to see if this structure is common to both mammograms. The *Radiologist* must then decide whether the clusters that exist or change in these clusters might indicate the presence of cancer, which disrupts ductwork within the mammary gland resulting in the aggregation of calcium. The *Radiologist's* diagnosis might rely on several cues or hints from the structure of the calcium buildup, including the number and arrangement of calcifications, as well as assessments provided by the *Diagnostic Aid.* Besides the technical information, other factors can influence the radiologist's recommendation, such as whether there is a family history of mammary gland cancer, the disruption a false positive diagnosis will have on Jane, and the consequence of not diagnosing potential cancer on Jane's life. The *Radiologist's* accuracy in diagnosing recent cases with similar calcifications can also affect their interpretation. In this case, the *Radiologist* decides that the presence of the calcifications in the mammogram warrants additional analysis. The *Radiologist* places their diagnosis in Jane's medical record. The Radiologist completes the examination of the remainder of the mammograms and decides that the cluster of calcium deposits is the only indication of potential malignancy. The system notifies Jane's *Primary Care Physician* who then contacts Jane and recommends additional tests but tells Jane not to worry as this happens frequently, and patients rarely receive cancer diagnoses.

Jane now contacts the *Diagnostic Center*, and the administrative assistant schedules her for a 3D mammogram. She repeats the process she did to have the original mammogram, except this time, a *3D Radiology Machine* captures multiple images from multiple points of view to permit a 3D reconstruction of the tissue. Again, once the *Radiology Technician* approves the images, she has Jane dress and return to her daily life.

Behind the scenes, the *Radiologist* reviews the original mammograms and notes from the first mammogram. The *Radiologist* also reviews the additional three-dimensional rendering of the tissue. In the original radiograph, the *Radiologist* can see a potential cluster of calcifications, but they cannot see whether these are clustered in one spot or if they exist at different layers in the tissue. Viewing the 3D radiograph, the *Radiologist* can use specialized viewing apparatus or rotate the image to see if the calcifications are all clustered in a small area in the tissue. Armed with this added information, the *Radiologist* can determine if the calcifications are in a group within one local area or if they exist at random locations in depth. Once again, the *Radiologist* must form a diagnosis. The *Radiologist* decides whether the pattern of calcification indicates the degradation of function in a localized area in the tissue as might occur in the presence of cancer. Furthermore, they must decide whether to simply monitor for future development, requiring a follow-up 3D mammogram after a few months, or request more invasive testing. The *Radiologist* decides that the appearance of the calcifications warrants additional evaluation and updates the *Electronic Patient Record* appropriately.

Jane's *Primary Care Physician* calls her, informs Jane of the findings, and discusses paths forward. In this case, Jane's tissue has not changed from previous mammograms, but clusters of calcifications are present and with new safer biopsy procedures available, the *Primary Care Physician* now believes the risk of undergoing the biopsy outweighs the risk associated with missing the presence of cancer. The *Primary Care Physician* instructs Jane on preparation, known risks, and follow-up procedures. Jane schedules the biopsy with the *Diagnostic Center.*

When she shows up for the biopsy, Jane is prepared and arranged in the operating space by a *Surgical Nurse*. The *Anesthesiologist* applies a local anesthetic. The *Radiologist* images the tissue once again to determine the location to biopsy. Guided by the *Radiologist*, a *Robotic System* inserts the needle into the tissue to approximately the correct location. The *Radiologist* again images the tissue. If the needle is in the correct place, the *Radiologist* takes the biopsy. If not, the needle is repositioned, and the *Radiologist* images the tissue again until it is placed properly at which time the *Radiologist* takes the biopsy. The *Radiologist* inserts a small piece of titanium into the tissue as a marker to indicate the location of the sample. The *Radiologist* withdraws the needle. The *Surgical Nurse* seals the wound and helps Jane prepare to leave.

Behind the scenes, an *Oncologist* examines the tissue biopsy and finds that it is not cancerous. The *Oncologist* updates Jane's records. The *Electronic Patient Record* system notifies Jane's *Primary Care Physician*. The *Primary Care Physician* contacts Jane a few days after the biopsy procedure and tells her the good news with any follow-up instructions.

Reviewing this stakeholder narrative, we see that there are multiple individuals and technological systems at play within this cancer screening process. We can also imagine areas where system designers can apply AI in this system. We might also recognize that AI is already applied in the *Diagnostic Aid*. From a purely system-centric point of view, we see that there are multiple technological systems at play, including the electronic patient records system, the screening, 2D radiology system, the 3D radiology system, the needle biopsy/imaging system, and whatever system the oncologist uses to support their diagnostic work. We could place our system boundary around any of these or around any combination of these systems. We also must recognize that we wish to design human–AI teams; therefore, we need to include humans and AI agents within our system boundary, not just the hardware and software systems. The more we include inside our system boundary, the more *complicated* our design becomes. We do not want to make our design more complicated than necessary. However, one of the inherent goals of human–AI teaming designers is to help address cognitive *complexity*, helping the humans in our system overcome complex interactions that make work difficult. We can define complexity in many ways, but one way to view complexity is through interactions, so we can ask ourselves "Where in this system do the most interactions exist which make decisions difficult?"

While we can develop methods to identify the degree of complexity, in this exercise, we simply observe that the number of interacting individuals increases within the 3D imaging and biopsy portions of this process. Further, some individuals take part in multiple portions of the process. Thus, these portions of the process must require common knowledge, skills, and abilities. Further, the fact that more individuals are present in the 3D imaging and biopsy portions of this process indicates that the cognitive work that the team performs within these portions of the process is highly complex. Therefore, let us assume an initial system boundary where the individuals and systems within the diagnosis center, beginning with the capture of the first mammogram through the collection of the biopsy sample, are within our system boundary, while all others are outside our boundary.

2.5 MODELING: "AS-IS" SYSTEM ARCHITECTURE AND BOUNDARY THROUGH BLOCK DEFINITION DIAGRAMS

2.5.1 Modeling Objective

To characterize the architecture of our "As-Is" system, the AOSM design method begins with constructing a set of three SysML Block Definition Diagrams (BDDs). BDDs often depict the components of a system in a hierarchical arrangement. We will depict our system first at the domain level where we show our system under design within the larger system and environment. Viewing the system at the domain level is important to help us understand external sources of variability that might impact the performance of our system. We will then depict the internal elements of our system via two internal BDDs, recognizing that some of these elements may inject variability into our system performance. These sources of variability are important to recognize as they can influence the performance of our system and the arrangement of our human–AI teams (Hollnagel, 2012). The first internal BDD is at the system level, and it shows the human–AI team as a single component of the system, along with the other people, subsystems, and equipment that comprise our system but are not part of the human–AI team. The final internal BDD displays the composition of the human–AI team, including the various humans and AI agents that comprise it.

2.5.2 Diagram Components and Composition

Figure 2.3 provides a domain level BDD containing generic elements. As shown, the first layer of the hierarchy in the domain level BDD is the domain or *System of*

Systems in which our *System of Interest* operates. The second layer consists of our *System of Interest* plus additional external *Interacting Systems, External Agents,* and *External Organizations* that meaningfully interact with our *System of Interest.* These interactions include providing our system with inputs, resources, and goals that impact the behavior or performance of our system and receive outputs from our system. These external systems or organizations may be decomposed into a third layer if a third layer is desired, or necessary, to capture interaction with the system.

Figure 2.4 shows a system level BDD containing generic elements. For this system level BDD, the first layer in the hierarchy is our *System of Interest.* If relevant, the second level depicts the *Subsystem(s)* that the *human–AI team* supports and that logically perform unique functions within the system. Then, the next level is a decomposition of the *System of Interest* with *Subsystem(s),* to include our *human–AI team,* other *non-HAT agents* that are part of the system (but not the HAT), and *Machines* (including non-agent software automations and hardware).

Figure 2.5 shows a BDD containing generic elements representing the HAT. In this HAT-level BDD, the first layer in the hierarchy is our *human–AI team*. The next level is a decomposition of the team into the *human agents, AI agents,* and non-agent automations, indicated by AI tool, which the designer could partially or fully replace with other agents.

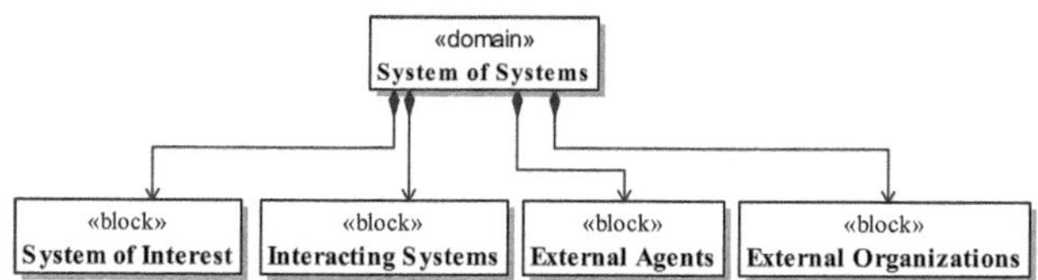

FIGURE 2.3 Generic Block Definition Diagram showing the Domain in which the System of Interest operates and external systems and agents that the System of Interest interacts with.

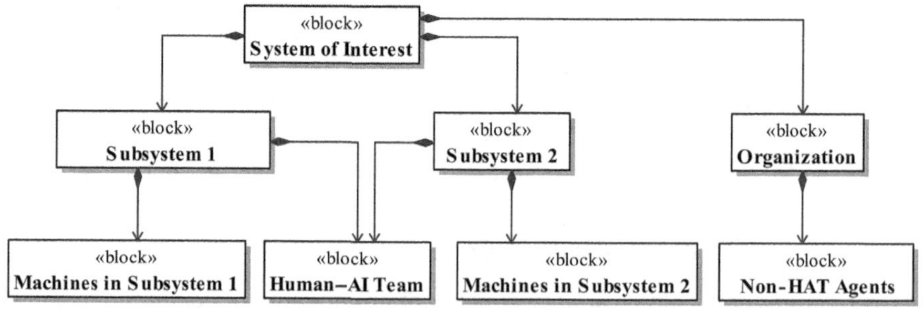

FIGURE 2.4 Block Definition Diagram showing the System of Interest and representations of the components it will typically contain.

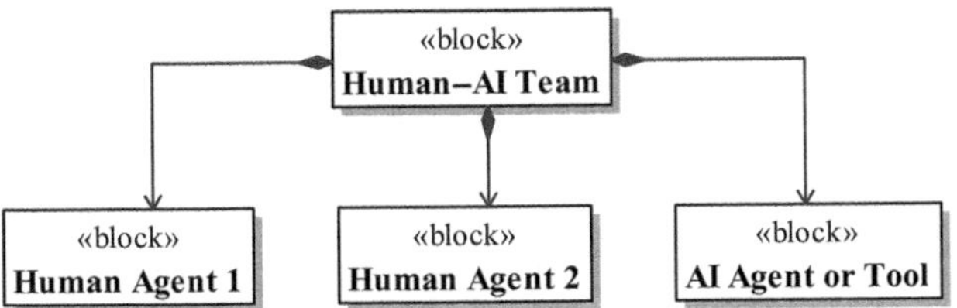

FIGURE 2.5 Block Definition Diagram showing the HAT and the types of components it will typically contain.

2.5.3 Considerations

While SysML blocks traditionally represented hardware systems and subsystems, AOSM captures humans and AI agents as blocks. This representation enables our model to capture additional characteristics and behaviors of these humans and AI agents that SysML and corresponding software modeling tools reserve for blocks. Additionally, as we construct our "As-Is" and eventually our "To-Be" systems, the modeler should capture the AI and other software algorithms separately from the hardware in which the AI algorithms reside. This will allow additional design flexibility later in the design process.

2.5.4 Example Application

Example Description: For context, our example system exists as part of a larger health care environment. The specific domain is the subset of the environment that deals with diagnosing mammary gland cancer. Our system, the *Diagnostic Center*, is part of this domain and interacts with other actors in the domain that are not part of our system. *Primary Care Physicians* refer *patients* to our system. The *Diagnostic Center* accesses the external electronic patient records and updates that record with our system's findings. Depending on the results of our system's detections, the Radiologist may refer the patient to Oncology.

Figure 2.6 shows the Domain level BDD. This diagram shows the *Diagnostic Center HAT Domain* at the top. The connectors shown between the domain and the underlying blocks indicate that the domain comprises the underlying blocks. This connector indicates that the *Diagnostic Center HAT Domain* comprises the elements indicated by blocks below it, including patient and *Primary Care Physician*. This diagram indicates people by their role. Thus, Jane does not appear in the domain, but the diagram depicts Jane's role as *Patient*. The diagram also depicts Jane's *Primary Care Physician* and the other actors that are outside the *Diagnostic Center*. The diagram depicts the system itself, the *Diagnostic Center*, as a single block.

Example Description: Within our *Diagnostic Center*, there are two main systems, the one that performs radiology and the one that performs biopsies. Both systems consist of machines, along with our human–AI team (which in the "As-Is" state does not actually contain any AI agents). Additionally, our system also includes administrative leadership, which provides the system with guidelines, policies, funding, staff support, and business goals. Finally, the figure shows two notes for future reference.

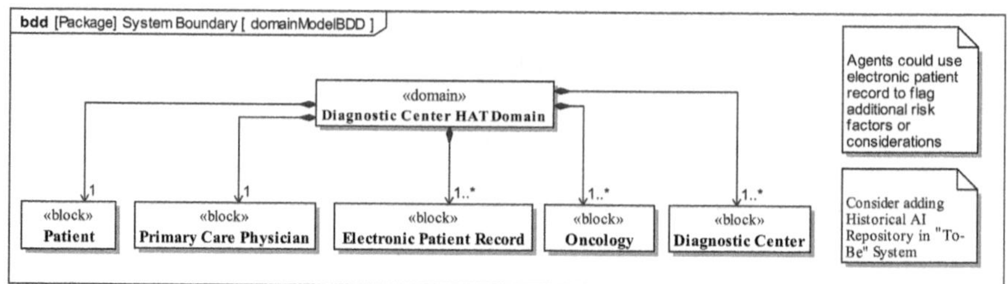

FIGURE 2.6 Block Definition Diagram showing the Diagnostic Center HAT Domain and its components.

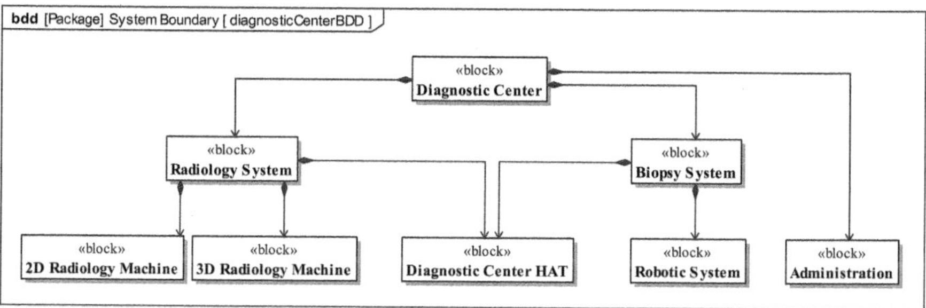

FIGURE 2.7 Block Definition Diagram showing the Diagnostic Center System and its components.

Figure 2.7, the System level BDD, shows the architecture for the "As-Is" *Diagnostic Center* system. Once again, the connectors show that in the "As-Is" configuration, the current version of the *Diagnostic Center* system comprises multiple elements, including our *Diagnosis Center HAT*, represented as a single block. Other elements include people that are not part of the team, such as the *Administration*; subsystems within the system, which may or may not include our HAT; and hardware, software, and automations that are part of the system but not part of our HAT.

Figure 2.8, the HAT level BDD, shows the composition of our "As-Is" human–AI team. The connectors show that the "As-Is" *Diagnostic Center HAT* comprises multiple elements, including multiple human agents along with the *Diagnostic aid.* In our example, *Diagnostic Aid* is not an intelligent agent; however, it has the potential to be replaced by an agent or multiple agents (human or AI), and, thus, we choose to represent it here as part of the team. Note that the current system does not yet have AI agents, although other "As-Is" systems might include AI agents. The only agents in our example are the people in the "As-Is" system.

While these diagrams do not include every element that might appear in the Stakeholder Narrative, they do include some important hardware and software elements that appeared. These diagrams at least begin to capture some elements in our architecture and may be iterated on as we progress through the AOSM method.

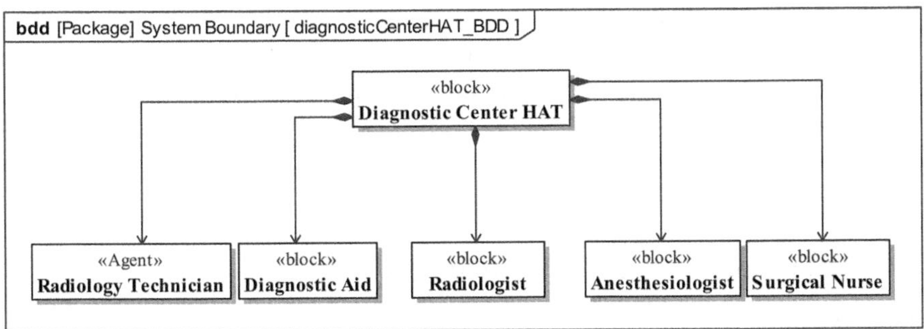

FIGURE 2.8 Block Definition Diagram showing the Diagnostic Center HAT and its components.

2.6 COGNITIVE TASK ANALYSIS

Task analysis is the process of learning and understanding how a user accomplishes a task through inquiry or observational analysis of the user or other subject matter experts who understand how the task is performed. Task analysis is part of the human factors engineering domain, and professionals in this domain have used this approach when designing work processes, redesigning work processes, when analyzing work to understand human skills in support of personnel selection, and when analyzing work to design training. *Cognitive task analysis* (CTA) is an extension of traditional task analysis. Cognitive systems engineering and psychologists emphasize cognitive task analysis to understand the cognitive aspects of task performance. As with traditional task analysis, analysts or designers perform cognitive task analysis through natural or controlled observation of the work as it is conducted or by applying structured interviews, questionnaires, focus groups, recorded artifact reviews, or mental walk-throughs. As it is often difficult to obtain domain experts to take part in cognitive task analysis, analysts or designers usually begin CTAs by reviewing available literature, such as training manuals, procedural documents, and the documentation of previous cognitive task analyses. Designers can construct initial models of the "As-Is" system from these documents prior to beginning interviews to help the researcher understand the vocabulary of the domain experts and establish a baseline understanding of the work that the domain experts perform. There are numerous resources available for detailing various cognitive task analysis methods and execution of conducting task analyses including *Working Minds* by Crandall and colleagues (Crandall et al., 2006).

Conducting the cognitive task analysis and the resultant construction of high-level Use Case Diagrams are iterative and simultaneous processes. Designers with limited system familiarity should conduct a cognitive task analysis to produce the necessary familiarity with the system prior to building the high-level Use Case Diagrams. However, if you as the designer are sufficiently familiar with the system, you can begin by building the Use Case Diagrams described in the next section and then conduct the cognitive task analysis focused on the primary processes identified in the Use Case Diagrams. Otherwise, it is recommended that you (1) review the available documentation, (2) conduct an initial task analysis with a user or customer of the system to develop a Stakeholder Narrative that provides a reference to a complete end-to-end experience with the system, and (3) conduct an initial task analysis with a primary agent or two within the human–AI team to understand the domain, processes, and systems involved.

2.7 MODELING USE CASE DIAGRAMS

2.7.1 Modeling Objective

Applying the information gathered from the cognitive task analysis, the AOSM language implements two high-level SysML Use Case Diagrams (UCDs). The first UCD is the Domain level UCD, which provides a high-level representation of the domain, showing the interactions between external actors and the domain level processes, or functions, performed or supported by the system. The designer can develop a UCD at multiple levels of abstraction, depending upon the required level of system detail.

The second UCD is the System level UCD, which provides a high-level representation of the system, showing the interactions between the internal processes performed by the system and the human–AI team members, other system actors, and internal and external systems, software, and hardware elements. You can develop the UCD at multiple levels of abstraction. However, the number of levels you should represent depends upon the system you are designing. At a minimum, the UCD should capture the primary process(es) that you seek to improve.

2.7.2 Diagram Components and Composition

A Use Case Diagram consists of multiple use cases or high-level processes, functions, or transactions performed by the system. The UCD represents each use case by an oval. The UCD labels each use case with a verb–noun phrase to indicate the name of the activity that the humans in the "As-Is" system are performing. The UCD places all use cases inside the diagram's system boundary. In the UCD, agents, systems, and key hardware and software elements, referred to as actors in SysML, are located outside the system boundary and connected to the use case ovals they interact with. UCDs often include three distinct types of use case elements. Primary use cases are use cases that the primary user of the system triggers. The UCD elaborates the primary use cases through "Include" or "Extend" use cases. Include use cases are subsets of the primary use case that the system always executes as part of the primary use case. In our system, the primary care physician might always update the patient's record, and, therefore, we might provide the use case of Update Patient's Record as an "Included Use Case". Extensions are use cases that occur by exception. For example, a Radiologist rarely orders a biopsy after a mammogram. Therefore, conducting a biopsy is an extension of the primary use case. The include and extend use cases are each attached to the primary use case by specialized connectors within SysML.

The Domain level UCD consists of the primary processes/activities in the environment that our system participates in or supports. Figure 2.9 shows a Domain

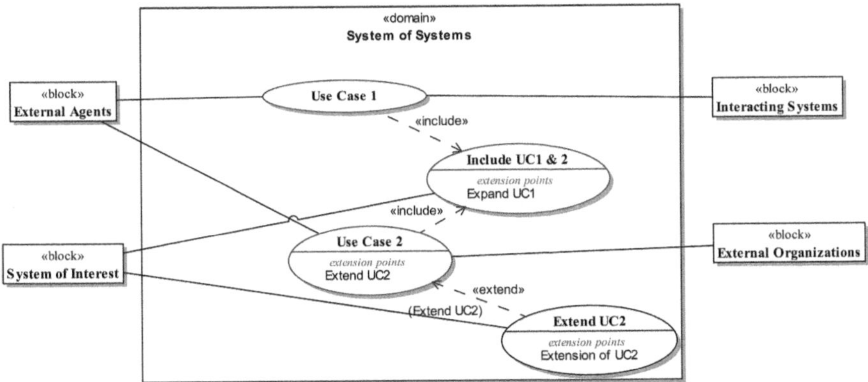

FIGURE 2.9 Generic Domain level Use Case Diagram showing the Domain Use Cases in which the System of Interest participates and external systems and agents that the System of Interest interacts with.

level UCD containing generic elements. As shown, this will include processes that are not the primary responsibility of our system, *Use Case 1,* and *Use Case 2,* or may not currently involve our system, but could benefit from the involvement of our system. In addition, this UCD will also show the processes our system has primary responsibility of performing, represented in this diagram by *Include UC1 & 2* and *Extend UC2*, which are inclusions/extensions of the domain use cases. Including the larger use cases of importance to our stakeholders as well as the use case, includes, or extends that our system contributes to, provides context for the application of our system. The presence of the domain use case and extensions or includes provided by our system illustrates the interface between our system and the external systems and will help to identify the information exchanged between our system and its domain as we progress through the design process. This concept of always capturing the structure and behavior of the system that encompasses the system under design is important to understand not only the context of the system under design but also the potential variability of the inputs to the system of interest from the external environment within open systems. The actors included in the Domain level UCD should be the same actors captured as blocks in the domain level BDD. We recommend placing primary actors (those actors who pursue their own goals and participate in performing the processes) on the left and secondary actors (those providing resources or receiving outputs from the processes) on the right.

The System level UCD consists of the primary processes/activities that our system is responsible for executing; these should be any processes with direct connections to our system in the Domain level UCD. Figure 2.10 shows a system level UCD containing generic elements. This diagram shows a more detailed decomposition of the relevant Use Cases from the Domain UCD, along with the relationships between the individual members of the human–AI team to the individual use cases for the system. The actors included in the System level UCD are each of the lowest-level

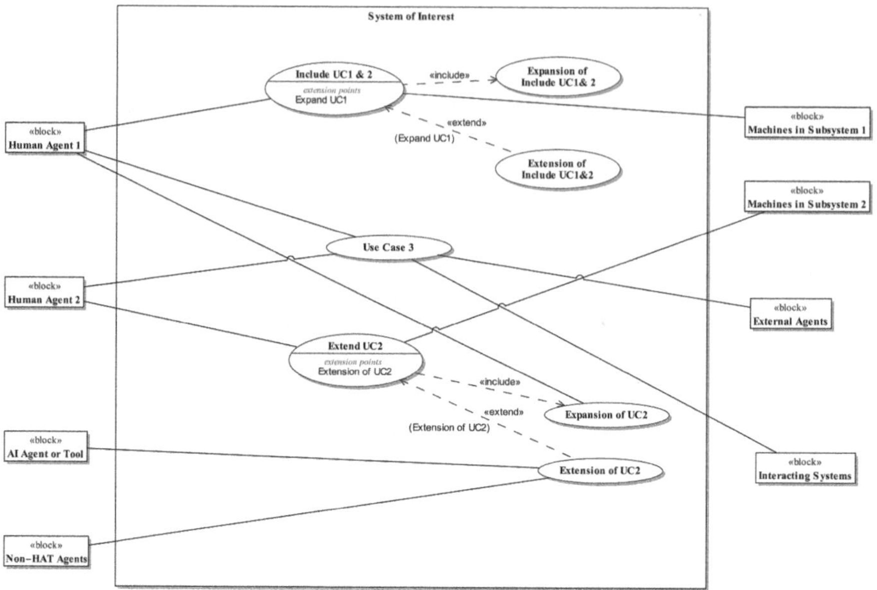

FIGURE 2.10 Generic System level Use Case Diagram showing the System of Interest Use Cases which the human–AI team is responsible for, in addition to external systems and agents that the System of Interest interacts with.

blocks identified in the Domain level BDD, System level BDD, and HAT level BDD. However, the UCD does not show consolidated system and HAT blocks as actors since the UCD represents these blocks in their decomposed form. We recommend placing HAT team members and external agents that support system goals on the left and the remaining actors on the right.

2.7.3 Considerations

UCDs traditionally capture humans as stick figures external to the system boundary, but as with the BDDs, the AOSM language captures humans and AI agents as blocks. This will enable our model to capture additional characteristics and behaviors of these humans and AI agents that SysML (and corresponding software modeling tools) reserve for blocks.

2.7.4 Example Application

Example Description: While our system's main function is to perform cancer screening, this activity is part of two larger functions within the domain: Monitoring mammary health and determining interventions. The patient monitors their mammary health in conjunction with their primary care physician. The primary care physician records the status of the monitoring process within the patient's electronic record on a reoccurring basis. Our system becomes involved either when the patient or their primary care physician detects an anomaly, or when the primary care physician recommends routine screening based on the patient's demographic information. If an abnormality occurs, the patient and the primary care physician will have to select a course of action. Of course, one course of action is to do nothing and continue to monitor. Our system is involved with providing information, and recommendations, to support that decision. If the radiologist in our system diagnoses the possible presence of cancer, oncology will also become involved.

Figure 2.11 shows the Domain level UCD. This diagram shows the *Patient*, *Primary Care Physician*, and *Diagnostic Center* as blocks on the left side of the diagram. The placement of these blocks on the left side of the system indicates that these agents are primary actors. The UCD shows other actors or external systems mentioned during the Stakeholder Narrative on the right as secondary actors. Inside the UCD domain boundary are two primary processes performed by the domain: *Monitor Mammary Health* and *Determine Interventions*. *Perform Mammary Cancer Screening* supports or extends both processes. In our example, our system, the *Diagnostic Center*, is not responsible for performing either of the primary processes; rather, our system performs the extending process *Perform Mammary Cancer Screening*. The "extends" relationship indicates that the extended use case is optional, occurring only when warranted.

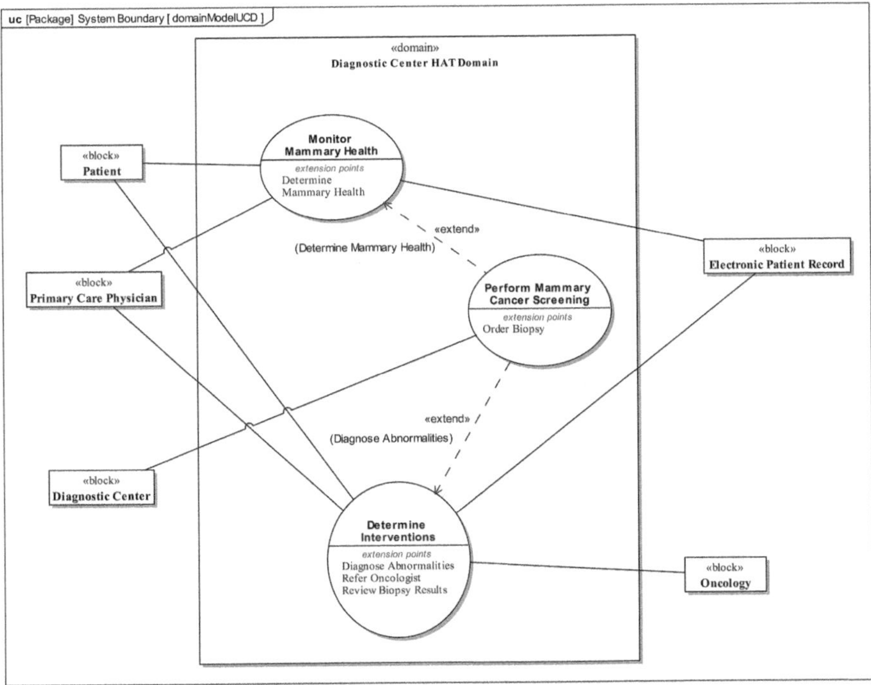

FIGURE 2.11 Use Case Diagram depicting Domain level use cases and corresponding actors.

Example Description: Our system's main function is to perform mammary cancer screening, which begins with mammography for initial and routine screenings. In our system, this begins with the lowest cost, simplest option of 2D mammography imaging. The radiology technician performs mammography on the patient using a 2D radiology machine. In cases where the analysis of the 2D mammogram indicates the need for additional imaging, the radiology technician will perform 3D mammography imaging, using a 3D radiology machine. Our "As-Is" system includes a Diagnostic Aid that analyzes the image to identify potential cancerous lesions, aiding the radiologist in detecting and identifying areas of concern within the mammogram. When abnormalities require further investigation, the diagnostic center HAT will perform a biopsy. The biopsy procedure involves the surgical nurse supporting the radiologist as they use a robotic system to take the cultures and an anesthesiologist to provide anesthesia for the procedure. If required, the surgical nurse will send the biopsy cultures to oncology for further examination. The patient and radiologist engage in all these activities, and the team provides feedback to the electronic patient record on the results of all activities, which alerts the primary care physician. The administration oversees the entire operation providing guidelines, policies, funding, staff support, and business goals.

Figure 2.12 shows the System level UCD. The diagram shows the diagnostic center human–AI team members as individual blocks on the left-hand side of the diagram. The blocks in Figure 2.12 are connected to the use cases that they individually support. The diagram also shows additional external actors that provide

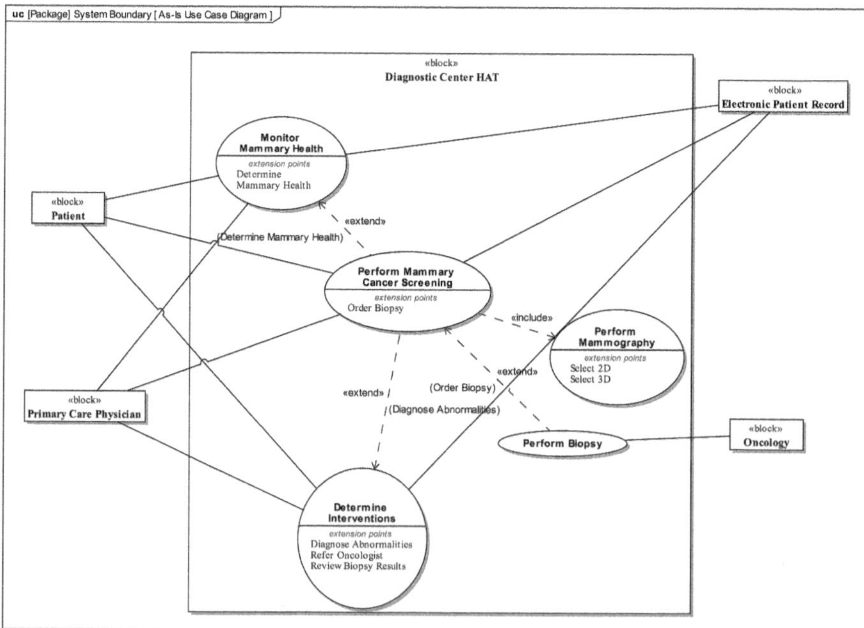

FIGURE 2.12 Use Case Diagram depicting System level use cases and corresponding actors.

goals to our system, including the *Patient,* and the *Administration*, on the left. The remaining actors appearing in the domain and system level BDDs appear on the right in Figure 2.12. Inside the UCD system boundary is the use case performed by our system, *Perform Mammary Cancer Screening.* The use case *Perform Mammary Cancer Screening* includes the use case *Perform Mammography.* The use case *Perform Biopsy* extends the use case *Perform Mammary Cancer Screening.* The diagram also shows that the use cases *Perform 2D Mammography* and *Perform 3D Mammography* extend the use case *Perform Mammography.*

2.8 TYPICAL AND ATYPICAL SCENARIO DEVELOPMENT

The AOSM method does not prescribe a specific cognitive task analysis methodology; however at a minimum, you should ensure that your analysis includes the evaluation of both Typical and Atypical Scenarios. The atypical scenarios should be those that stress the system and its actors. For our purposes, we are inspired by the Decision-Centered Design method described by Crandall, Klein, and Hoffman in *Working Minds: A Practitioner's Guide to Cognitive Task Analysis* (Crandall et al., 2006). Decision-Centered Design focuses on designing systems around the tough decisions that users face when executing work tasks, instead of designing around the normal workflows or around the latest technological advancement. By focusing on tough decisions, designers can minimize system brittleness and focus on the most critical portions of the cognitive work. Thus, as we conduct cognitive task analyses by interviews, observations, or mental walk-throughs, we focus on the key decision points within the work processes, especially those decisions that are difficult, complex, high-stake, or rare.

At this stage, the designer identifies the typical and atypical scenarios, documenting the scenarios in a list. Forming these scenarios supports meaningful data collection through cognitive task analysis. For each process identified in the UCDs, conduct a cognitive task analysis with the key people currently involved in accomplishing the tasks for both typical and atypical workflows. The Atypical scenario should be occasional in frequency but still likely to occur and should capture a situation where the "As-Is" system is known to be stressed. The designer should be sure to identify critical decisions that the people must make during task execution, key information the people need to make those decisions, and current team and system interactions.

For this example system, we might consider many circumstances that present challenges to the team. For example, initial mammograms that are poor in quality due to improper exposure, particularly dense tissue, or the inability to properly position the patient to obtain useful images. It might be difficult to compare current and previous mammograms due to changes in technology. Radiologists or technicians can be inexperienced. Calcifications exist in normal mammograms but can indicate cancer. Other individuals can disrupt people's timing by interrupting their work, placing pressure on the team to perform faster than normal. Equipment failures can occur. Each of these off-nominal or atypical conditions cause the team to adjust behavior. While it may not be possible for the design team to include all off-nominal conditions in their analysis, understanding a range of expected system variations can result in designs that are more robust to similar variation.

Currently, it is not necessary to diagram the workflows of the typical and atypical scenarios. The design team can decompose the workflows of the typical and atypical scenarios when modeling the "To-Be" system to ensure that the future system can appropriately manage stress points, critical interactions, and key decisions. While it is not necessary to create Activity Diagrams for these scenarios as this time, we provide Figure 2.13 for context and comprehension of the "typical" scenario for the Perform Mammary Cancer Screening Use Case. This figure begins from the patient's entry into the system for initial screening through to exiting the system either because results are normal or because the diagnosis results in transferring the patient to oncology.

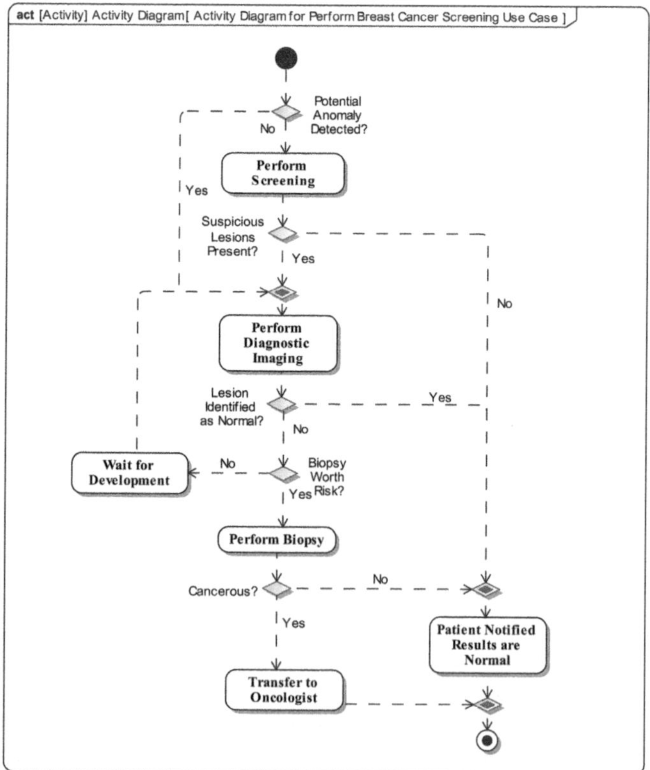

FIGURE 2.13 Activity Diagram depicting Domain level use cases and corresponding actors.

2.9 DESIGN OBJECTIVES

The final item we would like you as the designer to consider are design objectives. Based on the Stakeholder Narrative and the Typical/Atypical Scenarios, we might begin to envision potential design objectives for our example system. Certainly, as Jane proceeds through the process, we can imagine that her state of mind and stress levels are high during the multiple follow-up visits. Given the eventual negative determination, these stress levels were higher than warranted. It is certainly important that we keep the patient in this stressful state for as short a time as possible. Therefore, we might see the need to shorten the overall process time. We can also see the impact of decisions made throughout this process. If the radiologist had understood that there was no issue earlier in the process, this information would have reduced Jane's stress as she would not have had to proceed through this process. On the other hand, it is important that the radiologist identify cancer as early as possible as early detection is more likely to result in positive patient outcomes. Therefore, it is important to avoid both false positives (e.g., false alarms) and false negatives (e.g., misses). We might ask how to improve diagnosis to eliminate false negatives while maintaining or reducing the false positive rate. Either of these will help to reduce the stress on the patient. Interestingly, eliminating false positives also reduces the stress on the resources and staff within the system who must perform unnecessary diagnostic steps in response to these false positives. Finally, we might understand that, at least in countries with aging populations, the need for medical services is likely to outpace increases in staff, and therefore, reducing the required work or the number of staff required per patient served will have positive outcomes on the long-term sustainability of medical services. We can formulate design objectives based on these needs, and others, as we begin to reconsider redesigning this system.

False alarm rates during screening are approximately 15% in the United States; about 10% in England, Canada, and Australia; and less than 5% in Switzerland and the Netherlands (Mohd Norsuddin et al., 2015). While missing the presence of cancer can lead to loss of life, the financial burden due to false alarms in the United States exceed $4 billion per year for follow-up procedures when cancer is not present (Ong & Mandl, 2015).

2.10 CHAPTER SUMMARY

This chapter begins to implement the AOSM method by applying SysML modeling tools to capture the system boundary and system architecture of the "As-Is" system. It is important to capture the "As-Is" system so that we can begin to understand the richness of the cognitive work the teams perform in their current environment. Also important is the recognition that human–AI teams will form open and soft systems. Therefore, we need to constantly be thinking about and expanding upon the variability that the human–AI team will be exposed to from the domain in which the team operates and the variability that can arise internal to the system due to changes in the behavior, capacity, or state of individuals, AI agents, and system components within our system boundary.

Understanding the richness of the cognitive work occurs through conducting cognitive task analyses, enabling the development of a Stakeholder Narrative, which begins to capture the people and systems involved, the user's "pain points", the key decisions, and critical interactions between humans and AI agents. The development of Typical and Atypical Scenarios enables the systems modeler to understand typical workflows through the system and places where the system is stressed or is brittle under its current design. These places of potential system failure indicate potential focus areas for improvement in the "To-Be" system.

2.11 QUESTIONS AND DISCUSSION TOPICS

1. Which components of an AI agent does the meta-model discussed in this chapter explicitly capture? Why is it important to understand these elements of an AI agent early in the design cycle?
2. This chapter discusses the fact that this class of systems are open, soft systems. What do these terms imply about the class of systems to which HATs engage, and why does this fact make it important to define both the domain and the system during the initial stages of design?
3. Given the radiology example, discuss the reasons why you believe that the AI community has been unsuccessful in replacing the Radiologist or other members of this team.
4. This chapter suggests using both BDDs and UCDs when determining the system boundary; why is it necessary to consider both views of the system when defining the system boundary?
5. Why is it important to understand the atypical scenarios as discussed during the design of HATs?

2.12 PROJECT QUESTIONS AND ASSIGNMENTS

1. For your selected project, identify one to three stakeholders who provide goals for your system. Select one and develop a Stakeholder Narrative.
2. For your system, develop the Domain level Block Definition Diagram. Identify at least two sources of variability from items outside your system boundary, which have the potential to impact its performance.
3. For your "As-Is" system, develop the system level and HAT level Block Definition Diagrams. Identify at least two sources of variability from inside your system boundary, which have the potential to impact its performance.
4. For your system, conduct inquiry or observational cognitive task analyses. What are the tough decisions that the agents of your system face in executing their work?
5. For your system, develop the Domain level UCD. Which processes does your system perform or own? Which processes does your system support

but does not have primary responsibility for? Are there any processes that your system does not participate in currently, but it could participate in, in the future?

6. For your system, develop the System level UCD. Which actor(s) from the Domain level BDD provide design goals for your system?

REFERENCES

Crandall, B., Klein, G., & Hoffman, R. R. (2006). *Working minds: A practitioner's guide to cognitive task analysis*. MIT Press.

Hitchins, D. K. (1992). *Putting systems to work*. John Wiley & Sons.

Hollnagel, E. (2012). *FRAM: The functional resonance analysis method modelling complex socio-technical systems*. CRC Press.

Johnson, M., & Bradshaw, J. M. (2021). Chapter 16 – The role of interdependence in trust. In S. N. Chang & J. B. Lyons (Eds.), *Trust in human-robot interaction* (pp. 379–403). Elsevier Inc. https://doi.org/10.1016/b978-0-12-819472-0.00016-2

Johnson, M., Bradshaw, J. M., & Feltovich, P. J. (2018). Tomorrow's human–machine design tools: From levels of automation to interdependencies. *Journal of Cognitive Engineering and Decision Making*, *12*(1), 77–82. https://doi.org/10.1177/1555343417736462

Killingsworth, C. D., & Bohil, C. J. (2021). Breast tissue density influences tumor malignancy perception and decisions in mammography. *Journal of Applied Research in Memory and Cognition*, *10*(4), 642–655. https://doi.org/10.1016/j.jarmac.2021.07.005

Levy, J. (2015). *UX strategy: How to devise innovative digital products that people want* (1st ed.). O'Reilly Media, Inc.

Miller, M. E., McGuirl, J. M., Schneider, M. F., & Ford, T. C. (2020). Systems modeling language extension to support modeling of human-agent teams. *Systems Engineering*, *23*(5), 519–533. https://doi.org/10.1002/sys.21546

Mohd Norsuddin, N., Reed, W., Mello-Thoms, C., & Lewis, S. J. (2015). Understanding recall rates in screening mammography: A conceptual framework review of the literature. *Radiography*, *21*(4), 334–341. https://doi.org/10.1016/j.radi.2015.06.003

Ong, M.-S., & Mandl, K. D. (2015). National expenditure for false-positive mammograms and breast cancer overdiagnoses estimated at $4 billion a year. *Health Affairs*, *34*(4). https://doi.org/10.1377/hlthaff.2014.1087

Sterling, L. S., & Taveter, K. (2009). *The art of agent-oriented modeling*. MIT Press.

Zorpette, G. (2023). Just calm down about GPT-4 already. *IEEE Spectrum*. https://spectrum.ieee.org/gpt-4-calm-down

3 Goals and Responsibilities

3.1 UNDERSTANDING WORK IN THE "AS-IS" SYSTEM

Before we begin to redesign our system, it is important to understand the work that the current team conducts within the "As-Is" system. This and the subsequent chapter explain the process steps for understanding the "As-Is" system. Figure 3.1 shows these process steps. This figure shows the overall AOSM process on the left, with the second step highlighted. The right panel of Figure 3.1 shows the process steps discussed in this and the subsequent chapter. As shown, this step begins with developing a Goal Diagram, followed by Creating Concept Map(s), Building a Responsibility Diagram, and Characterizing the Team Type(s). We intend the goal and responsibility diagrams to decompose and provide a simplified depiction of the work while we intend the concept map to help to understand interactions and complexity of the work. Defining the team type or the types of subteams within our HAT will have a direct impact on our design as we go forward. Therefore, it is important to clearly understand and define the type of team or subteams within our HAT.

3.2 GOALS

Within the Use Case diagrams of the previous chapter, the primary actors pursue goals which drive the use cases shown in Chapter 2, in Figures 2.10 and 2.11 (Larman, 2004). ***Goals*** represent a desirable state that one or more agents are seeking to achieve. It is important that while goals express desired end states, it is not always possible for an agent to achieve their goals as the goals may be physically unrealizable or conflict with other goals, such that all goals cannot be achieved simultaneously. If you have project management experience, you may be familiar with the common example of desiring projects that are of high quality, delivered fast, and with low cost. The common reply to such a request is to pick two of these attributes as they compete, preventing the designer from achieving all three goals in a single project. This type of prioritization and trade-offs among competing goals is common. As such, the pursuit of goals often requires judgment and prioritization.

We are interested in goals since they describe the desired end state, such that the designer can specify what the team needs to achieve, without describing the process that an agent needs to complete to achieve the goals. Thus, during design of human–AI teams, this attribute of a goal provides us knowledge regarding what our human–AI team desires to achieve, without requiring us to quickly narrow to one solution for how the goal is to be achieved, providing a viable trade-space in which to explore diverse designs.

DOI: 10.1201/9781003428183-3

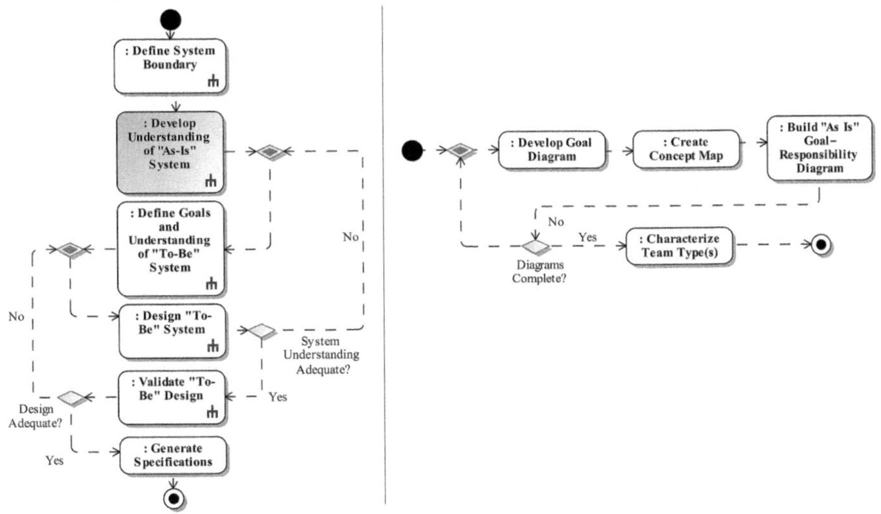

FIGURE 3.1 A depiction of the AOSM process on the left with the current sub-step highlighted. The right panel shows the sub-steps associated with Develop Understanding of "As-Is" System step.

While goals are not commonly captured or represented in typical software or system modeling languages, including UML or SysML, they are recognized in various human systems modeling frameworks (Elm et al., 2003; Geddes, 1989; Hollnagel, 2002), as well as in human–agent modeling frameworks (Deloach & García-Ojeda, 2010; Sterling & Taveter, 2009). A useful example is the *G*oals, *O*perators, *M*ethods, and *S*election rules (GOMS) framework. This framework models humans as goal-seeking entities who apply selection rules to select among various methods to achieve a desired goal (Card et al., 1983). This framework recognizes that an agent can utilize various methods to achieve a goal. Thus, the framework provides flexibility for the human to recall or improvise multiple methods to employ when pursuing their desired goal.

The literature recognizes that animals and other natural systems pursue goals through self-corrective *feedback*. Specifically, an entity selects an action which it predicts will achieve its goal, senses the result of the action, and uses this feedback to correct the predicted relationship between action and outcome before selecting the next action (Guttman, 1991). The application of feedback to adjust or steer actions through control and communication to achieve a selected goal is a central concept in *cybernetics* where control is viewed as a dynamic, iterative process in which a goal is selected, an entity decides upon an action after observing its environment, implements this action, observes the result of this action, compares the results of this action to the goal, and finally decides whether to continue with the current action or selects another action, which might be more effective in achieving the goal (Wiener, 1961). Figure 3.2 shows a typical action process to pursue an individual goal. As shown, this process involves goal setting, perceiving the state of the world, interpreting the

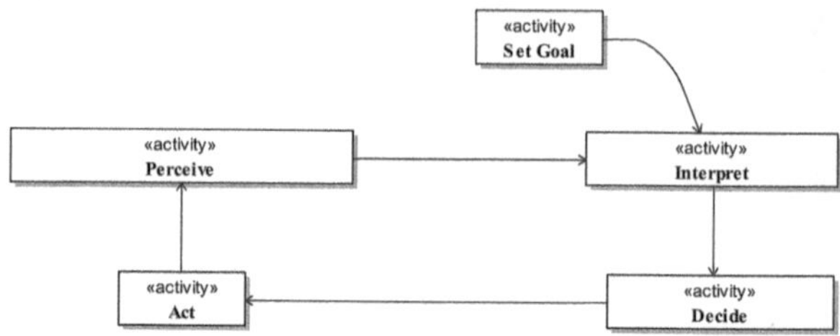

FIGURE 3.2 Typical action process required by an agent or team to achieve a desired goal.

state of the world by comparing it to the goal, deciding on a future action, taking the action, and repeating this cycle to achieve the desired goal.

A common example of this action process is that of steering a sailboat across a lake. We might have the goal of getting to a picnic site on the other side of the lake. Perceiving the water current, wind, and sailboat design, we might select sail and rudder placements. If we perceive over time that we are moving toward our desired location at a reasonable speed, then we may maintain the sail and rudder placement, but we will adjust if we are going off course or making inadequate headway to achieve our destination at a permissible time. These concepts of goal setting, perceiving, and monitoring progress; deciding upon actions; and implementing actions are important in human–AI teams. This process permits the agent or team to respond when variability in the environment or the actions of another intelligent agent in the environment creates circumstances where the selected action does not have the desired result. The sensors and actuators discussed in the previous chapters as components of intelligently designed AI agents permit the application of dynamic feedback processes, like the dynamic feedback processes in which humans engage.

Importantly, agents embed goals in higher-level goals or embed other goals within a goal. For example, we might want to sail to the picnic site, but this goal might support a higher-level goal of spending the afternoon at a picnic with a friend. If we cannot find the correct sail and rudder settings to move rapidly across the lake to the picnic site, we might decide to alter our landing location as it would be faster to land at another location and walk a while, which would permit us to achieve our higher-level goal of spending the afternoon at the picnic. This also illustrates that higher-level goals often require more time to complete and are more difficult to quantify, while lower-level goals persist over shorter periods of time and are more concrete (i.e., less abstract). However, monitoring progress toward higher-level goals may cause us to alter lower-level goals. As humans, we often consider feedback at multiple levels of abstraction to support decision-making.

Although we can discuss any individual agent pursuing their own goals, a defining attribute of a team is the fact that the team works together to pursue shared goals, which creates team cohesion (Filho et al., 2015). We will discuss the importance of shared goals and information within teams in more detail in the subsequent chapter. Teams of agents can share goals, particularly higher-level goals, across the entire team. Subsets of the team can also share goals, particularly mid-level goals. Finally, individuals within a team may pursue other goals in support of higher-level goals that are pursued by subsets of team members or the entire team.

Wiener and others recognized that cybernetic control is a cornerstone of intelligence (Wiener, 1961). As such, cybernetics can apply to biological, technical, or social systems. In the context of this book, we can consider individual humans, AI agents, and human–AI teams as cybernetic systems. As such, these systems pursue goals by selecting actions that elements or agents within the system believe are sufficient to achieve the goals but use feedback to adjust the selection of future actions. Consistent with embedded goals, we must recognize that the team will not pursue a single goal but instead will pursue higher-level goals and adjust lower-level goals as changes in the human, team, environment, or a competitor occur.

Because one can decompose goals in a manner similar to decomposing tasks, it is often difficult to distinguish goals from tasks. There are a few diagnostic qualities we can assess against to ensure that we are capturing goals and not tasks. First, a goal should be a desired end state rather than a means to an end state. Second, fulfilling a goal should have inherent value to the agent or team. Third, a goal should be achievable by multiple methods or means. Finally, goals *may* be an end state that an agent may not be able to accomplish, but the agent must be able to continually steer and progress toward the desired end state.

3.3 MODELING GOALS

The following text demonstrates one method for modeling goals. Specifically, we will demonstrate the use of a goal hierarchy, where we illustrate decomposing each higher-level goal into lower-level goals and each branch in the goal tree independent of one another. In this structure, no lower-level goal supports more than one higher-level goal. This is the most common structure for modeling goals within both the human–computer interface and AI literature. The human–computer interface literature acknowledges that a user must decide both whether the goal has been achieved and whether a change in action will assist in attaining the goal. Therefore, human decisions are associated with each goal, and the information necessary to support these decisions is important for representation within the user interface (Endsley & Jones, 2012; Humphrey & Adams, 2011). The AI literature also acknowledges that achieving lower-level goals are stepping stones toward achieving higher-level goals (Geddes, 1989; Rouse et al., 1987). Therefore, these frameworks often model goals in a hierarchical arrangement, where one or more decisions support each goal, and each decision is linked to multiple information requirements, which can be supported by multiple data elements, which is the case in the AOSM meta-model.

Other methods exist in the literature for modeling goals, including networked arrangements of Goals. For example, Functional Abstraction Networks (FANs) have been proposed that model goals as existing in a network, where a lower-level goal can support one or more higher-level goals (Elm et al., 2003; Potter et al., 2002). Therefore, FANs do not arrange the goals in a tree structure. Instead, FANs represent goals in a network. The literature in this area also acknowledges the need to understand information elements to understand goal attainment. This literature often terms information elements as commodities. FANs embed processes in each goal where the process converts raw data into these commodities (Duncan et al., 2023). We encourage the interested readers to explore this or other arrangements for modeling goals as an alternative method to the goal hierarchies we illustrate, if desired.

One other construct for modeling goals is useful to discuss. Thus far, we have discussed functional goals. Functional goals represent an end state that we wish to achieve, as we discussed earlier. Thus, we execute processes (i.e., functions) to achieve these goals, and we apply these goals to answer the question "What do we wish to achieve?" In their book on Agent-Oriented Design, Sterling and Taveter also discuss *Quality Goals*. Quality goals answer the question "How do we wish to achieve our goal?" (Sterling & Taveter, 2009). These authors suggest that quality goals include performance, safety, social, and security goals. In their view, it may be important to pursue a functional goal in a way to achieve maximum performance, for example, a robot loading boxes onto a palette may need to load boxes as fast as possible, so the user can select the quality goal of performance as the most important. Alternately, if this robot is working close to people, the safety of the people might be prioritized over performance, and so the designer can emphasize the safety of human teammates over the performance of the robot. The goal model can capture this priority. Further, you may capture multiple, competing quality goals in the model if you expect the members of the HAT to dynamically modulate the importance of these goals over time. For example, we might want the robot to prioritize safety when people are in its vicinity but prioritize performance otherwise. More recently, Schneider and colleagues have recognized that as we leave static environments, such as manufacturing environments, these quality goals can become dynamic and therefore become a source of variability in system operation (Schneider et al., 2023). For example, when operating an aircraft, there may be times where optimizing fuel efficiency is important to the pilot and other times where minimizing time to destination is important to the pilot. The fact that each of these quality goals can be important to the pilot can be captured in the model. This information permits the designer to consider the need for the system to pursue the functional goal of arriving at a destination under variations in the relative importance of these two quality goals during system design. Therefore, as we consider quality goals, we might be interested in the degree of quality associated with a quality goal (e.g., performance of the palletizing robot must be at least five boxes per minute), or we can consider the relative importance of quality goals at any moment (e.g., performance is twice as important for the palletizing robot when no people are within 50 feet of the robot, but safety is weighed as more important than performance when a person is within 50 feet of the palletizing robot). That is, each quality goal might have both a value and a relative importance as compared to other quality goals.

Capturing the goals of our system will enable us to prioritize system design in a manner that focuses on achieving the most important outcomes for the "As-Is" system. While we may decide to add or remove goals during the design of our "To-Be" system, analyzing, determining, and tracking goals for our "As-Is" system helps to ensure that the redesign does not inadvertently ignore important aspects of current work processes. As discussed in the AOSM meta-model section of Chapter 2, Roles fulfill goals, and Goals are achieved as agents accomplish assigned Responsibilities that contribute to a Role.

Roles are a logical collection or grouping of responsibilities that contribute to achieving a goal. Typically, roles are defined such that each role requires a distinct set of knowledge, skills, and abilities, such that humans who must play each role require similar training that is different than the training required for individuals who play a different role. Roles should not be confused with agents. As indicated in the meta-model, agents play a role by fulfilling the responsibilities associated with the goal. A single agent might play multiple roles; for example, the pilot in command of an aircraft might play both a role associated with controlling the aircraft and a role of team leader. Note that these two roles require different capabilities. Further, two agents can play the same role. For example, a copilot might also assume the role associated with controlling the aircraft but might not play the team leader role. Often, these same or similar roles will be useful in our "To-Be" system. Identification of roles in our "As-Is" system requires that we identify the responsibilities that our human–AI team must accomplish and then group them, based on the types of capabilities that are required to fulfill the responsibilities.

The goals captured in our goal hierarchy provide insight into these responsibilities, but, before we begin identifying them, we recommend constructing concept map(s) of the key outcomes the team is seeking to achieve. Combining the goal hierarchy and concept map products enables a deeper understanding of the responsibilities that do and should exist within our system of interest. The designer can create concept maps during discussions with representative HAT members from the "As-Is" system. You might consider either forming a different concept map with each category of human agents (e.g., Radiologist, Radiology Technician) or, if members work closely together, developing a single concept map with that portion of the team (e.g., Radiologist, Anesthesiologist, Surgical Nurse). However, once you develop concept maps for everyone, the concept maps can be synthesized into a single concept map. Once the goals are well understood, the designer should begin to identify responsibilities that each agent must fulfill to achieve each of the goals. We will demonstrate this through the use of a Goal–Role Diagram.

3.4 MODELING: GOAL HIERARCHIES IN A GOAL DIAGRAM—"AS-IS" SYSTEM

3.4.1 Modeling Objective

To model the goals of our "As-Is" system, the AOSM language constructs a ***goal hierarchy***, depicted within a Goal Diagram. This diagram organizes goals into broad high-level goals and then subdivides these goals into lower-level, more specific goals. The goal hierarchy enables the prioritization of goals, with higher-level goals trumping lower-level goals. Additionally, it enables us to identify where we have goal conflicts and need to make trades among goals that compete for system and environmental resources, including the cognitive resources available to our HAT. The goal hierarchy will enable us to identify, and assign, responsible agents for each of the goals. In the "As-Is" system, this may reveal that some goals have no agent that is currently responsible for achieving the goal, or it might indicate that the same agent is challenged to pursue numerous, possibly competing goals. Additionally, it can also demonstrate where multiple agents share goal responsibilities.

You may have noted the use of the words "Actor" as we discussed diagrams in Chapter 2 and the word "Agent" when discussing the Goal Diagram. In the previous chapter, we were discussing BDDs and UCDs, which are standard diagrams in the SysML language. This language uses the generic word "Actor" to refer to any entity that is acting. The word Actor may refer to people, artificial agents, other systems, organizations, or any other entity capable of executing an activity. In diagrams specific to AOSM such as the Goal Diagram discussed here, we are interested in designing for specific cognitive entities, and, therefore, we use the term agent to represent either AI agents or humans within all AOSM diagrams.

The goals in the goal hierarchy should be traceable to the system's primary agents and to the identified use cases. Based on the cognitive task analysis, and the modeling activities accomplished thus far, it may be useful to begin by brainstorming known goals in a free-form list before organizing and constructing a hierarchical diagram. The designer can also derive goals from interviews with the members of the "As-Is" HAT during a cognitive task analysis.

3.4.2 Diagram Components and Composition

Figure 3.3 illustrates a Goal Hierarchy containing generic elements. The diagram captures goals as blocks, applying a label that is a verb–noun phrase. As shown, the first layer of the hierarchy is the highest-level goal for our system of interest, *Achieve System of Interest Primary Goal.* The next layer presents goals for each of the primary actors that provide goals to the system, including the primary goal(s) of our human–AI team. Each actor presented on the left side of the System level UCD should have a goal represented at this level (e.g., *Achieve Goal for External Actor 1, Achieve Goal for External Actor 2, Achieve HAT Goal*). If not, then you should revisit the diagram to see if a goal is missing, or you may need to update to the System level UCD. For traceability, the designer should annotate the source of the goals by associating the goal with the actor in the diagram, as indicated by Actor 1 and Actor 2 within the diagram. The HAT can also be associated with a goal, if desired. Actors on the System level UCD, which do not have an associated goal, can indicate that you have missed a primary actor and you will need to capture their goals. Remember that the actors in this diagram include not only the members of the human–AI team but also any organization in which the HAT exists, as well as the external actors in the higher-level system (e.g., the patient and primary care physician in our example). These second-layer goals should then be subdivided into lower-level subgoals as shown. Agents who are responsible for these goals, Roles associated with

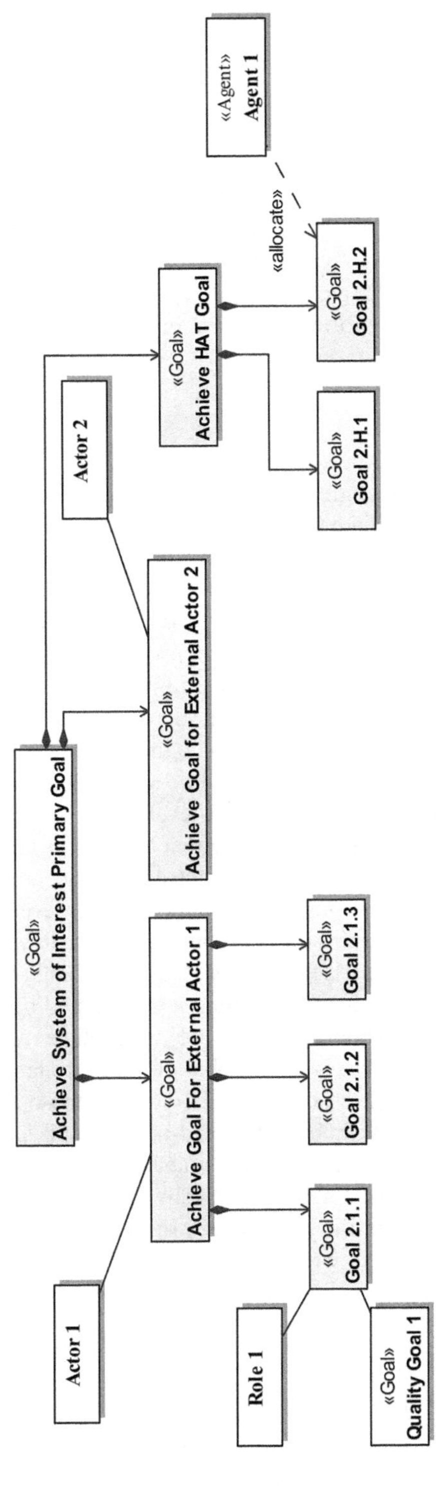

FIGURE 3.3 Generic goal hierarchy.

performing the goal, as well as quality goals, can also be associated with each goal within this diagram.

The AOSM extension to SysML includes custom diagrams. In this chapter, we introduce the Goal Diagram and the Goal–Responsibility Diagrams. Each of these diagrams are based on SysML Block Definition Diagrams. The method derives Goals, Roles, and Responsibilities from UML Activities, while the method derives Agents from SysML Blocks. Appendix A illustrates a profile for use in Cameo or Catia Magic System of Systems to create the implementation of these diagrams and entities, as shown in this text.

3.4.3 Considerations

How deep and detailed should the sub-goal division go? The designer should provide a diagram that is detailed enough for roles to emerge, particularly at a level that allows the designer to differentiate the roles that exist at the individual team member level of the "As-Is" system. That is, if your current team consists of three members, the goals should not be so high-level that they are all "team" responsibilities but rather should go deep enough so that they are (or could be) aligned to roles played by individual team members. However, stay focused on the goals that the members in the HAT are trying to achieve, and do not simply capture current roles and goals that exist in the system. Recall that with a goal focus, we may discover that there are some goals that exist for the team, but due to cognitive or other limitations, no one is currently performing. Similarly, there could be some roles in the "As-Is" system, which do not actually contribute to any higher-level goal.

One indication that the decomposition has gotten too deep is that the "goals" are not goals but rather tasks. Lower-level goals can contribute to accomplishing higher-level goals, but as you decompose these goals into lower and lower level goals, there is a tendency to begin to list tasks that contribute to accomplishing the lowest-level goals. Recall our guidance on differentiating goals from tasks: (1) A goal should be a desired end state rather than a means to an end state; (2) fulfilling a goal should have an inherent value to the agent or team; (3) a goal should be achievable by multiple methods or means; and (4) goals *may* be an end state that is not capable of being declared accomplished but rather must be continually progressed toward. Looking at those first two criteria, ask yourself, is this potential goal a desirable end in-and-of itself, or is it just means to another end? If it is a means, then it is a task. If it is a desirable state, then it is a goal.

For example, in the original Goal Hierarchy brainstorm for the Diagnostic Center Goal Hierarchy, we identified "Optimize Image Quality" as a potential goal. However, upon further reflection, we realized that quality images are not a desirable outcome on their own, but rather they are a means to achieving the goal of "Provide Correct Diagnosis".

3.4.4 Example Application

Example Description: Numerous entities provide goals to our system. The patient is looking to maintain an elevated level of health, which not only includes early detections and interventions but must also be balanced against their other needs, such as minimizing their out-of-pocket expenses, minimizing pain and discomfort, avoiding unnecessary and invasive procedures, and supporting their emotional health. The administration needs to ensure the longevity of the organization and thus is concerned with numerous business outcomes: Profit (revenues and costs), quality (i.e., system performance), legal liability, and reputation. Our human–AI team desires to be successful with high-performance outcomes, while also attending the needs of the individual team members and the cohesion of the team.

Figure 3.4 shows the Goal Hierarchy for our Diagnostic Center System. The first level is the overarching goal of the system *Optimize Diagnostic System.* This goal is decomposed into three lower-level goals: *Maximize Team Performance*, representing the human–AI team's goal for our system; *Achieve Optimal Patient Outcome,* representing the patient's goal for our system; and *Maximize Organizational Longevity*, representing the administration's goal for our system. Each of these actor goals are further subdivided into lower-level goals that contribute to the higher-level goal. For example, *Achieve Optimal Patient Outcome* involves both *Achieve Optimal Patient Health* and *Maximize Patient Comfort.* Some of the goals may compete with each other within a section of the hierarchy; for example, *Maximize Organizational Longevity* comprises both *Maximizing Reputation* (which includes costly activities) and *Maximizing Profitability* (which would seek to keep costs low and revenues high). Similarly, goals from different entities can also compete, such as the HAT's desire to *Detect All Anomalies* and the patient's desire to *Minimize Patient Financial Cost.*

3.5 MODELING: CONCEPT MAPS

3.5.1 Modeling Objective

A *Concept Map* is a diagram that shows the relationships between concepts, usually items, ideas, or information. For our context, the concept map displays the key activities, actors, and outcomes that fulfill a particular high-level goal. The connections between these concepts are "linking phrases" that describe the relationship between the concepts. Overall, the Concept Map(s) identifies: (1) The cognitive work required to achieve the key outcomes, (2) the outcomes that the system needs to achieve, and (3) the interactions and overlaps that occur between concept map entities.

You should focus the Concept Map(s) on the fulfillment of a higher-level goal and inform the content of the Concept Map with the cognitive task analysis. It is helpful to create the Concept Map through collaboration with member(s) of the current human–AI team, or at least to have member(s) review and provide feedback on the concept map(s).

> The systems engineering literature often employs a similar diagram to the Concept Map, which is referred to as a Systemigram. Systems engineers often use these diagrams to understand user needs in sociotechnical systems (Czachorowski et al., 2023)

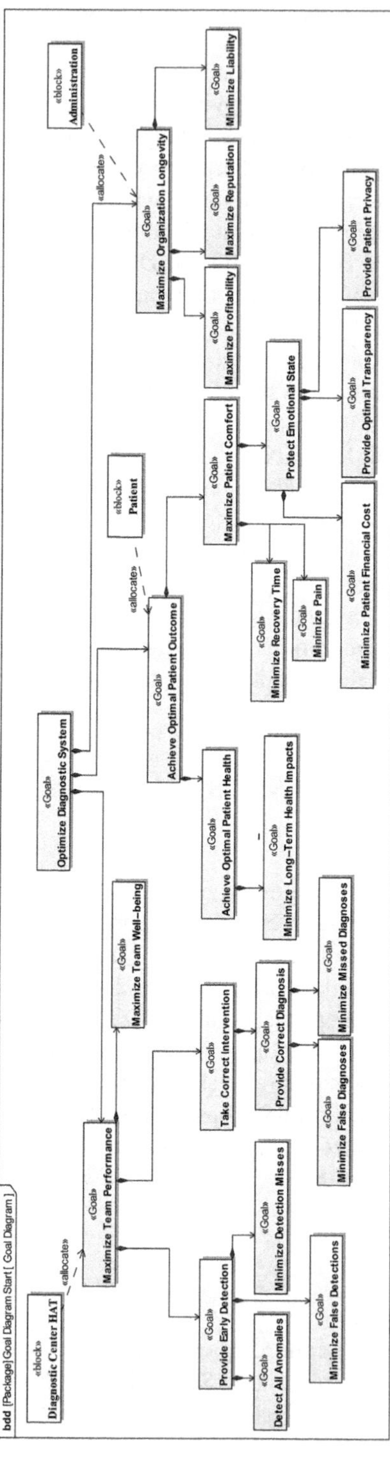

FIGURE 3.4 Goal Diagram for the Diagnostic Center System.

3.5.2 Diagram Components and Composition

Figure 3.5 illustrates a Concept Map containing generic elements. As shown, the Concept Map begins with a *Framing Question* on how to achieve one of the goals provided by a primary actor to the system (typically second layer of the Goal Hierarchy). Developing the Concept Map then proceeds through brainstorming the activities and intermediate products (i.e., concepts), which needs to occur to achieve this outcome, along with the *Relationships* between the *Concepts*. Many of these *Concepts* necessitate, include, or are supported by other *Concepts*. This diagram documents concepts as circular or rectangular blocks with labels that are either nouns or verb–noun phrases. The diagram captures all relevant relationships with connectors and the relevant "linking phrase". The diagram captures connectors as lines with directional arrows, with the linking phrase being consistent with the directional arrow. Linking phrases are typically verbs or verb-phrases. As shown, each concept may be related to other related concepts in a hierarchical fashion (indicated by the straight-line relationships in Figure 3.4). However, the diagram can also include interrelationships between related clusters of concepts, as indicated by the two curved lines in Figure 3.4. Although one can build Concept Maps within SysML tools, other tools are available (Institute for Human and Machine Cognition, 2023).

3.5.3 Considerations

Some of the concepts in the Concept Map may overlap with goals from the goal hierarchy, but the Concept Map should also include activities and the actors involved and thus should display relationships and interactions not captured in the goal hierarchy. To ensure completeness, it is helpful to refer to the portion of the goal hierarchy that corresponds to the Concept Map question to verify that you have captured all subgoals. It is also helpful to refer to the BDDs to ensure that you have captured all actors.

While Concept Maps are often organized in hierarchies for effective visualization, there will not be a direct mapping to the Goal Hierarchy, and there will likely be numerous links across the diagram. A robust concept map will resemble a web or network, rather than a hierarchy.

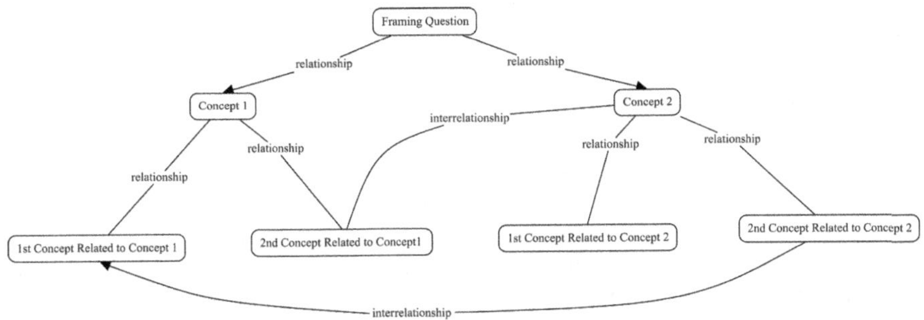

FIGURE 3.5 A generic concept map.

For a simple system, it may be possible to create a Concept Map for the system level goal of the goal hierarchy. For more complex systems, concept maps for second-level, or even lower, goals will be necessary. The example below provides a concept map for one second-level goal in the Diagnostic Center Goal Hierarchy. Due to the interconnectivity between this system's goals, this concept map includes many of the activities and outcomes that would support the other second-level goals.

3.5.4 Example Application

Example Description: The human–AI team's primary goal is to maximize team performance. They achieve this through an effective conduct of their core activities (performing early screening examinations, making an accurate diagnosis, and determining effective interventions), as well as through effective communication with each other and other system stakeholders, ensuring a positive customer experience, and prioritizing high-risk patients. Performing early examinations enables the team to identify (and resolve) anomalies early in their development, potentially increasing survival rates and enabling less invasive interventions. To conduct early examinations, the barriers to entry must be low, including the need to minimize the time, cost, invasiveness, and pain/discomfort caused by the examination. Making an accurate diagnosis includes detecting and accurately characterizing anomalies, while avoiding false alarms and misses. Quality radiology images and biopsy samples support these activities. Obtaining quality images is impacted by ensuring the patient is correctly positioned to get the correct angles, selecting the correct exposure for the image based on tissue density and other factors, the location of anomaly (if present), and the ability to evaluate if the image taken is suitable for anomaly detection and characterization. Similarly, obtaining quality biopsy samples involves numerous factors including the location of the sample, size of the sample, effective anesthesia administration, and correctly positioning the needle to take the sample. The diagnosis, combined with identifying potential interventions and understanding the relative impacts of these interventions, supports determining an effective intervention.

Effective communication between the human–AI team members enhances all of these activities. Accurate and timely documentation of findings in the electronic patient record further supports communication across the HAT. Clear communication with the team enables effective coordination of team activities. This communication also enhances the team members' knowledge, skills, and abilities—strengthening the individual contributions of each team member to the team. Many aspects of this communication center on protected health information, and thus the communication needs to occur in a manner that maintains the patient's privacy.

Whether or not the patient judges the team to be highly performing also depends upon their perceptions, and it is important that the team creates a positive customer experience. The timeliness of the team's activities influences the customer experience, as well as the ability of the team to manage expectations about the process and its outcomes, the attentiveness of the team members to the patient, and the ability of the team to create an enjoyable interaction, along with the environmental aesthetics.

All teams operate under limited resources (time, personnel, equipment, budget); thus, the ability to prioritize high-risk patients will enable the team to focus attention on those patients who need it most. This can influence: the timing and frequency of screening examinations, the identification of anomalies (including avoiding misses), and determine the most effective intervention. Accurately identifying these patients relies upon the communication between the actors in the system level BDD, as well as the medical history documented in the electronic patients record.

Figure 3.6 shows the Concept Map for the *Maximize Team Performance* goal of the Goal Hierarchy. The Concept Map begins with the question: "How does the team maximize performance?" The concept map provides numerous activities, actors, and outcomes. In this concept map, the primary activities identified are *Communication, Prioritizing High-Risk Patients, Perform Early Examinations, Make Accurate Diagnosis,* and *Determining Effective Interventions.* The actors include *HAT Members, PCM, Patient,* and *Oncologist.* Outcomes include *Enjoyable Interaction, Positive Customer Experience, Maintaining Privacy, Effective Coordination,* and *Maintain Low Barriers.* These outcomes identified in the Concept Map are suggestive of a goal for the system. The Goal Hierarchy Diagram may already capture these goals, such as *Maintaining Privacy.* Alternately, the outcomes may be indicators of quality goals (performance, safety, social, and security goals) such as *Enjoyable Interaction, Positive Customer Experience, Effective Coordination,* and *Maintain Low Barriers.* During the cognitive task analysis, the analyst or designer should identify the measures, if any, the team currently uses when assessing these outcomes to establish measurable quality goals. For example, the designer may measure *Positive Customer Experience* through a post-visit customer satisfaction survey, with a quality goal of achieving at least 90% positive ratings.

3.6 MODELING: GOAL–RESPONSIBILITY DIAGRAMS—"AS-IS" SYSTEM

3.6.1 Modeling Objective

Recall that team members achieve Goals by fulfilling Roles, which are logical groupings of Responsibilities. While Roles already exist in the "As-Is" system, we want to ensure that these Roles are deliberately defined in the "To-Be" system, based on an understanding of capabilities, teaming relationships (including coordination and communication), and interactions/interdependencies between Responsibilities. Clearly defining the Responsibilities required to accomplish the Goals of our system will inform the options and trade-space for grouping the Responsibilities into Roles. The AOSM Goal–Responsibility Diagram maps Goals to Responsibilities.

At this stage, we construct a Goal–Responsibility Diagram for our "As-Is" system. We will later update this diagram to reflect additional Goals and Responsibilities for our "To-Be" system. The objective is to identify the key Responsibilities that must be conducted to achieve the goals. These Responsibilities are actions and should be functionally oriented, without any presumed temporal sequence or implementation in mind.

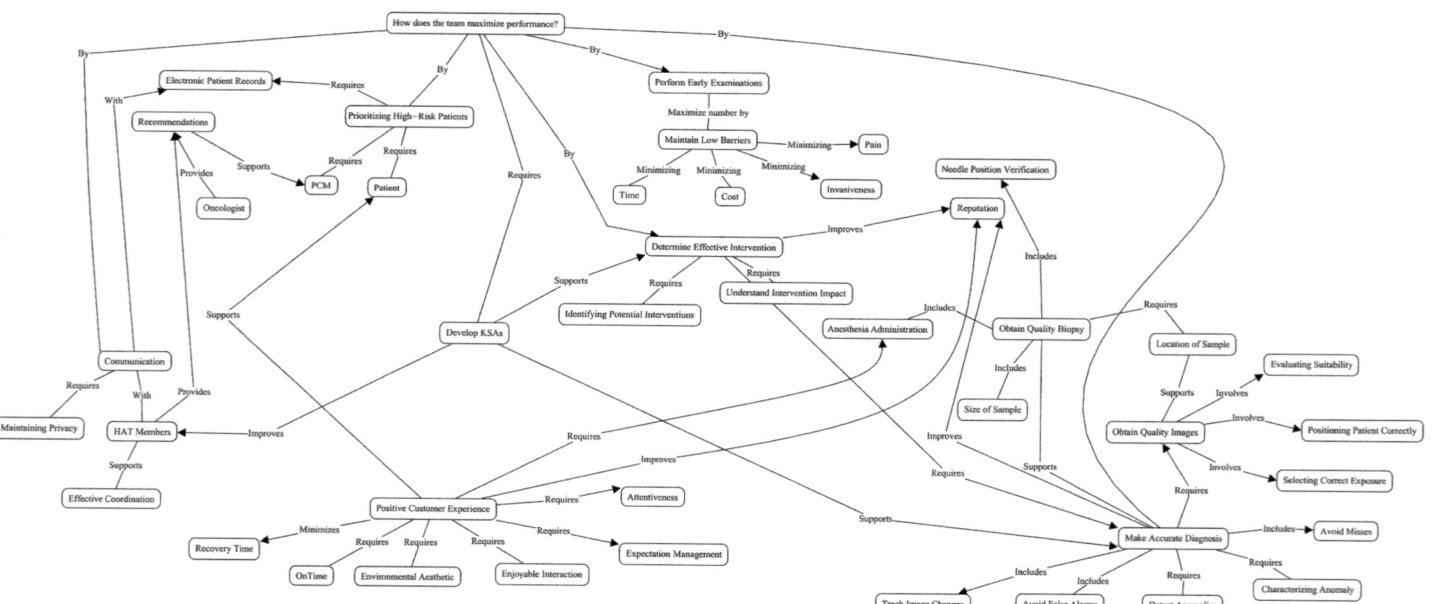

FIGURE 3.6 Concept map for "How Does the Team Maximize Performance?".

3.6.2 Diagram Components and Composition

Figure 3.7 shows a Goal–Responsibility Diagram containing generic elements. To prevent the diagram from becoming too cumbersome and unwieldy, we recommend constructing separate diagrams for each of the second-level goals in the Goal Hierarchy. Figure 3.7 contains the portions of the Goal Hierarchy in Figure 3.2 associated with External Actor 1: *Achieve Goal for External Actor 1*, *Goal 2.1.1*, *Goal 2.1.2*, and *Goal 2.1.3*. Each of these Goals are associated with the requisite Responsibilities to achieve that goal. For example, *Goal 2.1.1* is associated with *Responsibility 1*, *Responsibility 2*, and *Responsibility 3*. In addition, to responsibilities for each of the lowest-level goals in the hierarchy, there may be responsibilities associated with just the higher-level goals, such as *Responsibility 9* and *Responsibility 10*. Additionally, there may be responsibilities, such as *Responsibility 6*, which are associated with multiple goals, and thus these responsibilities should be associated with both goals.

3.6.3 Considerations

As mentioned, we recommend constructing separate diagrams for each of the second-level goals in the Goal Hierarchy. Begin by identifying responsibilities for each of the lowest-level goals in the hierarchy tree and then evaluate the higher-level goals to identify any responsibilities not fully covered by those identified for the lower levels.

To be comprehensive, it may be useful to consider the steps outlined in the Action Process in Figure 3.1: Perceive, Interpret, Decide, and Act. For each goal, consider what functional activities need to occur with respect to Perception, Interpretation,

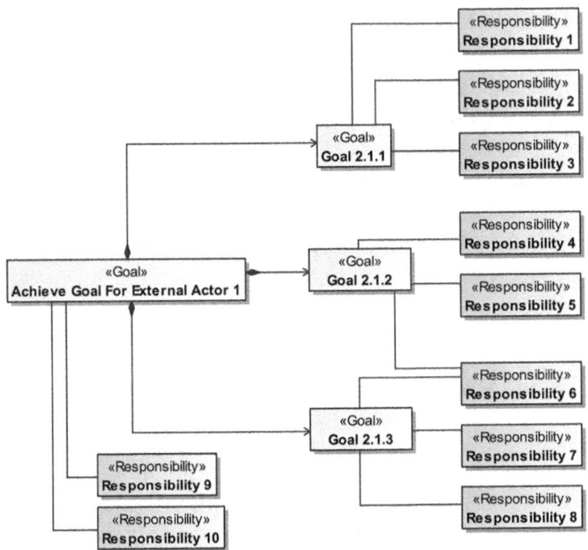

FIGURE 3.7 A generic Goal–Responsibility Diagram.

Decision, and Action. Some goals may have all four of these stages, whereas other goals may have only several, as the goal itself may represent just one or two of these stages in achieving a higher-level goal. At the highest level, a goal will include all of these, as this is the Action Process to achieve the goal. Additionally, a single goal may have multiple responsibilities of the same stage; for example, there may be multiple Perception responsibilities associated with a goal. It may also be helpful to refer to the Concept Map to identify activities that indicate responsibilities. Finally, discussions with end users who participate in your "As-Is" system team may lead you to see other responsibilities that members have within the team. However, as you transform these discussions to responsibilities, be careful to capture these responsibilities such that they are not specific to the current system.

3.6.4 Example Application

Figure 3.8 shows an example Goal–Responsibility Diagram for the *Maximize Team Performance* goal of the Goal Hierarchy, including the lower-level goals *Maximize Team Well-Being*, *Provide Early Detection*, *Take Corrective Action*, and the subgoals associated with each of these. Using the Perception, Interpretation, Decision, and Action framework, we identified responsibilities for each of the lowest-level branches of the goal tree. For example, *Provide Early Detection* contains the subgoals *Detect All Anomalies*, *Minimize Detection Misses*, and *Minimize False Detections*. The value in early detection is to identify tissue anomalies early enough in the cancer-growth process to be able to provide minimally invasive treatments that enable full recovery, with the lowest financial and emotional repercussions, and with overall patient health improvement. This requires detection before the patient, or their primary care physician, even knows that there is a problem. Detecting all anomalies provides Perception responsibilities to *Obtain Quality Images*, *Obtain Quality Biopsy*, *Obtain Complete Mammograph History*, and *Obtain Patient Health History*. Detection of anomalies also requires Interpretation responsibility to *Differentiate Normal from Abnormal* through the comparison of the images to previous mammography images, analysis of biopsy samples, and factoring in statistical likelihoods based on patient health history. The Decision and Action stages of the *Provide Early Detection* goal occur at its highest level, with the additional responsibility to *Recommend Further Diagnostics* if the data are inconclusive or *Alert to Anomaly* if the Radiologist detects an anomaly. Using this same process, we identified initial responsibilities for the remaining goals in the diagram.

3.7 CHAPTER SUMMARY

The AOSM method requires the designer to identify Goals to describe the desired end state for their system of interest without describing how the system or how team members within the system achieve these goals. This provides the design team knowledge regarding what the human–AI team desires to achieve, without requiring the designers to quickly narrow to one solution for how the HAT achieves the goal. This process provides a viable trade-space in which to explore assorted designs. There

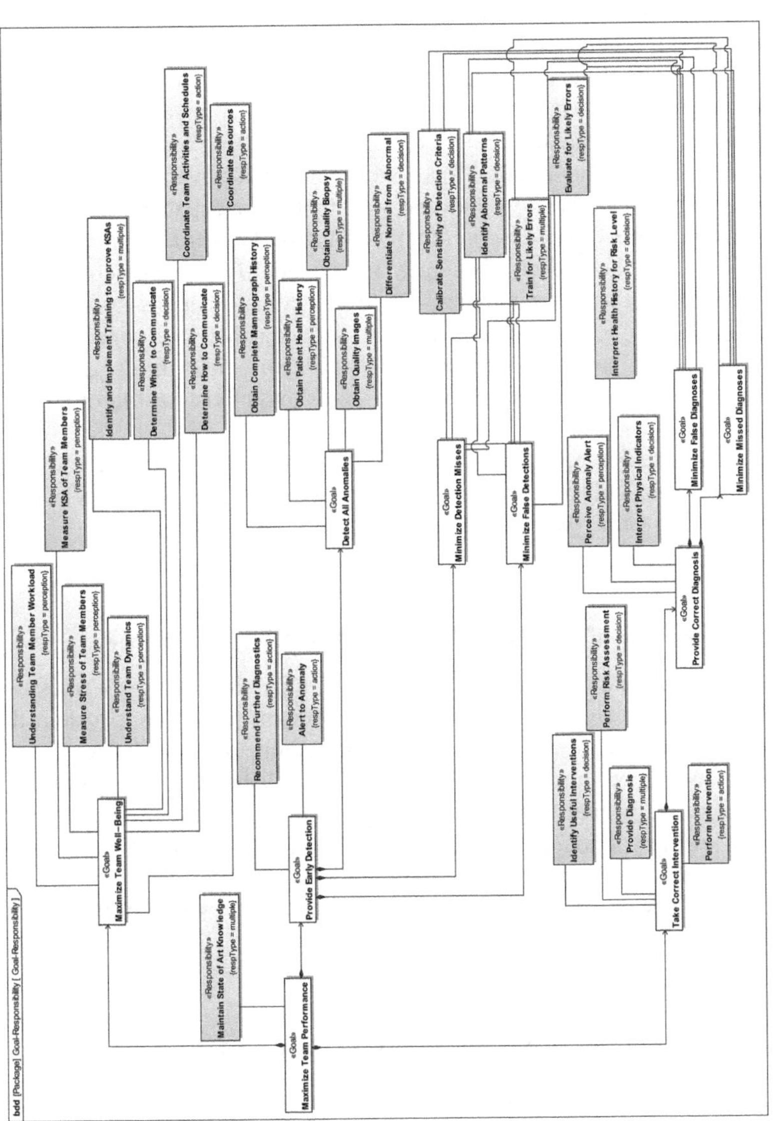

FIGURE 3.8 Goal–Responsibility Diagram for Maximize Team Performance.

are numerous methods and frameworks for capturing goals. AOSM uses the Goal Hierarchy, where the designer decomposes each higher-level goal into lower-level goals and designs each branch in the goal tree to be independent of one another. It enables us to identify where the HAT must address goal conflicts and the trades they need make, as goals may compete for system and environmental resources, including the cognitive resources available to our HAT.

When capturing goals, especially lower levels of the Goal Hierarchy, it can be common to accidentally capture tasks. To aid the designer in distinguishing tasks from goals, we provide four potential differentiators: (1) A goal should be a desired end state rather than a means to an end state, (2) fulfilling a goal should have an inherent value to the agent or team, (3) a goal should be achievable by multiple methods or means, and (4) goals *may* be an end state that is not capable of being declared accomplished but rather must be continually progressed toward.

We then discuss the use of Concept Maps to identify: (1) The cognitive work required to achieve the key outcomes, (2) the outcomes that the system needs to achieve, and (3) the interactions and overlaps that occur between concept map entities. Combining the Goal Hierarchy and Concept Map provides the designer a deeper understanding of the responsibilities that do and should exist within our system of interest.

Finally, using the Perception, Interpretation, Decision, and Action framework, we identify agent responsibilities, for each of the goals in the Goal Hierarchy of the "As-Is" system through Goal–Responsibility Diagrams.

3.8 QUESTIONS AND DISCUSSION TOPICS

1. This chapter discusses the vital role of feedback in achieving goals, as well as the concept of cybernetics for goal attainment. Describe a system that relies upon feedback and dynamic control systems to pursue its goal.
2. Select a common household task (e.g., washing dishes, cooking dinner) and identify a list of goals. Organize these goals into a goal hierarchy. Identify any tasks that you identified but did not include as goals due to not meeting the requirements for goals.
3. Why is it important to capture system Goals and Responsibilities before identifying system Roles?

3.9 PROJECT QUESTIONS AND ASSIGNMENTS

1. Construct an initial Goal Diagram for your system.
2. For your system, identify at least three quality goals. Which goal(s) are they associated with, and what is the standard of performance (i.e., the degree of quality)? Does the relative importance of any of these quality goals change according to the system states, external environment, or other contexts?
3. For one of the primary goals of your system, identify a framing question and construct a Concept Map. What concepts appear in the Concept Map,

which you have not identified in the Goal Diagram? Which concepts have the most interactions or overlapping relationships? Based on your cognitive task analysis, what are the overlaps, “pain points”, or “bottlenecks” of your system?

4. For your system, build out a Goal–Responsibility Diagram for the primary goal provided by the human–AI team. Indicate which responsibilities are Perception, Interpretation, Decision, and Action.

REFERENCES

Card, S. K., Moran, T. P., & Newell, A. (1983). *The psychology of human-computer interaction*. Lawrence Erlbaum Associates, Inc.

Czachorowski, K. V., Haskins, C., & Mansouri, M. (2023). Minding the gap between the front and back offices: A systemic analysis of the offshore oil and gas upstream supply chain for framing digital transformation. *Systems Engineering*, *26*(3), 241–256. https://doi.org/10.1002/sys.21652

Deloach, S. A., & García-Ojeda, J. C. (2010). O-MaSE: A customisable approach to designing and building complex, adaptive multi-agent systems. *International Journal of Agent-Oriented Software Engineering*, *4*(3), 244–280. http://citeseerx.ist.psu.edu/viewdoc/download?doi=10.1.1.460.7287&rep=rep1&type=pdf

Duncan, M. C., Miller, M. E., & Borghetti, B. J. (2023). Analysis and requirement generation for defense intelligence search: Addressing data overload through human—AI agent system design for ambient awareness. *MDPI Systems*, *11*, 561. https://doi.org/10.3390/systems11120561

Elm, W. C., Potter, S. S., Gualtieri, J., Easter, J., & Roth, E. M. (2003). Applied cognitive work analysis: A pragmatic methodology for designing revolutionary cognitive affordances. In E. Hollnagel (Ed.), *Handbook of cognitive task design* (pp. 357–382). CRC Press.

Endsley, M. R., & Jones, D. G. (2012). *Designing for situation awareness: An approach to user-centered design*. CRC Press, Inc.

Filho, E., Tenenbaum, G., & Yang, Y. (2015). Cohesion, team mental models, and collective efficacy: Towards an integrated framework of team dynamics in sport. *Journal of Sports Sciences*, *33*(6), 641–653. https://doi.org/10.1080/02640414.2014.957714

Geddes, N. D. (1989). *Understanding human operators' intentions in complex systems*. Georgia Institute of Technology.

Guttman, H. A. (1991). Systems theory, cybernetics, and epistemology. In A. S. Gurman & D. P. Kniskern (Eds.), *Handbook of family therapy* (pp 41–62). Routledge/Taylor & Francis Group.

Hollnagel, Erik. (2002). Cognition as control: A pragmatic approach to the modelling of joint cognitive systems. *IEEE Journal of Systems, Man and Cybernetics*, *10*, 1–25.

Humphrey, C. M., & Adams, J. A. (2011). Analysis of complex team-based systems: Augmentations to goal-directed task analysis and cognitive work analysis. *Theoretical Issues in Ergonomics Science*, *12*(2), 149–175. https://doi.org/10.1080/14639221003602473

Institute for Human and Machine Cognition. (2023). *Cmap*. https://cmap.ihmc.us/

Larman, C. (2004). *Applying UML and patterns: An introduction to object-oriented analysis and design and iterative development* (3rd ed.). Prentice Hall.

Potter, S. S., Elm, W. C., Roth, E. M., Gualtieri, J., & Easter, J. (2002). Bridging the gap between cognitive analysis and cognitive engineering. In M. D. McNeese & M. A. Vidulich (Eds.), *State of the art report (SOAR): Cognitive systems engineering in military aviation environments: Avoiding cogminutia fragmentosa!* (pp. 137–168). papers2://publication/uuid/065A545D-3064–4BA3-BE50-FA90D68DECCE

Rouse, W. B., Geddes, N. D., & Curry, R. E. (1987). An architecture for intelligent Interfaces: Outline of an approach to supporting operators of complex systems. In *Human-computer interaction* (pp. 87–122). Lawrence Erlbaum Associates, Inc.

Schneider, M. F., Miller, M. E., & McGuirl, J. M. (2023). Assessing quality goal rankings as a method for communicating operator intent. *Journal of Cognitive Engineering and Decision Making*, *17*(1), 26–48. https://doi.org/10.1177/15553434221131665

Sterling, L. S., & Taveter, K. (2009). *The art of agent-oriented modeling*. MIT Press.

Wiener, N. (1961). *Cybernetics: Or control and communication in the animal and the machine* (2nd ed.). MIT Press.

4 What Is Teaming?

4.1 WHY FOCUS ON HUMAN TEAMING?

Before we discuss designing and modeling human–AI teams, it is useful to examine knowledge of human–human teams. The design philosophy discussed in Chapter 1 includes examining the existing or "As-Is" systems to inspire the application of automation. Some might view the word "system" to represent an engineered system, such as an aircraft or an automobile. This text applies a broader definition of system to include sociotechnical systems. Sociotechnical systems attempt to understand the integration of humans with hardware and software as systems, instead of reductionist approaches that consider the design of the hardware and software components separately from the human components of a system (Behymer & Flach, 2016).

Applying this design philosophy, it is important to understand principles from human–human teaming to understand how human–agent team design might influence human and team performance. This philosophy may be particularly relevant when humans may not have experienced interaction with AI agents in systems and expect AI agents to respond like human teammates.

4.2 WHY DO WE TEAM?

First, we might ask, what is a team? As mentioned in Chapter 1, a *team* can be defined as two or more agents who train and work together interdependently, with diversified knowledge, skills, and abilities to accomplish work that is complex or critical enough that it requires greater knowledge, skill, ability, capacity, or redundancy than a single agent can provide in the available time frame (Cooke et al., 2013; Salas et al., 1992). This single sentence illustrates a team's ability to overcome three separate constraints. These constraints included limited time, limited knowledge, and error tolerance. For reference, Figure 4.1 shows the structure of a human team. As shown, the team consists of two or more humans where each human has unique knowledge, skills, abilities, and capacity. Each human additionally forms a mental model of the world. Also shown are team mental models and transactive memory systems, which we will discuss later in this chapter.

An example of a time constraint occurs in a typical *Remotely Piloted Aircraft* (RPA) where a pilot controls the aircraft while the sensor operator controls the sensor. The pilot must manipulate separate controls in real time as they cannot stop the RPA in midflight to permit the pilot to pan, zoom, and otherwise control the sensor. A single individual, no matter their knowledge, skills, or abilities, does not have the capacity to visually monitor and physically track both the aircraft path and the camera controls simultaneously. However, two individuals can synchronize their actions,

DOI: 10.1201/9781003428183-4

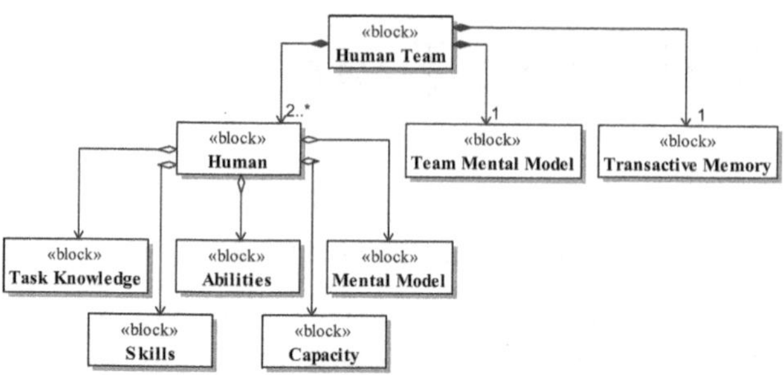

FIGURE 4.1 A block definition diagram showing relevant the components of a human–human team and each human in the team.

to provide real-time control of both the aircraft and the sensors. Thus, this team has more capacity than a single individual, permitting it to overcome the time constraint.

A typical hardware design team can overcome the limitations in knowledge of individual team members to achieve complex designs. An example hardware design team might include a mechanical engineer, an electrical engineer, and a program manager. The mechanical engineer has specialized knowledge regarding the strength of materials, physical configuration on critical forces within the final assembly, and manufacturing or assembly techniques relevant to building the structural part of the product. The electrical engineer understands circuit and power design necessary to control actuators within the product. The project manager has specialized skills in organizing the project, facilitating communication among stakeholders, and holding the engineering team accountable for completing the project on time and on budget. Importantly, each of these team members have specialized knowledge which permits them to understand certain aspects of the project and to integrate this knowledge to develop a quality product. Further, training a single individual to accomplish all these activities would require extensive time and effort. By dividing the knowledge across individuals, they synchronize their knowledge to provide a better understanding than any one of them could obtain on their own, while making training of individuals who are experts in each specialized area manageable.

Teaming to reduce error rate is best illustrated through commercial aviation where both a Captain and a First Officer are always in the cockpit. In this case, should one of them become incapacitated or make a perceptual or cognitive error that could endanger the aircraft, the second individual is available to help recognize and overcome the potential issue. Are both individuals necessary in the cockpit? They have similar knowledge, and, under normal circumstances either one of them could complete all tasks necessary to fly the aircraft. However, redundancy increases system *resilience* to error.

It is also important that while each of these reasons can motivate the use of teams, it is common for these human–human teams, once they are formed, to address multiple concerns. For example, while the First Officer is in the cockpit primarily to overcome errors, during times of high workload, this individual will perform activities synchronously with the captain to reduce the captain's workload. Therefore, in addition to providing system resilience to error, this individual also helps to overcome time constraints. This adaptive behavior of the team provides additional resilience within the system.

Forming a team with a sufficient number and diversity of teammates provides adequate cognitive and physical *capacity* to overcome the complexity and criticality challenges in complex, time, or safety-critical environments. The team members not only bring diversified knowledge, skills, and abilities to the team but also train to work together interdependently (Delise et al., 2010). The result is a team comprising individuals with varied perspectives on the task, which are encoded in each operator's *mental models* of the task at hand, as well as their mental models of their teammate's ability to respond to task demands (van den Bossche et al., 2011).

Just as with human–human teams, human–AI agent teams (HATs) also can overcome the constraints of limited time, limited knowledge, and error tolerance. HATs also have the potential to overcome cognitive capacity limitations of human teammates, providing skills and abilities beyond human capacity. The humans' mental models of each other and the AI agent teammates shape the effectiveness of the HAT. Prior experience with comparable human–human teams will influence an individual's mental model and their expectations of both their human and AI agent teammates.

4.3 TYPES OF TEAMS

We can characterize teams along multiple dimensions. For example, Sundstrom and colleagues discuss four types of teams which are differentiated by the team member characteristics, the integration of the team with the work environment, and the work cycles, or timing (Sundstrom et al., 1990). This framework includes four types of teams, including *Action*, *Production*, *Project*, and *Advice* teams. *Action* and *Production teams* are highly integrated with their environment. As such, the environment often dictates the pace of the work, and therefore their responses are time critical. Differences in their environments drive further differences in these group types, where Action teams perform in highly unpredictable environments, Production teams work in predictable environments. This difference drives Action teams to require a broad range of specializations to counter the environmental variability. However, the stability of the environment permits Production teams to achieve high performance without the need for the same broad range of specializations. Action teams might include sports teams, entertainment groups, surgery teams, cockpit crews, or military groups. Production teams include teams such as assembly, manufacturing, mining, flight attendants, or maintenance groups. *Project* and *Advice teams* are not highly integrated with their environment. As such, their environment does not dictate the pace of their work, providing team members time to respond. These groups are differentiated by team specialization with Project teams having diverse team membership, while members of Advice teams have knowledge that is focused in narrow areas of expertise. Each of these types of teams can impose different requirements on teaming, particularly regarding time constraints and knowledge sharing. Project teams might include research groups, engineering and development teams, and task forces while Advice teams might include committees, review panels, or advisory councils.

Just as the type of human–human team influences communication, specialization, and timing of team member interactions, the type of HAT will equally pose a unique set of teaming requirements. When allocating roles and responsibilities to the agents in a HAT, the designer should consider the diversity of responsibilities, level of integration, and pace of the work cycle.

4.4 TIMING OF COMMUNICATION DURING TEAMING

As timing and communication are particularly important in action teams, significant research has focused on communication within this team type. High-performing Action teams comprise a group of diverse members who are committed to increasing their performance toward common outcomes. The team members collectively possess the skills necessary to address the task at hand, have the interpersonal skills (e.g., social sensitivity, emotional engagement, and communication patterns) necessary to perform as a team, and have the training to understand when and willingness to play specific *roles* within a team. As such, high performing teams will select talented (Noe et al., 2011) and adaptable operators (Cox & Szajnfarber, 2018) from diverse disciplines (Kearney et al., 2009) and train them intensively (Delise et al., 2010) to focus on clear objectives (McComb et al., 1999).

High performing human–human Action teams take part in planning, execution, and feedback processes, often referred to as *transition*, *action*, and *interpersonal processes* (Marks et al., 2001). Table 4.1 defines these three processes. Further, this table shows the specific activities in which these teams engage. For example, when conducting transition processes, teams analyze, formulate, and plan the mission; specify specific goals; and formulate strategies for mission execution. These activities define the high-level direction the team intends to execute during the action phase.

It is useful to separate the processes shown in Table 4.1 from the phase of activity in which they are conducted. In many environments, the team's activities might be decomposed into three phases. In the first phase, which we will call the *preparation phase*, the team engages primarily in transition processes in preparation for the second phase. During the second phase, which we will refer to as the *action phase*, the team engages primarily in action processes. However, the team must also adapt their plans to environmental changes. Thus, they will periodically reengage in transition processes. Additionally, teams are likely to engage in interpersonal processes during the action phase to support one another. In the third and final stage, which we will refer to as the *after-action phase*, the team will engage transition and interpersonal processes in which they examine their behavior during the action phase to create future-looking plans, goals, strategies, and methods to address interpersonal conflict in preparation for future activities.

TABLE 4.1
Taxonomy of Team Communications, Adapted from (Marks et al., 2001).

Process	Definition	Specific Activities
Transition	Process steps in which teams focus primarily on evaluating and planning activities to guide the accomplishment of a team goal or objective	Mission analysis, formulation, and planning Goal specification Strategy formation
Action	Process steps where teams conduct activities to accomplish the goal or objective	Monitoring progress toward goals Systems monitoring Team monitoring and backup behavior Coordination
Interpersonal	Process steps teams use to manage interpersonal relationships. These processes can occur during transition and action phases.	Conflict management Motivation and confidence building Affect management

Teams exchange information and plans during the transition and interpersonal processes. Teams that effectively engage in robust transition and interpersonal processes during the preparation and after-action phases reduce the need for communication and explicit coordination during the action phase. The limited time available to Action and Production teams makes the transfer of communication into the preparation and after-action phases and out of the action phase important. Therefore, it is often the case that high performing teams will not only develop detailed plans prior to activity execution but often practice the steps in this plan prior to execution. These planning and practice activities help to clarify the roles and responsibilities of each team member and contribute to the construction of team mental models so that team members are not required to transfer specific instructions or other communications to synchronize team member activities during activity execution. As an example, commercial aviation crews practice emergency scenarios so that the Captain and First Officer clearly understand their roles prior to flight. Therefore, the pair can respond rapidly with limited communication if the known and practiced emergency scenario occurs during flight. The crew's engagement in planning and practice increases their likelihood of success when required to act to overcome the emergency.

Communication is the conveyance of information between two or more entities and involves actions to both send and receive information. As we discuss communication and, by extension, coordination of actions within human–human teams, the required effort can be used to classify these actions. *Explicit communication* is the purposeful exchange of a discrete message (e.g., sensor operator might say to the pilot: "Target is moving down the highway"). *Explicit coordination* is an explicit communication for synchronizing actions among team members (e.g., pilot tells the sensor operator: "We cannot lose them. Switch to high contrast mode and zoom in to make sure we have got the right bike".). *Implicit communication* is observable behaviors, other than explicit communication, which convey information between teammates (e.g., seeing the pilot take over manual control, the sensor operator begins increasing camera zoom to compensate for the increased distance of the target from the aircraft). Finally, *implicit coordination* involves explicit or implicit communication for a purpose other than synchronization of action. A significant reason for engaging in the preparation and after-action phases is to allow teammates to utilize explicit communication outside the action phase to create common knowledge among teammates. This permits the teammates to communicate and coordinate implicitly during the action phase, thus reducing the need for communication during this phase.

The focus of high performing teams often extends beyond the team members themselves to include examination and modification of policies and processes that impact their performance (Dickson et al., 2009). In the commercial aviation example, the crew has procedures that they apply during an emergency. Further, high performing teams will find ways to modify hardware and software systems to improve performance (Cox & Szajnfarber, 2018). When employing supporting hardware and software, these teams often utilize operator creativity to take advantage of detailed control of the systems to extend the software and hardware functions beyond the designed system capacity with minimal design changes (Jacques & Strouble, 2011). As a result, they can adapt the system at a pace that outstrips the system development cycle (Cox & Szajnfarber, 2018). For example, Cox and Szajnfarmber provide examples where military pilots control the power to aircraft engines in a specialized fashion, and these actions by the pilots permit the aircraft to fly lower and slower than the aircraft's designers intended for it to fly. This modification to the control process permitted the aircrew to perform specialized missions that were not otherwise possible.

These high performing teams also often create domain-specific vocabulary and gestures (Woods & Hollnagel, 2006). *Domain-specific vocabulary* permits team members to express commonly occurring complex concepts with a single word or short phrase. Sometimes, this vocabulary takes the form of acronyms, jargon, or slang where human team members invent words to represent exceptional circumstances. This abbreviated form of communication reduces the time and effort for communication during activity execution.

Based on this overview of human–human teaming, we might ask: "Can these concepts apply to the design of HATs?" AI agents certainly can possess knowledge and skills which extend beyond those of the human operator, and thus these entities may be desirable members of future high performing teams. However, the ability to coordinate individual adaptations toward a common goal is a key attribute for members of high performing teams and is a capability that AI agents currently often lack. Thus, it is important that we decide whether to view AI agents as adaptable hardware and software systems to be modified by the human team members or whether we seek to develop true HATs in which the AI agents can coordinate their behavior with other team members in response to the environment and the actions of teammates. The latter will require AI agents that can adapt to the needs of the human, as well as the environment, rapidly and reliably (Sycara & Lewis, 2004). Although this type of adaptation is an advanced function for AI agents to accomplish, we will discuss the need for and ability of HATs to adapt in this way later in this text.

In this text, we will assume the need to design integrated user interfaces and AI agent transparency to improve the operator's knowledge of the AI agents, which will support coordination (Stowers et al., 2016). However, the currently available AI agents typically lack insight into the operator's understanding of the situation and task. Therefore, current AI agents require explicit communication to coordinate with their human teammates (Klein et al., 2004). Typically, designers will design this type of explicit coordination to occur during the mission. Thus, current design methods often limit human communication with AI agents exclusively to explicit communication. This differs from coordination activities in high performing human–human teams where these teams deliberately shift the cost of coordination from the action phase to less critical time periods, including the preparation and after-action phases. HATs are often not designed to take part in or benefit from human activities in these less critical time periods (Chen et al., 2018). As a result, there are frequent references within the HAT literature to the need to design AI agents to support implicit communication or coordination and intent inferencing (Espinosa et al., 2004; Riley, 1989). However, the literature rarely provides precise definitions of these terms or methods to achieve robust implicit coordination. If we seek to design AI agents for well-coordinated HATs, we must understand coordination. However, more fundamentally, we must first understand the human use of mental models and memory processes that support team coordination.

Timing is also a significant issue in HATs. In the previous chapter, we discussed the steps in goal-directed behavior for an individual or a team. However, when we examine the members of a team, it is apparent that each member of a successful team executes this process in a synchronized fashion. John Boyd, in the study of military strategy, labeled this loop with the four stages of observe, orient, decide, and act and pointed out that each cognitive agent must complete this loop (Boyd, 1976). He further pointed out that if one agent can execute this loop faster than their adversary, such that the agent can take their next action faster than the adversary can orient, decide, and act, the agent can disrupt any plans or strategies that the adversary can employ, providing the agent a decisive advantage in a military engagement. While this is true in adversarial relationships, it is also true in teaming relationships. Figure 4.2 uses the terminology we introduced for goal-directed work in the previous chapter, rather

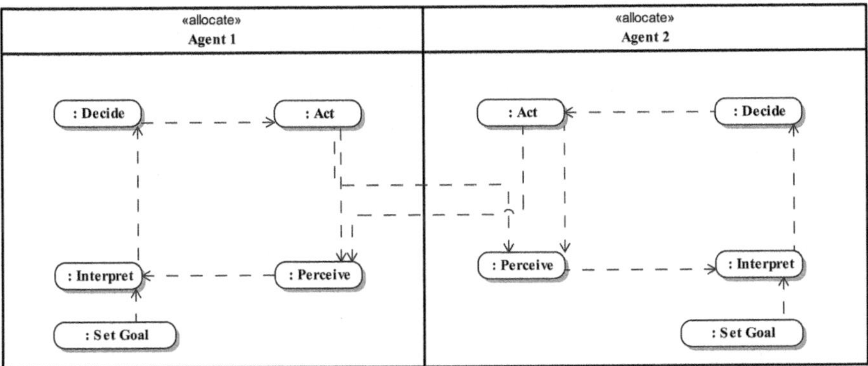

FIGURE 4.2 Depiction of the Temporal Control Loops for a pair of agents during interleaved interaction.

than Boyd's observe, orient, decide, and act terminology, although these elements are evident within our terminology. This figure shows the cognitive loops for two interdependent agents performing time sequential work where one agent completes their task, and the other agent must perceive and understand this work before taking their action. Although this type of turn-taking is only one type of potential interaction between agents, the time synchronization is evident in this type of interaction. As AI agents are often capable of completing this loop faster than humans, if the interface between a human and an AI agent is designed to permit the AI agent to respond to the human faster than the human can complete their loop, the human is left simply responding to the agent and not considering all information in the environment. Existing literature has demonstrated that the timing of AI agents in an environment has a significant influence on human behavior as this observation would suggest (Goodman et al., 2017; Rouse, 1976). Thus, it is important to consider not only the actions and communication among teammates in HATs but also the timing of these interactions.

4.5 MENTAL MODELS AND EFFECT ON COMMUNICATION

As we will discuss in more detail in the next chapter, humans employ both feedback and feedforward approaches when selecting actions to take in the world. In a feedforward process, we use knowledge from the past to predict how an action might influence something in our world before we take an action. Therefore, we anticipate the effect that each action we take in the world will have upon other objects in the world. We then monitor changes in the world after we take an action and use this feedback to determine whether the desired effect occurred, adjusting our actions if the original action did not have the desired effect. The feedforward processes require that we form *predictive models* which permit us to anticipate the world. These models are referred to as *mental models*. Similarly, teamwork requires that team members share certain parts of their models so that individual team members coordinate their actions to have the desired effects. In this context, we will discuss *shared mental models*. Finally, it might be that we do not have the knowledge to make predictions ourselves, but we might know that others in our team might have this knowledge. *Transactive memory systems* refer to information structures that permit team members to understand which team member holds specific information. As depicted in Figure 4.1, each human maintains their own mental models. However, team mental models, a specific type of shared mental model, and transactive memory are specific to each team and represent knowledge relevant to the function of each team. In the cognitive science literature, creating and maintaining shared mental constructs are often referred to as constructing *common ground* (Clark, 2005; Clark & Brenan, 1991), and updating task-relevant information to understand current and future states is often referred to as building *Situation Awareness* (Endsley, 1995).

4.5.1 MENTAL MODELS

A *mental model* is an internal mental representation of elements of the external reality (Johnson-Laird, 1980). When applied to the use of manufactured systems, mental models permit us to understand the purpose, form, state, and function of a system, as well as to predict the behavior of the system (Rouse et al., 1992). However, humans can apply mental models to any object or process in the world. They permit us to describe the characteristics of the object, how it will function or act within the environment, and how it will affect us or our goals in the world. Note, this is an essential element in goal-seeking agents as the mental model permits the agent to predict how one or more potential actions will affect the world, permitting the agent to predict whether an action is likely to support goal attainment.

It is important that we construct mental models as we observe, perceive, and build understanding, that is, situation awareness, of the world. These models often change over time, are based on perceived information, and are not necessarily correct. We are constantly constructing these models, using them to make predictions about future actions of objects in the world, and adjusting our models when the objects or their behavior do not match our expectations.

Errors in constructing mental models often occur because we observe actions, not the internal workings of objects in the world. Thus, we assume internal functions of objects based on the behavior of similar, well-understood objects through analogy. Furthermore, because we do not understand the internal function of many objects, we can assume causation for correlated events. For instance, I recently hit a piece of debris on the road and later noticed that my car shook during breaking. Therefore, I assumed that that car was damaged when I struck the road debris. In fact, the brakes had simply worn with time, which had nothing to do with the road debris. I simply became more aware of the shaking after striking the road debris and falsely associated the encounter with the road debris with the shaking of the car during breaking.

Importantly, while we are not aware, there is significant evidence that we are constantly applying our knowledge to mentally simulate or predict the behavior of objects in our environment (Friston et al., 2016; Sajid et al., 2021). This constant prediction permits us to direct attention to maximize information gain and allows us to learn by comparing observations in the world to predictions as surprise observations inform the need for model improvements. Referring to the automotive example in the last paragraph, my car was shaking during breaking before it encountered the road debris. However, as I did not anticipate a change in the behavior of my car, I did not focus my attention on the car's behavior before hitting the road debris and therefore did not perceive the shaking. However, after encountering the road debris, I anticipated the change in state of my car, focused my attention on its behavior, and suddenly became aware of the shaking. This result occurred because my mental model of my car permitted me to anticipate that my car's behavior may have departed from normal after encountering the road debris. However, my mental model did not include the fact that my breaks had not been serviced for the last 75,000 miles, providing a potential explanation for my car's failure.

4.5.2 Shared Mental Models

Team members can share elements of mental models (Mohammed, 2010). Individuals can construct the shared elements of mental models by observing a common environment and teammates as they work in the common environment, as well as through dialogue. These observations and communication can occur during training, prior to beginning the operation during the preparation phase, during the action phase, or through interpersonal processes during the after-action phase (Eccles & Tenenbaum, 2004; Entin & Serfaty, 1999; Jones et al., 2011).

Literature differentiates shared mental models, that is, models shared between dyads of individuals, and team mental models, that is, models shared across an entire team (Langan-Fox et al., 2004). Researchers have studied shared mental models in the context of teamwork (Mathieu et al., 2000; Rouse et al., 1992). Team mental models expand traditional mental models, including information to enable team coordination (Eccles & Tenenbaum, 2004).

A meta-analysis of team cognition indicated that improving the similarity in structure and content of teammates' mental models improves team behavior process and performance (DeChurch & Mesmer-Magnus, 2010). More recent research has confirmed that improved shared mental models and mutual understanding of team-centric content correlate with team performance (Mesmer-Magnus et al., 2017). This indicates that the ability of teams to build consistent shared mental models regarding team cognition improves team performance.

Leaders within teams can enhance the consistency of shared mental models. For example, Entin and Serfaty found that training leaders to periodically communicate key elements of their mental model to their team members improved team performance and teamwork scores (Entin & Serfaty, 1999). This training additionally led to an increased use of implicit coordination mechanisms as the communication permitted the team members to synchronize their mental models with the mental model of the leader. By having the team leader communicate their mental model to their team members, the team members apply this mental model to estimate the team leader's intent and to guide their own actions to enhance team coordination. Humans do not require a single, monolithic mental model and will employ a variety of models during execution (Endsley, 1988). While researchers have made efforts to extend the concept of shared mental models to HATs (Carpenter & Zachary, 2017; Howard & Cambria, 2013; Lapso et al., 2023; Scheutz et al., 2017; Yen et al., 2006), this effort has met with limited success to date. However, this is an ongoing area of research.

Rouse proposed a taxonomy of training knowledge and skill requirements (Rouse et al., 1991) and extended this taxonomy with colleagues to include teaming knowledge and skills (Rouse et al., 1992). This taxonomy illustrates the types of information within mental models, which are necessary for successful team execution. This training taxonomy was divided into three categories: System, Task, and Team. Although this taxonomy was designed to inform the training content for new systems, it can also be useful to consider the types of knowledge that are necessary for an agent (human or AI) to form robust mental models. In a more naturalistic environment, it is important for humans to construct mental models of objects beyond the system they are interacting with. One can either replace or augment the task knowledge category to include other environmental artifacts, such as weather and individuals (e.g., other drivers, adversaries, or others) beyond team members. This taxonomy further decomposes each category into three types of knowledge: "What", "How", and "Why". Each type of knowledge has a continuum of levels ranging from detailed to abstract, which is valuable because the human's ability to switch among these levels of abstraction during troubleshooting or planning permits the use of abstract thought to draw analogy and thought about details of operation to develop detailed plans.

Table 4.2 expands on Rouse and colleagues' taxonomy to provide a revised taxonomy. As stated earlier, this taxonomy provides further insight into the types of knowledge an agent requires to support their mental models, enabling them to interact effectively within a team. The table expands on the three categories of system, task, and team to include environment and adversaries. It is important that the environment and non-team members, referred to as adversaries, play a role that constrains and shapes the performance of the system, team, and task environment. As such, humans form mental models of these entities to project the effect of these entities upon team performance.

As a reader, you might observe that there is a large amount of information within complex scenarios, which affects the performance of any team. Rasmussen suggested that humans employ a skills, rules, knowledge framework to address this information. According to this framework, in novel situations, team members rely on knowledge to attempt to reason about behaviors to employ. As they gather information, they develop rules that are employed to shortcut or subset the decision space. With additional experience, situations become recognizable and the users are able to match a current situation with a past situation and employ previously learned, skilled behavior to respond.

4.5.3 Transactive Memory Systems

Shared and team mental models are most important in Action and Production teams, where it is important for team members to highly couple their activities with the environment. However, shared and team mental models do not represent the disparate or unique knowledge between or among team members.

Transactive memory systems map where knowledge resides among team members, the credibility of the individual who holds the knowledge, and who can acquire essential information (Wegner, 1986). This knowledge is particularly useful in both Action Teams and Project Teams where the individuals within the team each have specialized knowledge that contributes to the performance of the team in complex and unpredictable environments. This transactive memory can differ or be shared across team members and is constructed as communication occurs within a group, thus it is often not robust in short-lived groups (Yan et al., 2021). This knowledge does not reside with any individual, particularly in larger groups. Instead, group members construct transactive memory structures across the collective memory of the group members. The AI agent literature has explored the embedding of transactive memory in AI agents and the use of combined mental models and transactive memory in AI agents (Lapso et al., 2023).

4.6 DEFINING INTERDEPENDENCE AND COORDINATION

Earlier, we defined a team as two or more entities who train and work together interdependently, with diversified knowledge, skills, and abilities to accomplish work that

TABLE 4.2
Training Knowledge Taxonomy.

Level	Knowledge Type		
	What	**How**	**Why**
		System Knowledge	
Detailed	Characteristics of system elements	Functioning of system elements	Requirements fulfilled
	Relationships among equipment elements	Co-functioning of equipment elements	Objectives supported
Abstract	Temporal patterns of system response	Overall mechanism of system response	Physical principles and theories
		Environmental Knowledge	
Detailed	Relevant environmental characteristics	Environmental function	Constraints and opportunities imposed
	Relationship to team function	Function of team under environmental constraints	Objectives permitted
Abstract	Temporal patterns of environment	Mechanism of environmental changes	Physical principles and theories
		Task Knowledge	
Detailed	Situations	Procedures	Operational basis
	Criteria, how to assess performance	Strategies	Logical basis
Abstract	Analogies	Methodologies for assessing alternatives	Mathematical principles and theories
		Team Knowledge	
Detailed	Team member roles	Functioning of team members	Requirements fulfilled
	Relationships among team members	Co-functioning of team members	Objective supported
Abstract	Temporal patterns of team performance	Overall mechanism of team performance, intent projection	Behavioral principles and theories
		Adversarial Knowledge	
Detailed	Adversary roles	Function of adversary team members	Requirements seeking to fulfill
	Relationships among adversary team members	Co-functioning of adversarial team members	Objectives seeking to fulfill
Abstract	Temporal patterns among adversarial team members	Mechanism of team performance, intent projection	Behavioral principles and theories

Source: Adapted from Rouse et al. (1992)

is complex or critical enough so that it requires greater knowledge, skill, ability, capacity, or redundancy than a single entity can provide (Delise et al., 2010; Salas et al., 1992; Schneider et al., 2022). Our definition of HATs in Chapter 1 specifies that these teams must be composed of at least one human and one AI agent who perform work interdependently where the interdependent performance requires a coordination of activities, usually facilitated by a form of communication. This section further defines the two terms: Interdependent and coordination.

Interdependence occurs when two or more team members depend upon one another to complete a goal. That is, neither team member completes all the tasks to complete the goal, instead the team fulfills the goal through their interaction. Johnson and colleagues have pointed out that interdependence can be discussed as either hard or soft (Johnson et al., 2019). *Hard dependence* occurs when a single team member is unable to complete the goal without any aid from one or more teammates. *Soft dependence* occurs when a team member can accomplish the goal on their own, but the team can achieve their goal to greater effect, for example, more accurately, faster, or safer, when one or more teammates contribute toward completing the goal. A common example of soft dependence is when a passenger in a car serves as navigator to the driver in novel locations. Drivers commonly perform this navigation function, but it is safer and easier for the driver to permit the passenger to assume this role when a capable passenger is present.

Sheridan and Verplank suggest that team members either "share" or "trade" work (Sheridan & Verplank, 1978). Through sharing, team members either extend each other's capabilities (i.e., the team performs work that could not be performed by a single individual) or relieve teammates from some tasks (i.e., take some portion of the work away from a teammate to make the work easier or safer). When trading work, the teammate can back up the individual or replace the individual, and this sharing permits the individual to attend to other duties or diversifies the work to reduce monotony and errors that might occur due to lapses in attention that can occur as individuals perform repeated work.

The coupling of interdependence is also important and can be described as pooled, sequential, or reciprocal (Thompson, 2003). *Pooled interdependence* does not necessarily require communication as each entity performs separate functions that are not time dependent. In *sequential interdependence*, one individual must perform an activity before another can perform their activity. This time dependence creates the need for communication so that the subsequent teammate knows that it is their turn to perform their task. One might think of an assembly line as an example where one individual must perform a first set of tasks before the next individual can perform their task. When the work is repetitious, team members can rely on cues such as elapsed time to gain sufficient information to support interdependence. The presence of these cues reduces the need for explicit communication. In *reciprocal interdependence,* the team members must each perform multiple interleaved tasks, with the tasks performed by one or more other individuals. In this case, teammates must communicate the order and timing of tasks, as well as any issues with task completion, to enable the back-and-forth exchange among team members.

Coordination can be defined as a cyclical communication process that enables synchronized actions of teammates who are working on interdependent tasks (Crowston,

1991; Schneider et al., 2022). Based on this definition, while communication (which supports coordination) may occur within the preparation or after-action phases, coordination occurs during the action phase. Teammates can plan and practice coordination such that they do not require explicit coordination. However, communication is often important in systems where (1) the times and occurrences of tasks may be dependent on external events, (2) the task to be performed is dependent upon tasks performed by other team members, or (3) the tasks to be performed to achieve a specific goal may have consequences for the actions of other team members who are pursuing a separate team goal (March & Simon, 1958).

For effective coordination, each agent must be able to observe and predict the actions of team members. Prediction occurs as members of the team construct shared and team mental models, as well as transactive memory systems of the team. Observation is important as it permits the agent to compare the observed actions of team members to the predicted behavior.

4.7 COMMUNICATION AND COORDINATION

We defined explicit and implicit communication and coordination earlier. As indicated earlier, explicit coordination requires communication to be directly focused on managing dependencies and synchronizing actions. Therefore, it requires explicit communication. Implicit coordination is described as dependency management without dedicated or purposeful communication regarding synchronization (Butchibabu et al., 2016). Implicit coordination involves either implicit communication or explicit communication. Implicit coordination provides context that implies the team activities necessary for coordination. Implicit coordination driven by implicit communication might occur when one team member is handed an artifact to work on. This handoff implies that the team members completed the previous tasks. Therefore, there is no need for explicit communication as the individual receiving the artifact relies on the shared mental model of the task to assume they may begin their task as they receive the artifact. Teams can drive implicit coordination using explicit communication. For example, when baseball coach tells a player that they are "on-deck", this terse explicit communication tells the player that they should begin stretching, move closer to home plate, pick out a bat, and mentally prepare to bat once the current batter completes their turn at bat. This use of domain-specific vocabulary is explicit communication as the coach is clearly conveying information. However, they do not provide precise instructions to the player on what he or she should do. By relying on the shared mental model, agreed upon by the definition of this shared terminology, the player knows to get their bat, advance to the on-deck circle, and begin to stretch or warm up in preparation for batting. It is important that the shared or team mental models enable efficient communication and perception to be extended to support coordination as well as convey other relevant information through communication (Espinosa et al., 2004).

Based on the previous discussion, it might also be clear that communication can vary from abstract concepts to concrete data. For instance, the coach might use the domain-specific language "on-deck" with an experienced player but might need to instruct a beginner using extremely specific language. Therefore, a single communication potentially implies numerous coordination activities. It is also relevant that to provide explicit commands or commands at a low level of abstraction requires the sender (e.g., coach) to have more knowledge of the situation than they require to provide a command at a higher level of abstraction. For instance, the coach cannot tell a child to do all actions associated with being "on-deck" at one time. Instead, they need to monitor the child, giving them each command only after they have completed the earlier command.

Coordination in human teams shifts from being highly explicit toward implicit as teams become familiar with each other and the task (Butchibabu et al., 2016; Espinosa et al., 2004; Mathieu et al., 2000; Rico et al., 2019). Thus, the team constructs the knowledge structures, shared mental models, and transactive memory system as the team explicitly coordinates, eventually enabling implicit coordination. Importantly, this learning is specific to a set of conditions, but the team will extend this knowledge to analogous situations. Humans adapt their coordination type depending on their perception of the team's cognition. If the team is managing interdependencies well and the individual teammates perceive the team behaving consistent with their expectations, implicit mechanisms are more prevalent. Conversely, if the team's actions are unsynchronized or the team's actions do not match an individual teammate's expectations, the teammate will revert to explicit coordination mechanisms. This reversion improves their ability to manage interdependencies and refine team cognition. Importantly, each team member applies this knowledge of coordinating activity to select a set of coordinated actions. The team applies this knowledge without any leader or team member issuing specific and highly detailed commands to each team member to facilitate coordination. This coordinating activity is different from general broadcast methods often employed by AI agents where these agents send messages without regard to who the recipients are or what they need to synchronize their actions.

4.8 TYPES AND COST OF COORDINATION IN TEAMWORK

It is important to recognize that "nothing is free". Instead, teamwork processes require cognitive resources and time to execute. The effort required to support coordination increases as the size of the team increases. Therefore, Klinger and Klein suggest that the cost of coordination increases with the addition of team members such that the marginal value of adding additional team members decreases with each increase in team size (Klinger & Klein, 1999). As this cost is not readily apparent when viewing humans working in teams unless one specifically examines these coordination processes (Macmillan et al., 2004), it is important to understand this cost and the effect that adding additional team members has on coordination costs.

Coordination costs can be classified into four categories (Klein & Bradshaw, 2005; Schaefer et al., 2016):

1) ***Synchronization overhead***, the time one team member spends waiting for another team member to complete a prerequisite task before beginning their task;

2) ***Communication overhead***, the effort required to manage a handoff or to initiate a required activity by another team member;
3) ***Redirection overhead***, the time team members spend following an out-of-date plan after a team member signals a new plan but before all team members understand the change; and
4) ***Diagnosis overhead***, the time spent diagnosing a problem that occurs because of interrelated activity.

Klein and colleagues list activities required to support coordination, including communication, monitoring, and feedback activities. Some activities might be the responsibility of a frequent sender, such as a team leader, and some are clearly delegated to frequent receivers, that is, the team members. Each of these costs builds upon the work of Clark and Brenan who initially provided a list of costs for constructing common ground (Clark & Brenan, 1991). The sender bears the formulation and production costs, as well as the cost associated with formulating and conducting the steps in a plan. The receiver bears other costs such as reception, understanding, and delay costs. Finally, team members share many of the costs discussed by Clark and Brenan, including start-up, directing attention, asynchrony (e.g., interruptions), change, display, fault, and repair costs.

Coordination can occur across a spectrum from being fully explicit (e.g., all coordination is explicit, and messages must be complete) to being fully implicit (e.g., a sender and receiver conduct activities without explicit communication). Further, the cost of any given coordination exchange may vary for different entities within a team, depending on whether communication and coordination are implicit or explicit. To illustrate, let us examine two cases; in the first case, the team relies on a team leader to communicate using fully explicit coordination, while in a second case, the team leverages highly implicit coordination.

4.8.1 Fully Explicit Coordination

In the case of fully explicit coordination, the team leader (i.e., sender) is responsible for formulating all activities for each team member, formulating the communication of these activities, and producing this communication. Further, the team leader must monitor each team member (i.e., receiver), as well as formulate and communicate any required changes. The team members bear the cost of receiving and understanding the communication, costs associated with delays or interruptions as they receive the communication, and costs with detecting and repairing faults in the steps they receive from the team leader.

To relate to this behavior, you might consider an interaction you have had with a young child the first time they perform a task. A task as simple as making a peanut butter and jelly sandwich requires you to specify to the child that they must get a plate, two slices of bread, peanut butter, the jelly, a knife, and a spoon. You might have to further explain where each of these is located as well. Once you observe that they obtained these items, you must explain using the knife to scrape the proper amount of peanut butter from a jar and then to spread it on the bread. Then place the knife in the sink to avoid putting peanut butter on the counter, then to use the spoon to obtain the proper amount of jelly and how to spread it over the peanut butter or on the second slice of bread, placing the spoon in the sink. Finally, you must provide instructions for putting everything away and covering the sandwich to prepare it for eating. In the case of explicit communication, you must formulate each of the steps, consider the order of the steps, then communicate this order at an appropriate time, monitor performance of each task, formulate, and communicate appropriate feedback or error correction information and then communicate the subsequent task.

In fully explicit coordination, the team member receives this communication, understands the activities, executes the activities, and processes any feedback. They do have to wait for this communication or interrupt other activities to receive this communication, and they may detect and correct errors on their own during the process. In our peanut butter and jelly example, the child does not have to perform any cognitive processing to formulate the plan and communication or have the knowledge of the plan or process prior to execution.

Based on this discussion and example, at the extreme, fully explicit coordination requires the team leader to formulate each team member's sequence of activities at the lowest possible, that is, most detailed, level of control. Further, the team leader must formulate and transmit this communication at extreme detail, placing most of the cognitive burden on the team member. Additionally, formulating and communicating add tasks to the team leader that they would not have to perform if they were to perform the task themselves. The team member must receive the full set of communication placing the full burden on the team member for this action, but as the team leader fully specifies the information, the team member does not have to spend significant resources to understand the communication. Instead, the team member's only requirement is to translate the communicated steps directly to action.

4.8.2 Implicit Coordination

Fully explicit coordination is exceedingly rare in human–human communication, which is important as this communication is extremely burdensome for the team leader, consuming a substantial portion of their cognitive resources. Even in this example, most of us can make a peanut butter sandwich without being consciously aware of the individual steps we are performing, allowing us to think about other activities while we are making this sandwich. However, to formulate and communicate this task to someone else in an explicit manner requires the majority of our cognitive resources. Because consciously formulating the entire plan to enable us to convey the steps in the appropriate order prior to execution, as well as monitoring and providing feedback during execution use many of our conscious mental resources. Note that even this tedious example is not completely explicit coordination, as we did not have to explain what a plate, fork, or knife is or how the individual should use these utensils. Further, we did not need to explain peanut butter or jelly, and their properties that affect our ability to spread them, etc. This is knowledge an agent will require to complete the task and that an AI agent will need to gain before it can perform the task.

As the coordination becomes more implicit, the team leader can begin to formulate and communicate at a more abstract level, for example, by communicating the use of a plan. At the fully implicit extreme, the team leader only formulates and communicates information critical to coordination. In our example, this might be as simple as instructing the child to make themselves a peanut butter and jelly sandwich. Once the child has an appropriate mental model of the task, they can formulate and execute the plan themselves without further communication from you. Therefore, the team leader can be less precise, reducing the burden for communication formulation, and abbreviate communication. As there is less information to receive, the team member incurs only a part of the cost of reception when compared to the full cost they incurred during explicit communication. As the team reduces the level of detail in coordinating communication, the team member utilizes context information and knowledge from their mental model of the task and team cognition to understand the communication and its implications for actions. The team members must then translate their understanding of the communication and environmental context to formulate their own activities. As such, the demands for understanding communication and formulating activities increase for the team member.

Returning to the earlier discussion of the lists of costs associated with coordination from the literature, each cost focuses primarily upon the costs associated with communication or actions taken by other teammates. However, Clark and Brennan include the cost of directing attention, which is associated with ensuring that teammates are aware of changes in the environment—cues that the team must use to trigger changes in future activities (Clark & Brenan, 1991). It is rational to partially exclude the cost of perceiving environmental cues from the cost of coordination as individuals bear this cost as they perform tasks in an environment. However, plans or other coordinating information can include environmental events to trigger changes, and individual team members often perceive the same environmental cues to trigger appropriate activities. Thus, the activity required to observe these environmental events plays a significant role in coordination. These cues affect the coordination of the team and are critical when trying to understand implicit communication within teams.

To coordinate effectively, the team leader must understand the team member's understanding of the team member's situation. In our simple example, the parent must understand the cleanliness of the child's hands to understand whether they should require additional steps, such as washing hands. As communication moves toward implicit communication, team members must develop an understanding of the team leader's situation. For example, for a couple of adults, the response to make yourself peanut butter and jelly might change if you understand whether the other individual might also be hungry. The fact that a team member must not only form their own understanding of their and their teammate's situation but also project their team member's understanding of these situations may not be initially obvious.

The need for the latter is discussed in an incident reviewed by Marstall and colleagues (Marstall et al., 2016). In this example, two individuals, sitting fore and aft in a two-person cockpit such that the individuals could not see each other, pilot a fighter aircraft. During the flight, a coded communication came in. The individual not actively flying the aircraft stated "I've got it" and accidentally tapped the control stick with their knee when they reached for a pad to begin decoding the message. The individual flying the aircraft interpreted the pressure changes on the control stick to mean that the copilot was taking control of the aircraft. As such, he also began to decode the message. In this instance, their projection of the other's situation was that the other individual was flying the aircraft. Fortunately, they were flying in formation with another aircraft, whose pilot noticed their aircraft slowly nosing over and beginning a dive toward the ground while both pilots were busy decoding the message. As the other aircraft pilot knew they were flying in a transit phase where the planes should be maintaining constant altitude, he alerted the two pilots to the diving aircraft, permitting them to take control of the aircraft and to avoid a ground collision. This near accident originated because each pilot in the first aircraft incorrectly projected the other's understanding of their activity.

When operating in a fully explicit paradigm, the team leader must remain aware of their own and their teammates' situations to guide activity formulation as they must decide which team member is in the best situation to perform each required activity. Additionally, they must be aware of their teammate's understanding of their own situation to understand if their teammate is likely to execute the actions. As communication becomes more implicit, responsibility for situation awareness shifts to the receivers. As the receivers must now understand communication from the sender, they must be aware of the sender's state and their own situation. Although this shift reduces the cognitive load on the team leader when they are coordinating with a single team member, this change in load increases when the team leader is coordinating with multiple team members.

To summarize, explicit coordination typically consumes the cognitive resources of the sender who must formulate, as well as communicate, an explicit command or request to each team member (Woods & Hollnagel, 2006). Implicit coordination distributes cognitive demand to the receivers. The receivers must then incur the obvious perceptual channel load (verbal, auditory, haptic, or visual) to gather information that makes them aware of the sender's as well as their own situation. Further, they must execute cognitive processes to abstract the available information, understand the utility as it relates to the context, and formulate the appropriate coordinating activity (Endsley & Kiris, 1995). Implicit coordination reduces real-time communication load on the team by multi-purposing a communication event from a sender to permit each teammate to select actions that they coordinate with the actions of other teammates based on their understanding of each teammate's situation and their assessment of each team member's understanding of their current situation. Importantly, as teams develop the knowledge required to understand situations and the implications of these situations for coordinating activity, they often develop domain-specific terminology which encapsulates the description of system states with accepted plans (Woods & Hollnagel, 2006). The use of these terms then encapsulates or abstracts the situation and plans to a higher level, facilitating the use of less explicit communication. For example, to avoid situations such as described by Marstall and colleagues, language is typically formalized in fighter aircraft such that the copilot should have clearly stated they were decoding the message, and, failing this communication, the pilot should have been clear that they were relinquishing control of the aircraft to the copilot. Had either of these communication events occurred, as trained, the pilots would have avoided this incident. However, this formalized communication again requires time and effort and can be a shortcut, especially when team members are weary or overburdened.

The teaming literature discusses the utility of explicit communication to improve team effectiveness. For example, a meta-analysis of 150 team communication studies found that team communication positively influences performance (Lacerenza et al., 2017). Communication to support explicit coordination becomes an issue in time-pressured environments where the pace of execution and volume of information restrict the time available to explicitly communicate coordinating information (Entin & Serfaty, 1999). In such environments, the highest performing teams demonstrate well-coordinated behavior with limited or no communication by governing their actions from team cognition, which allow implicit coordination such that everyone can anticipate the actions of their teammates (Burke et al., 2004; Lacerenza et al., 2017; Stout et al., 1999).

To emulate human teaming, AI agents will need to constantly monitor and predict the behavior of its team members. To perform this prediction, the AI agents must be able to participate in transition and interpersonal phases of teaming. These capabilities are important if the AI agent is to participate in an implicit coordination. Further, AI agents will need to understand when team members behave differently than expected to modulate communication between explicit and implicit communication. These behaviors are not necessary for teaming, but they are necessary for the team to successfully reduce communication and minimize the cost of coordination during teaming.

4.9 MODELING: UPDATING GOAL HIERARCHY AND GOAL–RESPONSIBILITY DIAGRAM—"TO-BE" SYSTEM

4.9.1 Modeling Objective

We developed a Goal Hierarchy and Goal–Responsibility Diagram within Chapter 3. However, before considering teaming, it is likely that we focused our goal and responsibility identification on the action part of the process, without considering the teamwork responsibilities. Furthermore, we did not consider the type of team we were designing and its implications for our goals and responsibilities. These same limitations often occur during the design of AI agents, with designers focusing on their design for the action process and the related task work, placing less emphasis on designing the AI agents to participate in the transition and interpersonal processes and teaming behaviors.

The team type varies primarily by diversity and environmental integration. Each of these variables also present different challenges to a team. The more diverse the team is, the less likely they are to have robust team mental models and transactive memory systems prior to working together. Human team members construct these structures during training, planning, and rehearsal. Therefore, it can be beneficial to consider how your team effectively builds these structures. The more highly integrated the team is with the environment, the more likely the environment will set the timing of activities that must be performed, imposing considerable time pressure on the team during the action process. To overcome this time pressure, it is important to move as much activity related to transition and interpersonal processes to times outside the time that the action process occurs. In human teams, this often occurs during planning, briefing, and rehearsal prior to the action process, as well as debriefing after the action process. Given this information, if your team is an action or production team, you may find it useful to develop goals and responsibilities specific to the preparation or after-action stages. By focusing on transition processes during these phases, you can determine the activities that are performed to build team mental models and that enable the team to leverage implicit communication and coordination. If your team is an action or project team, you may want to revisit your goal and goal–responsibilities diagram with an effort to understand any responsibilities that you might have missed that are necessary to build transactive memory.

4.9.2 Diagram Components and Composition

The diagram components and composition for both the updated Goal and the updated Goal–Responsibility Diagram are largely the same as for the original Goal and Goal–Responsibility Diagrams. These diagrams are intended to capture the same information as the original Goal and goal Responsibility Diagrams with the exception that we are expanding the goals to include any activities completed by the team during the preparation and after-action phases of the work in addition to the action phase of the work. Additionally, the responsibilities should be enhanced to capture the responsibilities that are necessary to support teaming, focusing on areas of the work where interdependence is necessary. Although not required or demonstrated in this text, the goals and responsibilities could be annotated to indicate the phase of activity where the work occurs as well as individual responsibilities that exist to support task or teamwork.

4.9.3 Considerations

Consider revisiting your team to understand the work they perform during both the preparation and after-action phases of their work. In some instances, these phases might not be conducted in a formal fashion, especially in production teams, but similar processes might be completed at shift changes. For example, during shift changes, nurses within an intensive care unit might brief incoming nurses on the status of each patient, plans that the physician has stated regarding each patient, and rules for engaging with the physician on duty or the primary care physician. You should develop an understanding of goals that the team attempts to complete during the preparation and after-action phases of their work, as well as the responsibilities associated with each goal. You should then capture these goals and responsibilities within the updated Goal and Goal–Responsibility Diagrams.

Further, it is important to understand the teaming behaviors that are conducted during the action phase of the work. You should determine which team members have shared goals and thus formed subteams within the larger team. For each subteam, you should then determine the responsibilities for each team member with respect to these shared goals. Finally, you should update your goal–responsibility diagram to include these shared goals.

4.9.4 Example Application

For our Diagnostic Center HAT, the subteam performing mammographic activities has specialized skills and little shared work during the action process. Further, although the need to maintain a schedule may exert time pressure, the environment is not imposing strict time limits. Therefore, we might classify this subteam as a project team. In this subteam, the radiology technician understands how to position patients properly for imaging and has detailed knowledge of the imaging system, while the radiologist has specialized knowledge associated with examining and interpreting the mammograms. Their shared work occurs through the mammogram itself. Therefore, they must agree upon what constitutes a quality mammogram including the content and exposure of the mammogram. In fact, research has shown that having radiologists and radiology technicians each independently rank and communicate the quality of each radiograph leads to better detection performance (Santner et al., 2021). Therefore, we added the responsibility of *Develop Shared Understanding of Mammogram Quality* to the diagram in Figure 4.3.

The second subteam performs the biopsy. Given the diversity of the team members and their need to integrate the actions necessary to administer anesthesia to the patient, this subteam can be classified as an Action Team. To account for the transition processes, we can add responsibilities for *Develop Agreed Biopsy Plan*, *Project Likely Biopsy Challenges and Plan Alternatives*, *Agree on Team Member Biopsy Responsibilities*, and *Prepare Necessary Resources*. For the after-action interpersonal processes, we can add the responsibilities of *Provide Performance Feedback* and *Agree on Changes to Future Behavior.*

4.10 CHAPTER SUMMARY

The topics of this chapter span from the rationale for why humans team with one another to the costs associated with teaming. Along the journey, we briefly discussed the memory structures that we utilize during teaming, including mental models, shared mental models, and transactive memory systems. As we team, we can communicate with one another to partition work among us. However, in team activities, individual tasks we share may not be independent of one another. Instead, we often need to coordinate our activities so that the tasks, the order of those tasks, and the timing of each task are coordinated so that we successfully achieve our goals.

The coordination of these interdependent tasks can require planning and communication. In human teams, we often rely upon plans, prior experiences, and training to permit us to understand how fellow team members will respond to the environment and our actions. As such, we leverage this knowledge during the action phase to reduce the communication that the team requires for coordination. This reduction can occur through the development of domain-specific language and gestures which permit us to rapidly express complex information, through learning of sequences

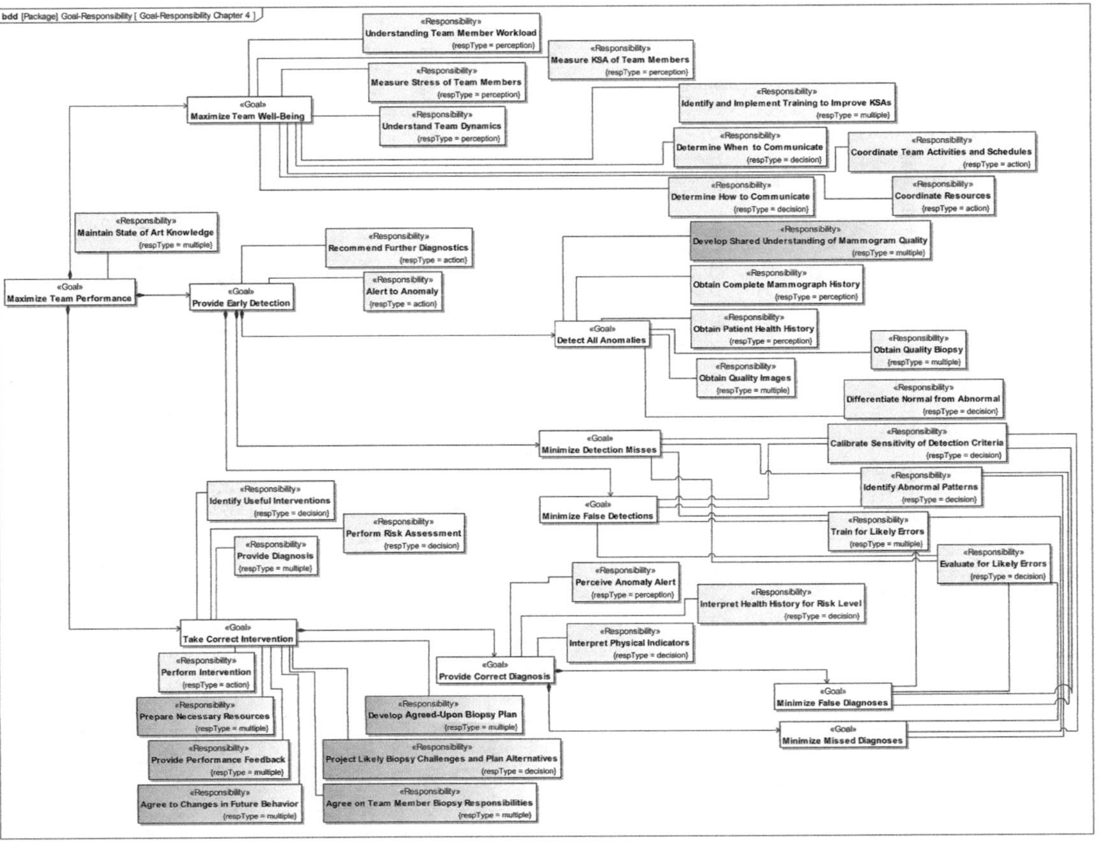

FIGURE 4.3 Revised Goal–Responsibility Diagram. Note the new responsibilities are darkly shaded.

and knowledge that our teammates will apply to execute the right steps at the right time during the action phase, and through the expression of commands or requests to others to execute high-level goals which relieve us from developing detailed goals and plans to achieve those goals. Although these mechanisms that aid coordination are common in human teams, the implementation of these behaviors requires the design of AI agents and HATs that have the intelligence to use similar mechanisms. We can divide these mechanisms into two categories. First, the human must be able to form a dependable mental model of the agent that permits the human to observe, predict, and direct the behavior of AI agents. Second, whenever possible, we should design AI agents to observe, predict, and direct the human through appropriate communication that facilitates coordination. Although these behaviors are not necessary in human–agent teaming, teaming may appear more natural to humans if the design team designs AI agents to take part in these activities.

4.11 QUESTIONS AND DISCUSSION TOPICS

1. Describe a team you have been a part of with an example of both implicit and explicit coordination that took place during the function of your team.
2. For the team you just described, which of the categories discussed by Sundstrom and colleagues do you believe this team falls into? Justify your answer.
3. Provide an example of processes your team performed prior to the action phase, and how these activities improved team performance during the action phase.
4. Provide an example of information that your team includes in their shared mental model and transactive memory. Justify why you provided information in each category.
5. For your team, describe both the advantages and costs associated with teaming.

4.12 PROJECT QUESTIONS AND ASSIGNMENTS

1. For your project, categorize your team using the categories discussed by Sundstrom and colleagues and provide a justification for your categorization.
2. For the human-based team you would like to redesign, describe the information your team exchanges during the transition process.
3. For each element of information your team exchanges during the transition process, how do you think it streamlines their processes during the action phase?
4. For the human team you are trying to redesign, what responsibilities and expectations exist among team members?
5. For the human team you are trying to redesign, what tasks does your team share or trade? What activities have pooled, sequential, or reciprocal interdependence?

6. For the human team you are trying to redesign, how do shared mental models contribute to implicit coordination?
7. For the human team you are trying to redesign, what are examples of advantages of the human team versus individuals, and can you list any shortcomings of the current team?
8. Provide at least three examples of the costs of coordination in the human team you are seeking to redesign.

REFERENCES

Behymer, K. J., & Flach, J. M. (2016). From autonomous systems to sociotechnical systems: Designing effective collaborations. *She Ji: The Journal of Design, Economics, and Innovation*, *2*(2), 105–114. https://doi.org/10.1016/j.sheji.2016.09.001

Boyd, J. R. (1976). *Destruction and creation*. US Army Command and General Staff College. https://doi.org/10.1108/pr.1998.27.2.158.2

Burke, C. S., Salas, E., Wilson-Donnelly, K., & Priest, H. (2004). How to turn a team of experts into an expert medical team: Guidance from the aviation and military communities. *Quality and Safety in Health Care*, *13*(Suppl. 1), 96–104. https://doi.org/10.1136/qshc.2004.009829

Butchibabu, A., Sparano-Huiban, C., Sonenberg, L., & Shah, J. (2016). Implicit coordination strategies for effective team communication. *Human Factors*, *58*(4), 595–610. https://doi.org/10.1177/0018720816639712

Carpenter, T. J., & Zachary, W. W. (2017). Using context and robot-human communication to resolve unexpected situational conflicts. *2017 IEEE Conference on Cognitive and Computational Aspects of Situation Management (CogSIMA)*, 1–7. https://doi.org/10.1109/COGSIMA.2017.7929596

Chen, J. Y. C., Lakhmani, S. G., Stowers, K., Selkowitz, A. R., Wright, J. L., & Barnes, M. (2018). Situation awareness-based agent transparency and human-autonomy teaming effectiveness. *Theoretical Issues in Ergonomics Science*, *19*(3), 259–282. https://doi.org/10.1080/1463922X.2017.1315750

Clark, H. H. (2005). Coordinating with each other in a material world. *Discourse Studies*, *7*(4–5), 507–525.

Clark, H. H., & Brenan, S. E. (1991). Grounding in communication. In L. B. Resnick, J. M. Levine, & S. D. Teasley (Eds.), *Perspectives on socially shared cognition* (pp. 127–149). American Psychological Association. https://doi.org/10.1201/b17574

Cooke, N. J., Gorman, J. C., Myers, C. W., & Duran, J. L. (2013). Interactive team cognition. *Cognitive Science*, *37*(2), 255–285. https://doi.org/10.1111/cogs.12009

Cox, A., & Szajnfarber, Z. (2018). System user pathways to change. In A. M. Madni, B. Boehm, R. G. Ghanem, D. Erwin, & M. J. Wheaton (Eds.), *Disciplinary convergence in systems engineering research* (pp. 617–634). Springer International Publishing.

Crowston, K. G. (1991). Towards a coordination cookbook: Recipes for multi-agent action. In *Alfred P. Sloan school of management: Vol. Doctorate*. Massachusetts Institute of Technology.

DeChurch, L. A., & Mesmer-Magnus, J. R. (2010). The cognitive underpinnings of effective teamwork: A meta-analysis. *Journal of Applied Psychology*, *95*(1), 32–53. https://doi.org/10.1037/a0017328

Delise, L. A., Gorman, C. A., Brooks, A. M., Rentsch, J. R., & Steele-Johnson, D. (2010). The effects of team training on team outcomes: A meta-analysis. *Performance Improvement*, *22*(4), 53–80. https://doi.org/10.1002/piq

Dickson, E. W., Singh, S., Cheung, D. S., Wyatt, C. C., & Nugent, A. S. (2009). Application of lean manufacturing techniques in the emergency department. *The Journal of Emergency Medicine*, *37*(2), 177–182. https://doi.org/10.1016/J.JEMERMED.2007.11.108

Eccles, D. W., & Tenenbaum, G. (2004). Why an expert team is more than a team of experts: A social-cognitive conceptualization of team coordination and communication in sport. *Journal of Sport and Exercise Psychology*, *26*(4), 542–560. https://doi.org/10.1123/jsep.26.4.542

Endsley, M. R. (1988). Situation awareness global assessment technique (SAGAT). *Proceedings of the IEEE 1988 National Aerospace and Electronics Conference*, 789–795. https://doi.org/10.1109/NAECON.1988.195097

Endsley, M. R. (1995). Toward a theory of situation awareness in dynamic systems. *Human Factors: The Journal of the Human Factors and Ergonomics Society*, *37*(1), 32–64. https://doi.org/10.1518/001872095779049543

Endsley, M. R., & Kiris, E. O. (1995). The out-of-the-loop performance problem and level of control in automation. *Human Factors*, *37*(2), 381–394. https://doi.org/10.1518/001872095779064555

Entin, E. E., & Serfaty, D. (1999). Adaptive team coordination. *Human Factors: The Journal of the Human Factors and Ergonomics Society*, *41*(2), 312–325. https://doi.org/10.1518/001872099779591196

Espinosa, A., Lerch, J., & Kraut, R. (2004). Explicit vs. implicit coordination mechanisms and task dependencies: One size does not fit all. In E. Salas & S. M. Fiore (Eds.), *Team cognition: Understanding the factors that drive process and performance* (1st ed., pp. 107–129). American Psychological Association.

Friston, K., FitzGerald, T., Rigoli, F., Schwartenbeck, P., O'Doherty, J., & Pezzulo, G. (2016). Active inference and learning. *Neuroscience and Biobehavioral Reviews*, *68*, 862–879. https://doi.org/10.1016/j.neubiorev.2016.06.022

Goodman, T. J., Miller, M. E., Rusnock, C. F., & Bindewald, J. M. (2017). Effects of agent timing on the human-agent team. *Cognitive Systems Research*, *46*, 40–51. https://doi.org/10.1016/j.cogsys.2017.02.007

Howard, N., & Cambria, E. (2013). Intention awareness: Improving upon situation awareness in human-centric environments. *Human-Centric Computing and Information Sciences*, *3*(1), 9. https://doi.org/10.1186/2192-1962-3-9

Jacques, D. R., & Strouble, D. D. (2011). *A-10 Thunderbolt II (Warthog) systems engineering case study* (No. ADA530838). https://scholar.afit.edu/docs

Johnson, M., Vignati, M., & Duran, D. (2019). Understanding human-autonomy teaming through interdependence analysis. In J. Kerstholt, J. Barnhoorn, T. Hueting, & L. Schuilenborg (Eds.), *Automation as an intelligent teammate: Social psychological implications* (pp. 1–20). North Atlantic Treaty Organization Science and Technology Organization.

Jones, N. A., Ross, H., Lynam, T., Perez, P., & Leitch, A. (2011). Mental model: An interdisciplinary synthesis of theory and methods. *Ecology and Society*, *16*(1), 46–46. https://doi.org/46

Kearney, E., Gebert, D., & Voelpel, S. C. (2009). When and how diversity benefits teams: The Importance of team members' need for cognition. *Academy of Management Journal*, *52*(3), 581–598. https://doi.org/10.5465/amj.2009.41331431

Klein, G., & Bradshaw, J. M. (2005). Common ground and coordination in joint activity. In W. B. Rouse & K. R. Boff (Eds.), *Organizational simulation* (pp. 139–184). Wiley-Interscience.

Klein, G., Woods, D. D., Bradshaw, J. M., Hoffman, R. R., & Feltovich, P. J. (2004). Ten challenges for making automation a "team player" in joint human-agent activity. *IEEE Intelligent Systems*, *19*(6), 91–95. https://doi.org/10.1109/MIS.2004.74

Klinger, D., & Klein, G. (1999). An "accident" waiting to happen? *Ergonomics in design*, *7*(3), 20–25. https://doi.org/10.1177/106480469300100312

Lacerenza, C. N., Salas, E., Burke, C. S., Marlow, S. L., & Paoletti, J. (2017). Does team communication represent a one-size-fits-all approach? A meta-analysis of team communication and performance. *Organizational Behavior and Human Decision Processes*, *144*(February 2016), 145–170. https://doi.org/10.1016/j.obhdp.2017.08.001

Langan-Fox, J., Anglim, J., & Wilson, J. R. (2004). Mental models, team mental models, and performance: Process, development and future directions. *Human Factors and Ergonomics in Manufacturing*, *14*(4), 331–352. https://doi.org/10.1002/hfm.20004

Lapso, J. A., Peterson, G. L., & Miller, M. E. (2023, February). A hybrid cognitive model for machine agents. *Cognitive Systems Research*, *81*, 1–10. https://doi.org/10.1016/j.cogsys.2023.02.007

Macmillan, J., Entin, E. E., & Serfaty, D. (2004). Communication overhead: The hidden cost of team cognition. In E. Salas & S. M. Fiore (Eds.), *Team cognition: Understanding the factors that drive process and performance* (pp. 61–82).American Psychological Association. https://doi.org/10.1080/03637759309376288.

March, J. G., & Simon, H. A. (1958). *Organizations*. John Wiley & Sons.

Marks, M. A., Mathieu, J. E., & Zaccaro, S. J. (2001). Remember me a temporally based framework and taxonomy of team processes. *Academy of Management Review*, *26*(3), 356–376.

Marstall, J., Miller, M. E., & Poisson Iii, R. J. (2016). Collaboration in the cockpit: Human–system interaction beyond the autopilot. *Ergonomics in Design*, *24*(1), 4–8. https://doi.org/10.1177/1064804615572628

Mathieu, J. E., Heffner, T. S., Goodwin, G. F., Salas, E., & Cannon-Bowers, J. A. (2000). The influence of shared mental models on team process and performance. *Applied Psychology*, *85*(2), 273–283.

McComb, S. A., Green, S. G., & Compton, W. D. (1999). Project goals, team performance, and shared understanding. *Engineering Management Journal*, *11*(3), 7–12.

Mesmer-Magnus, J., Niler, A. A., Plummer, G., Larson, L. E., & DeChurch, L. A. (2017). The cognitive underpinnings of effective teamwork: A continuation. *Career Development International*, *22*(5), 507–519. https://doi.org/10.1108/CDI-08-2017-0140

Mohammed, S. (2010). Metaphor no more: A 15-year review of the team mental model construct. *Journal of Management*, *36*(4), 876–910.

Noe, R. A., Mcconnell Dachner, A., Saxton, B., & Keeton, K. E. (2011). Team training for long-duration missions in isolated and confined environments: A literature review, an operational assessment, and recommendations for practice and research. *Researchgate. Net*, *October*, 44.

Rico, R., Gibson, C. B., Sánchez-Manzanares, M., & Clark, M. A. (2019). Building team effectiveness through adaptation: Team knowledge and implicit and explicit coordination. *Organizational Psychology Review*, *9*(2–3), 71–98. https://doi.org/10.1177/2041386619869972

Riley, V. (1989). A general model of mixed-initiative human-machine systems. *Proceedings of the Human Factors and Ergonomics Society Annual Meeting*, *33*(2), 124–128. https://doi.org/10.1177/154193128903300227

Rouse, W. B. (1976). *Adaptive allocation of decision making responsibility between supervisor and computer* (T. B. Sheridan & G. Johannsen, Eds.). Springer US. https://doi.org/10.1007/978-1-4684-2523-9

Rouse, W. B., Cannon-Bowers, J. A., & Salas, E. (1992). The role of mental models in team performance in complex systems. *IEEE Transactions on Systems, Man, and Cybernetics*, *22*(6), 1296–1308. https://doi.org/10.1109/21.199457

Rouse, W. B., Cody, W. R., & Boff, K. R. (1991). The human factors of system design: Understanding and enhancing the role of human factors engineering. *International Journal of Human Factors in Manufacturing*, *1*(1), 87–104. https://doi.org/10.1002/hfm.4530010108

Sajid, N., Ball, P. J., Parr, T., & Friston, K. J. (2021). Active inference: Demystified and compared. *Neural Computation*, *33*(3), 674–712. https://doi.org/10.1162/neco_a_01357

Salas, E., Dickinson, T. L., Converse, S. A., & Tannenbaum, S. I. (1992). Toward an understanding on team performance and training. In R. W. Swezey & E. Salas (Eds.), *Teams: Their training and performance* (pp. 3–29). Ablex.

Santner, T., Santner, W., & Gutzeit, A. (2021). Effect of image quality and motivation of radiographer teams in mammography after dedicated training and the use of an evaluation tool like PGMI. *Radiography*, *27*(4), 1124–1129. https://doi.org/10.1016/j.radi.2021.05.006

Schaefer, K. E., Straub, E. R., Chen, J. Y. C., Putney, J., & Evans, A. W. (2016). Communicating intent to develop shared situation awareness and engender trust in human-agent teams. *Cognitive Systems Research*, *46*, 26–39. https://doi.org/10.1016/j.cogsys.2017.02.002

Scheutz, M., DeLoach, S. A., & Adams, J. A. (2017). A framework for developing and using shared mental models in human-agent teams. *Journal of Cognitive Engineering and Decision Making*, *11*(3), 203–224. https://doi.org/10.1177/1555343416682891

Schneider, M. F., Miller, M. E., Ford, T. C., Peterson, G., & Jacques, D. (2022). Intent integration for human-agent teaming. *Systems Engineering*, *24*(4), 291–303. http://doi.org/10.1002/sys.21616

Sheridan, T. B., & Verplank, W. L. (1978). Human and computer control of undersea teleoperators. *DTIC*. http://oai.dtic.mil/oai/oai?verb=getRecord&metadataPrefix=html&identifier=ADA057655

Stout, R. J., Cannon-Bowers, J. A., Salas, E., & Milanovich, D. M. (1999). Planning, shared mental models, and coordinated performance: An empirical link is established. *Human Factors: The Journal of the Human Factors and Ergonomics Society*, *41*(1), 61–71. https://doi.org/10.1518/001872099779577273

Stowers, K., Kasdaglis, N., Newton, O., Lakhmani, S., Wohleber, R., & Chen, J. (2016). Intelligent agent transparency: The design and evaluation of an interface to facilitate human and intelligent agent collaboration. *Proceedings of the Human Factors and Ergonomics Society*, *60*(1), 1706–1710. https://doi.org/10.1177/1541931213601392

Sundstrom, E., De Meuse, K. P., & Futrell, D. (1990). Work teams: Applications and effectiveness. *American Psychologist*, *45(2)*, 120–133.

Sycara, K., & Lewis, M. (2004). Integrating intelligent agents into human teams. In E. Salas & S. M. Fiore (Eds.), *Team cognition: Understanding the factors that drive process and performance* (1st ed., pp. 203–231). American Psychological Association.

Thompson, J. D. (2003). *Organizations in action: Social science bases of administrative theory*. Transaction Publishers.

van den Bossche, P., Gijselaers, W., Segers, M., Woltjer, G., & Kirschner, P. (2011). Team learning: Building shared mental models. *Instructional Science*, *39*(3). https://doi.org/10.1007/s11251-010-9128-3

Wegner, D. M. (1986). Transactive memory: A contemporary analysis of the group mind. In B. Mullen & G. R. Goethals (Eds.), *Theories of group behavior* (pp. 185–205). Springer-Verlag.

Woods, D. D., & Hollnagel, E. (2006). *Joint cognitive systems: Patterns in cognitive systems engineering*. CRC Press.

Yan, B., Hollingshead, A. B., Alexander, K. S., Cruz, I., & Shaikh, S. J. (2021). Communication in transactive memory systems: A review and multidimensional network perspective. *Small Group Research*, *52*(1), 3–32. https://doi.org/10.1177/1046496420967764

Yen, J., Fan, X., Sun, S., Hanratty, T., & Dumer, J. (2006). Agents with shared mental models for enhancing team decision making. *Decision Support Systems*, *41*(3), 634–653. https://doi.org/10.1016/j.dss.2004.06.008

5 Automation and System Redesign

5.1 DEFINING GOALS AND UNDERSTANDING OF THE "TO-BE" SYSTEM

To begin to redesign the "To-Be" system, it is important to understand the types of improvements that AI can bring to our "As-Is" system. This chapter describes the general process for considering and documenting the goals of your "To-Be" system. Figure 5.1 shows the process steps involved. The left panel in this figure shows the overall AOSM process, with the third step highlighted. The right panel of Figure 5.1 shows the process steps discussed in this chapter. As shown, this step begins with identifying the redesign objectives, creating the "To-Be" Goal Diagram and the "To-Be" Goal–Responsibility Diagram. We continue these process steps in the next chapter where we discuss derivation of capabilities.

5.2 WHY AUTOMATE?

To redesign the HAT, it is important for you to form a clear vision for including AI agents in the larger sociotechnical system you seek to redesign. This begins by understanding the objectives of your "To-Be" design and shaping the design process to achieve these objectives. Traditionally, technologists have suggested two primary premises for the use of automation and artificial intelligence. These include improving human efficiency to reduce cost, usually labor costs, and improving system safety. It is useful to briefly explore each of these. However, there are many other reasons we may want to employ human–agent teams. Thus, it is important to understand the potential benefit of AI agents beyond these two objectives, as well as variations on approaching these two objectives.

5.2.1 Promise of Cost Reduction

Systems employing human operators have an ongoing cost associated with the cost of recruiting, training, and sustaining the workforce. The organization bears these costs throughout the entire system's life cycle, and these costs rise over time. Therefore, labor costs can account for a considerable proportion of the total life cycle cost of a system. Further, in safety-critical systems, humans are susceptible to harm during operations. While direct monetary cost of labor can be significant over the life cycle of a system, the potential loss of life can have larger indirect costs. This is particularly relevant in western societies where the publicity associated with loss of an individual's life can influence public opinion against operations and force early closure or

DOI: 10.1201/9781003428183-5

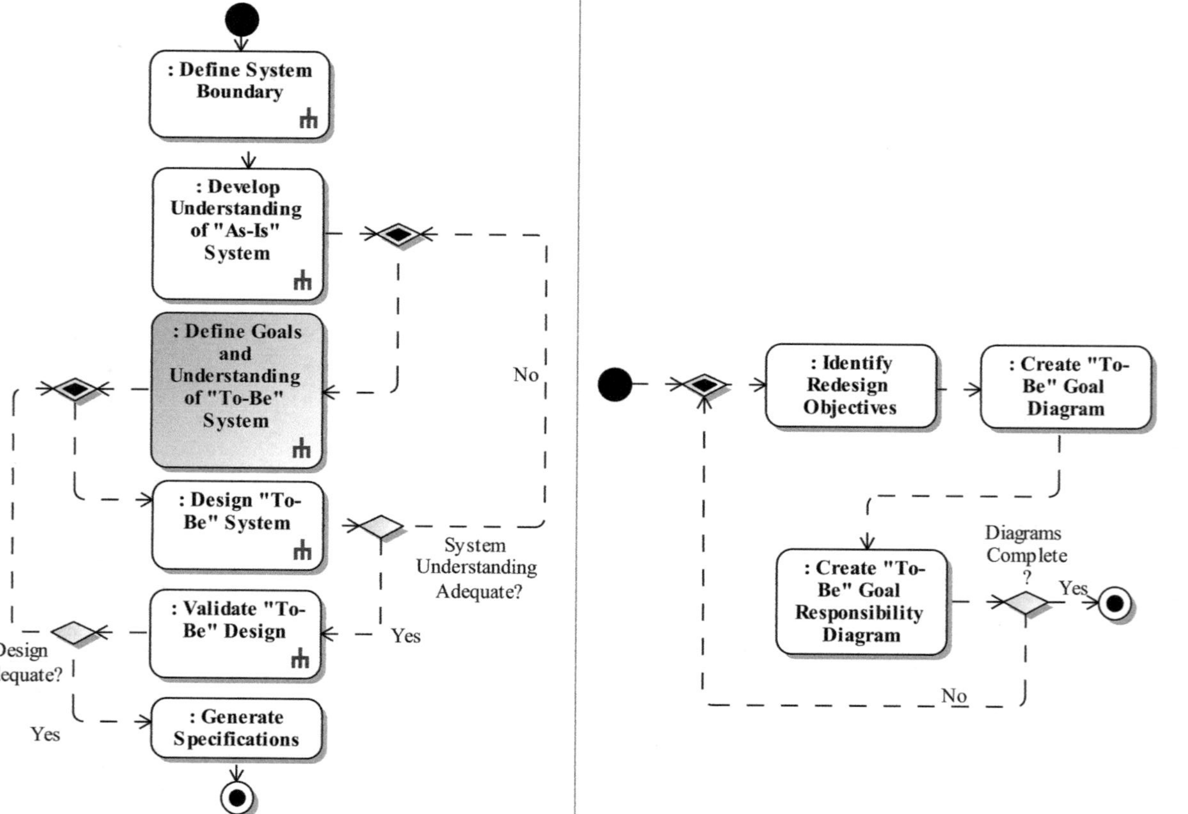

FIGURE 5.1 A depiction of the AOSM process on the left with the current sub-step highlighted. The right panel shows the sub-steps associated with Define Goals and Understanding of "To-Be" System steps.

discontinuation of facilities or operations and loss of investment. The replacement of humans with AI agents has the potential to alleviate both the monetary and publicity costs associated with employing humans to control systems.

Developers often assume that AI agents will replace humans, reducing the direct monetary cost of system operation. Historically, the introduction of AI or AI agents has reduced the required number of human operators in many applications. For instance, many large aircraft constructed around the middle of the 20th century included personnel with roles such as "Engineer" or "Navigator". In the intervening decades, advancements in avionics have eliminated the need for humans to perform each of these roles, reducing the direct personnel costs, including the salaries, benefits, and other costs associated with these individuals. Further, as each aircraft includes fewer individuals whose lives might be lost with the loss of an aircraft, the inclusion of AI reduces the indirect costs associated with loss of life. As a result, these aircraft have often become more capable with the introduction of automation, permitting shorter flight paths, increased comfort by avoiding weather, and lower air fares in commercial aviation. The advantage is larger in military aviation as this technology permits one to achieve similar or greater effects with far fewer aircraft. Thus, the introduction of automation has certainly reduced the labor required to operate most large aircraft. Superficially, these changes reduce the associated direct and indirect costs in these and many other industries.

However, the adoption of AI has introduced significant complexity. This increase in complexity increases the skills and knowledge required of the remaining crew and the skills and knowledge required of the support personnel. This increase in required skill translates directly to requirements for longer, more intensive, and often more frequent training to permit the humans to understand the utility and operation of the AI. Failure to deliver this training can have catastrophic consequences. A recent example is the loss of two Boeing 737 Max aircraft in separate incidents, resulting in the loss of 346 lives. Although many factors contributed to these incidents, a key factor was the decision not to provide adequate training for the pilots of these aircraft. It has been reported that airlines provided incentives to the aircraft designer to provide an aircraft that did not require simulator-based training to avoid the costs of providing this training, which would entail removing pilots from service to permit them to participate in simulator-based training (Canellas & Haga, 2020; Gates & Baker, 2019). Therefore, when a particular AI feature was engaged, the pilots did not understand its behavior or their role when interacting with the AI, leading to the loss of these aircraft.

Increases in system complexity also often require increases in the number and skills of support personnel. For example, the number of support personnel for certain unmanned aerial vehicles (UAVs) is dramatically higher than traditional aircraft platforms. Therefore, it is common for automation to increase both the direct costs associated with recruiting and training capable operators and the direct costs associated with increasing numbers and required capabilities of support personnel. Among these support personnel are the software engineers needed to support the life-cycle updates to the AI as well as an increased number of test engineers necessary to verify and validate that the repeated updates to the AI meet the design requirements. Therefore, it is important to develop tools which permit the designer of systems employing AI to understand not only the benefits associated with automating human activity but also the costs associated with the introduction of AI. For individuals associated with safety-critical industries, these tools need to provide insight into the direct costs and benefits of incorporating AI as well as the indirect costs and benefits.

Understanding the actual cost or benefit of applying AI and mechanical automation requires one to consider both direct and indirect costs. However, well-designed automation, mechanical or cognitive, can produce significant benefits. As an example, coal mining used to require sending hundreds of people underground into a dangerous and dirty environment, where many died from mining collapses as well as from long-term exposure effects to coal dust. Today, automated machines accomplish much of mining, requiring few individuals to enter the dangerous and dirty environment. Thus, designers should not assume that cost reduction and other benefits will occur anytime AI is introduced to enhance or replace human operators. Instead, designers should evaluate both the direct and indirect costs, as well as the benefits of automation.

5.2.2 System Safety, Decision-Making, and Resilience

The human operator is often described as the least predictable element within any system (Schutte, 2017). Importantly, individuals often intend this statement as derogatory toward human performance, implying that systems should perform in a standard fashion. To improve the predictability of the human element in safety-critical systems, system designers routinely provide standard procedures and checklists and then require the human operator to follow these procedures. Aircraft accident investigations often conclude that a significant cause in a substantial percentage of accidents is human error, often associated with the human's failure to follow the appropriate procedure or checklist. Therefore, a frequent refrain from automation designers includes the promise of increased system safety through the adoption of more predictable automation.

In fact, humans often do not have the mental capacity or knowledge to perform certain required tasks, which truly result in human error, resulting in system failures. For instance, thousands of people die in highway accidents every year that involve humans who are driving under the influence of alcohol or drugs that degrade their mental capacity, leading to deadly accidents. We can also observe that accidents are more likely to occur for younger drivers who have less experience and knowledge to handle complex highway environments. As discussed earlier, one method to increase human cognitive capacity is to include multiple operators, such as the two pilot configurations in commercial aviation, or requiring experienced drivers to aid novice drivers during driver education. AI agents, when designed appropriately, can provide similar backup behaviors to secondary humans to reduce the burden on the human, particularly, when the cognitive load on the human is undesirable, reducing the likelihood of accidents.

It is also important to recognize that the human controlling the system is not the only source of variability. Even if the designer could remove variability from the response of the system, the fact remains that most real-world operating environments contain a large amount of variability, and the system must respond appropriately to this variability to maintain safe performance. Weather can impose a significant amount of variability, as can less common real-world conditions. Further, unexpected system behavior occurs as well, often born from hardware or software malfunctions. AI, which works a substantial portion of the time (e.g., automatic lane keeping algorithms which track the location of a vehicle inside edge lines on a highway), often has difficulty adapting to less common conditions (e.g., road construction zones without edge lines or gravel roads). The variability in the environment produces circumstances which designers must fully anticipate and compensate for during the design of AI systems. However, the overall system must be capable of adapting to these situations to maintain safe performance; thus, even automobiles with advanced AI require the human to keep their hands on the steering wheel and remain engaged with the environment to "save the day" during automation failure.

Many of these issues were described by Bainbridge when discussing the automation of manufacturing processes in the 1980s (Bainbridge, 1983). Her landmark paper points out the ironies of automation. These ironies include the fact that designer errors can be a major source of operating problems, but these errors are rarely reported. A second irony is that designers typically cannot automate all operator tasks, which leaves the operator to perform a set of arbitrary tasks and the designers are often focused only on implementing the automation, without consideration for providing aids to structure or support the human in completion of the remaining tasks. Often, these tasks include monitoring the automation for failures and recovering the system after automation failure. As we will discuss later, humans are poor at monitoring systems and are often unable to recognize failures when they are not participating in control of the system. Further, without training and practice in operating the system, they are unlikely to have the physical skills and knowledge that is necessary to recover the system, The human-centered research has echoed these persistent concerns since (Endsley, 2023; Strauch, 2017).

An intelligent adversary can enhance unanticipated variability as they create uncommon situations to which the AI must respond. Once an intelligent adversary identifies these vulnerabilities, they can readily repeat this exploit to significant effect. Additionally, an intelligent adversary can deploy deceit to represent uncommon situations as if they were common. Such circumstances require the system to be capable of both detecting and responding to an altered environment, to provide an altered response, even when the alterations are subtle, masked, and difficult to detect. It is also important that the designer cannot anticipate all "intelligent adversaries". For example, activists upset with the behavior of autonomous vehicles discovered that they could stop these vehicles from moving by placing a traffic cone on the car's hood (Gordon, 2023). This behavior was repeated to disable individual vehicles. In another example, individuals aimed lasers into lidar sensors on autonomous vehicles, causing the vehicle to detect and react to nonexistent obstacles (Marks, 2020). Each of these behaviors is certainly adversarial to the adoption of AI embedded in self-driving automobiles.

The need to respond to unforeseen circumstances leads to the discussion of decision-making. Scientists have made significant advances in AI, which have permitted AI systems to rival human performance in specific tasks. For example, designers have constructed AI that can exceed human performance when performing object and pattern recognition and making decisions in controlled, predictable environments, such as playing Chess or the game Go. This success has resulted from the AI learning responses to specific patterns within a deterministic environment, where the environment constrains responses to prescribed rule sets. For the psychologist or human factors engineer, these systems are using processes that are somewhat analogous to the intuitive system, sometimes referred to as System 1, in the human decision-making literature (Patterson & Eggleston, 2017). In employing this process, researchers believe that the human recognizes a situation from experience and selects actions based solely on the association with this experience. In games like Chess or Go, AI systems can rapidly simulate thousands of games to learn moves which effectively emulate effective strategies (Hubert et al., 2017). This rapid learning is advantageous in a deterministic (i.e., highly predictable) game environment, giving artificial intelligence the ability to perform well in these environments.

As humans, we often form strategies where we devise a plan of future action that we intend to use to shape the world, including the plans or strategies of others, with the intent of enabling us to achieve our goals, given our current resource and capability limitations. Plans and strategies are distinct from one another. A *plan* is an arrangement of process steps that we intend to result in a desired outcome. A *strategy* is a blueprint or idea used to accomplish a goal that is open for adaptation and change when needed; thus, a strategy may include several plans which are adopted on the basis of changes in the environment or the behavior of others in the environment. Strategies might include plans that include shifts in cooperation with others or deceit, particularly when we are competing with others in an adversarial relationship where it is not possible for everyone to achieve their desired outcomes. Strategies once again require the projection of future activity. Today's deep learning systems do not create strategies but simply learn moves which were effective under similar situations. As we will discuss later, through adversarial learning, this decision space can become quite complex. AI agents pursue these complex, learned relationships. Importantly, as humans, we are unable to devise and understand the patterns the AI agents are following, which result in this complex behavior. As a result, we tend to anthropomorphize this behavior and assess that the AI is pursuing "strategies". However, there is a distinction between the ability to devise and pursue novel strategies as opposed to following learned patterns to achieve results based purely upon historical evidence.

We understand, however, that some decisions, particularly decisions required in new circumstances, require different reasoning processes. The psychology literature on human decision-making suggests that the analytic system, sometimes referred to as System 2, is applied in these circumstances. Researchers believe that these processes rely upon analogy to determine responses to circumstances that we have not previously experienced. These processes often include determining circumstances which are analogous to our present situation along one or more dimensions, abstracting rules from that context, and applying these rules in our current context to derive one or more potential courses of actions. Given time, we can then employ mental simulation of these potential courses of action to determine the course of action which fulfills some criteria before deciding upon an action (Klein, 1993). Mental simulation may include some aspect of estimating the potential risk associated with a given course of action, and we are likely to employ a course of action we believe will have "acceptable" risk. The literature supports the concept that we are biased toward response, that is, taking action without fully analyzing the decision space, especially in time-critical domains. In fact, people often adopt a strategy referred to as *satisficing* in which the decision-maker adopts the first potential response which is believed to provide the desirable resources to achieve our goal with "acceptable" risk (Simon, 1956).

While the AI community has developed approaches that provide rapid training and application of structures that respond to known circumstances, the search for ***general*** *intelligence*, which can innovate to determine solutions to a broad range of unknown problems, remains elusive. Current artificial intelligence consists of *specific* or *narrow intelligence*, in which the AI is highly proficient at a specific task(s) or in a limited domain(s). While these specific intelligences can be employed to search parameter spaces in efficient fashion, these systems are only applicable when the parameter space is well known, and the outcome of the system response based upon the modification of the parameter space is well characterized. Therefore, these systems are only useful under constrained circumstances within most real-world environments as they often do not, or cannot, consider all valuable information. Further, these systems often do not have the ability to anticipate future circumstances and often fail to include valuable information within their decision criteria. It is the human's ability to anticipate and innovate, due to our general intelligence, which separates and elevates our cognitive assets as compared to modern AI systems. While general intelligence certainly contributes to the potential for unexpected and variable responses, these abilities are necessary if the system is to adapt successfully to unforeseen circumstances to maintain safe operation.

To return to our earlier discussion of human error leading to accidents, the literature indicates that many of these accidents occur after unexpected or "failed" system response. Under these circumstances, designers expect the human operator to assume control of and recover the system. As the human often cannot anticipate these failures and may not have knowledge of recovery processes, this system behavior often imposes time pressures or circumstances that require the human operator to depart from their standard checklist and participate in unexpected behaviors. It is not surprising that incidents lead to accidents. However, incidents occur relatively frequently during operations with little fanfare or consequence. In fact, Holbrook has suggested that over a ten-year period in aviation, permitting the automation to function without a human would have reduced the number of worldwide accidents by 310. However, more than 48 million additional accidents might have occurred without direct human intervention to recover aircraft after automation failures (Holbrook, 2021)! Of course, as there is a bias toward AI adoption, designers often do not report or even record this later number. Therefore, this number is only an estimate.

Humans can certainly contribute to errors. Humans often commit these errors when they are fatigued, bored, or are experiencing cognitive overload. Teaming can also help to overcome many of these situations. Human teammates can and do often share work, relieving teammates as they become fatigued or bored when possible. Under other circumstances, human teammates inject cognitive activity as teammates become bored. This injection can include casual conversation, among other activities. Further, these teammates can provide cognitive support to overcome cognitive overload. Therefore, effective teammates can help reduce human error.

5.2.3 Increasing Factual Knowledge

When discussing teaming, we discussed including teammates with diversified knowledge, skills, and abilities. Over the past several decades, developers have constructed computers to support human needs, including providing access to precise facts or performing mathematical operations. Even though many of us know certain numbers precisely, we have limited ability to remember these numbers. For instance, we might know that the value of Pi is 3.14 or 3.14159, but rarely would a person provide the value of PI as 3.141592653589793238. Yet, that is the response an internet search might provide for PI. Thus, humanity has constructed computers to store and aid the recall of precise facts and to rapidly perform precise, accurate calculations, specifically because these are skills that we do not possess. Also, with the internet, computers can share precise facts, permitting the team to distribute knowledge without human effort. As a result, AI agents have the ability to enhance human cognition in these areas.

It is also important to recognize that humans employ heuristics in environments where they lack exact knowledge to make decisions, and these heuristics can introduce bias into decisions (Kahneman, 1973). While these heuristics will often lead to correct decisions in natural environments (Lopes, 1991; Patterson et al., 2010), some biases, such as anchoring on early information, can introduce error in environments where counter information is available. It is worth considering how a designer could design AI agents to overcome these decision biases. This may include providing and supporting contra-evidence and hypotheses in complex decision-making environments, as we will discuss in more detail in Chapter 9.

5.2.4 Improving Organization Performance

In this chapter, we have touched on the primary reasons given for employing automation, as well as misperceptions regarding these rationales. To review, the primary reasons given for employing automation often include:

1) Reduction of human effort and the expense that results from employing humans;
2) Removing humans from dirty, dangerous environments to protect their well-being;
3) Adding to the predictability of system response, resulting in an increased system reliability in predictable situations; and
4) Enhancing human knowledge, computation, and completeness of option consideration during decision-making.

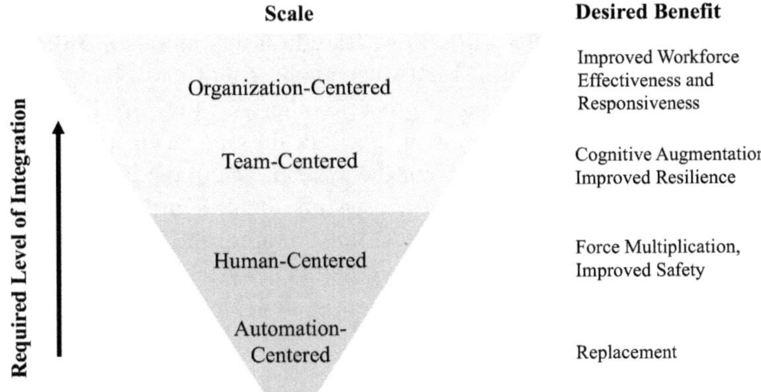

FIGURE 5.2 A hierarchy of enterprise level objectives and stakeholder values as they relate to a hierarchy of human-centric automation systems.

Figure 5.2 shows what might be a broader view of automation. As shown in this figure, we can consider automation at different scales. At the lowest scale, we develop systems that are automation-centered. In this case, we focus on the design of the AI itself, with little concern for how it integrates with other entities in the world. One might consider this type of automation, dating back thousands of years to Archimedes who considered the construction of simple systems to replace human physical effort. However, many systems, including systems motivated by the reasons given earlier, often view the problem within a similar context where AI is intended to replace the human, often to reduce the recurring cost of a human operator or to free the human from "mundane" tasks. Ideally, such a replacement would permit humans to perform other, more enjoyable tasks.

The term Human-Centered Automation has been in place since Billings proposed what he termed the "first principles of human-centered aviation automation" (Billings, 1996). In these principles, Billings recognized that the pilot in these systems is "responsible for safety of flight" and must command the aircraft and any AI within the aircraft. Therefore, AI designed from a human-centered viewpoint should improve system performance or safety through a process that requires the AI to communicate with a human in the system to achieve higher-level system goals. This design viewpoint often focuses on the design of the system with an individual human combined with AI to provide improved safety and serve as a force multiplier. AI at this level permits the human to offload less critical or difficult tasks to the AI to enable the human to focus on tasks that the AI cannot support and for the human to "back the automation up" or assume control when the AI is incapable of a successful performance.

Billing's discussion of human-centered automation noted that humans must be responsible in safety-critical systems. Specifically, pilots are responsible for the safety of flight, while air traffic controllers are responsible for maintaining aircraft separation and safe traffic flow. Based on this premise, Billings suggests the following principles. First, the human (pilot and controller) must be actively involved in the control process, otherwise they are likely to lose awareness of the state of the system. Second, the human operators must be "adequately informed" of the state of the system. Third, the human operators must be able to monitor the AI that is assisting them. Fourth, the behavior of the AI must be predictable. Fifth, the AI must monitor the human operators to support decision-making. Finally, every intelligent system, human or AI, must understand the intent of the other intelligent system elements.

Note that one of the purposes of Figure 5.2 is to provide a distinction between automation-centered or human-centered approaches and team or organization-centered approaches. The former usually considers technology to impact the work of a single human operator. The latter considers the impact of AI on a larger group of humans and focuses on the performance of an overall team of humans and AI agents. Although a designer can use the approach provided in this text to aid the design of human-centered automation, especially when multiple AI agents are designed to interact with the human, the approach is more important as the team size increases. These systems require a higher level of integration and include a larger design space than human–AI agent teams with smaller numbers of entities.

The differences between the approaches to automate system design depicted in Figure 5.2 are often a matter of focus. For example, past efforts at cost reduction have often resulted in systems that are more efficient, faster under nominal conditions, and result in operator cost reductions. Unfortunately, they are often more brittle and less adaptable to unforeseen circumstances (Endsley, 2015; Shively et al., 2016). Also, there are often tasks within an operator's role that AI cannot perform, such as physically activating an external system or monitoring for inputs that the designer failed to design the AI agent to detect (e.g., excessive engine vibration). As the remainder of an individual's work is automated, the designer often transfers these tasks, knowingly or not, to another human in the original team. This individual must then perform these tasks in addition to their regular duties. By reallocating these tasks to the second human, the designer removes slack time that the human may have had for planning or reacting when unforeseen circumstances arise. As a result, the system can become less resilient to unexpected external perturbations. Placing an emphasis on human-centered automation implies that the designer considers the secondary effects of AI implementation on the operators of the system, such that the designer considers their performance and system resilience during the design process. However, placing emphasis on the design of the team or organization often requires rearchitecting the arrangement and interaction of the human and AI agents to enhance the workflow, potentially increasing the ability of the system to respond to external influences in a positive or negative fashion. A primary objective of human–AI agent teaming is often to aid the design of systems that not only respond faster but also are more resilient to external perturbations.

5.3 IDENTIFY AND DETAIL PURPOSE OF REDESIGN

If this text has been successful, you are beginning to understand the substitution myth, that is, the mistaken belief that by introducing an AI agent into a system, a designer only needs to replace the everyday task work performed by the person. Inherent in this discussion is the fact that humans have goals, which often stem from the belief that human life is valuable, and therefore we must preserve our life and the lives of those who we cherish. As a result, we devise lower-level goals which support this overall belief system. While we can design AI to pursue goals that are consistent with human goals, AI agents do not inherently have goals. Further, these AI agents cannot be held accountable for their actions as there is no true means to reward or punish them for their actions (Inagaki, 2003). Philosophically, all goals are inherently human and although automation might pursue near-term goals, at some level of these goals, humans must impart these goals to the AI agent. Therefore, unless the designer defines an AI agent narrowly, such that its goals never change, humans must be able to interact with the AI agents to help them understand the goals they are to achieve. Subsequently, while AI agents might replace elements of a human's activities, they will need to interact with humans to remain in synchronization with other members of the team.

Related to the substitution myth, Bradshaw and colleagues have made a series of five observations about the typical assumptions that accompany the substitution myth (Bradshaw et al., 2013). First, they observed that AI agents are not a discrete component that a designer can simply "plug" into a system. Instead, the AI agents usually bring new or enhanced capabilities to a team where these capabilities require the AI agent to interact with the environment in a particular situation through a machine. Further, this cyber-physical environment in which the AI agent operates will include humans with which it will need to interact. Thus, the AI agent will need to not only perform task work but also interact with humans and other AI agents in this cyber-physical environment. Second, no entity, including humans, is truly autonomous. We may behave autonomously when performing certain actions, but as social beings we, as humans, will interact with others who will impose upon our autonomy, and the same restrictions will inherently apply to AI agents. We hope the exercise of answering the Chapter 1 questions illustrated this fact. Third, humans often conduct work in dynamic, fast-paced environments that require multiple agents to respond to complex tasks with goal priorities that shift in response to environmental constraints. An AI agent cannot contribute meaningfully to this team without having the ability to interact with members of the team. Additionally, if the AI agent cannot communicate with the humans in the team, the humans will lose awareness of the portions of the environment the AI agent might be monitoring and the work it is performing. As such, they cannot coordinate their work with the AI agent. Fourth, the suitability of an AI agent or a human to complete a set of task work is situationally dependent. For humans, this is obvious, as our capability and capacity change as we undergo training, become fatigued, consume certain substances, experience different lighting and noise environments, and experience numerous other effects. Similar effects occur for AI agents, as their sensors may or may not be as effective as human senses under some circumstances, and they may or may not have the knowledge to respond appropriately in an environment under some circumstances. Finally, these authors point out that adding semiautonomous AI agents to a system typically changes the work that the remaining humans in the team must perform. If designed poorly, the AI agents can increase the workload of the humans and have other effects, such as causing the humans to cognitively decouple from the work, such that they cannot contribute effectively to the environment.

Importantly, thinking about the use of AI to replace humans implies that you are expecting to complete the same amount of work with fewer people, effectively making every human more efficient or effective. Does it change the problem and the solutions we might envision if we approach the problem from the viewpoint of making humans more effective and efficient, rather than replacing a single human or type of human? In the project you are considering, it is likely that you have already thought an AI agent could readily do elements of the work that the humans find boring, tedious, dangerous, or difficult. You may have also seen areas, perhaps some of the same areas, where the humans are unable to perform the work either reliably or effectively. The question becomes then: How do we then integrate the AI agents into the current workflow in a way that significantly augments the humans' abilities to perform work? This augmentation can have many different benefits, including increased efficiency or cost reduction, improved performance or utility, error or accident avoidance, increased response time, decreased human workload, increased human engagement, or increased creativity. As we begin to think about designing our human–AI agent team, it is important to consider which of these benefits will be valued by the stakeholders in your system and should be adopted as design criteria as we begin to switch focus from understanding our "As-Is" system and begin to consider our "To-Be" system.

It is useful to consider potential design goals for our "To-Be" system. Table 5.1 provides a list of concepts and examples that we can consider when thinking about our design goals. This table provides titles for several implementation examples with a quick description of the implementation, application examples for where similar AI implementations exist, design goals this implementation can support, some design implications regarding requirements for the AI agent, and finally potential negative consequences for adopting each implementation. A few of these implementation examples are described in detail in Table 5.1.

5.3.1 Asynchronous Communication Support

Recalling the coordination costs from Chapter 4, a significant cost is synchronization overhead which refers to the time that one individual spends waiting for a team member to complete a task and provide information that is necessary for further progress. Unlike humans, AI agents do not need to rest and can be active continuously. Therefore, if the AI agent has knowledge regarding work that is, like a human, it can provide this information to another anytime, day or night. In geographically distributed teams, we might be able to request information from an AI assistant while the human they support is not available. This would require technology like the chatbot technology that is present in customer support aids for e-commerce websites or first-line telephone support systems. These systems can provide this information, day or

TABLE 5.1
Example AI Agent Implementations.

Implementation Examples	Description	Application Examples	Design Goal	Design Implications	Implementation Consequence
Asynchronous Communication Support	AI agent enables communication between parties at different points in time	– AI-enabled email – AI-enabled messaging system – Automated user-help chatbot on e-commerce website	– Cost reduction – Geographic/Chrono separation Delay reduction – Workload reduction – Increased consistency	AI must know what information it can be share and with whom	– May increase communication quantity and quality – May impact customer/user experience
Accident Avoidance	AI agent monitors environment, intervenes to prevent mishaps	– Automatic braking – Automatic ground collision avoidance systems	– Error reduction – Improved safety – Faster response time	AI must reliably be able to determine hazards	– May reduce human situation awareness or risk perception – May interfere with risky but necessary activities
Human Error Reduction	AI agent determines times that human's performance is degraded and aids to reduce the error likelihood	Automotive drowsiness detection and warnings	Error reduction Improved safety	AI must provide appropriate countermeasures	Humans may over rely on technology
Directed Attention	AI agent makes selected portion(s) of the user interface more salient or augment items in an environment	– Highlighted text displaying errors – Flashing/color-changing portions of user display – Graphical depiction of a high-speed vehicle approaching an intersection before it is visible	– Error reduction – Safety – Improved decision-making – Faster response time	AI must be capable of determining when something is important enough to redirect human attention from their current focus to the item of interest	May disrupt ongoing human activities

(Continued)

TABLE 5.1 *(Continued)*
Example AI Agent Implementations

Implementation Examples	Description	Application Examples	Design Goal	Design Implications	Implementation Consequence
Task-Load Balancing	AI agent takes over tasks during high-workload situations; AI agent hands off tasks to the human during times of low workload	– Adaptive automation systems based on user performance or physiology measures	– Error reduction – Workload management – Faster response time	AI must be able to reliably understand when workload is sufficiently out of balance	May reduce human situation awareness and increase handoffs and task-switching costs
Reminder System	AI agent alerts human to prevent an error due to an act of omission	– Automated re-ordering reminder system	– Error reduction – Workload reduction	AI must know required actions and determine when humans skip steps	May disrupt ongoing human activities
Recommender System Decision Aid	AI agent recommends selection based on established criteria	Product recommendations on e-commerce websites	– Improved Decision-Making – Workload reduction – Faster response time	Robust recommendation requires the AI agent to know the criteria the human would employ in the situation	May reduce the quality of decisions
Action Implementation System	AI agent executes actions	– Automated re-ordering system – Tsunami warning system	– Workload reduction – Faster response time – Always available	Must have ability and authorization to act	May reduce human situation awareness
Understanding Complex Information Spaces	AI agent analyzes complex datasets to support inferences	Weather prediction systems	– Improved decision-making – Identifying complex relationships	Must have access to relevant data	Not all correlations indicate causation, false attribution
Option Evaluator	AI automatically calculates the values for multiple decision criteria	– Navigation aids show estimated travel time to destination – Aircraft route planning, estimates regarding ETA, fuel, passenger comfort	– Improved decision-making – Workload reduction – Increased safety	AI must communicate options and be able to estimate values of important human criteria for decision-making	Adds communication to convey options and requires time for the human to evaluate options and understand recommendations
Option Generator	AI generates options to support decision-making	– Automotive navigation systems – Engineering design tools	– Improved decision-making – Reduced bias	AI must understand option space and important criteria	Requires time to consider and decide among increased number of options

night, regardless of human availability. Further, they can provide consistency as they can offer only specific information to specific requests. The AI agents, however, need to know what they can provide, and the technology has the potential to increase communication quantity and reduce its quality as an individual using an AI assistant will likely need to know what its assistant told team members to maintain an appropriate team mental model and perhaps correct certain responses.

5.3.2 Accident Avoidance

In recent years, accident-avoidance systems have been developed where an AI agent detects potential collision hazards and reacts to avoid these collisions (Norris, 2016). In automobiles, this is referred to as Automatic Braking systems; and in aircraft, similar Ground Collision Avoidance Systems have been developed (Bice et al., 1990). These systems either sense obstacles in the environment or use environmental knowledge, like terrain maps, to determine potential obstacles and then consider the vehicle's path, velocity, and acceleration to project whether a collision is imminent. To avoid confusion, these systems typically determine the point of no return, where an action such as braking or pulling up the nose of the aircraft will not avoid the collision, and take this action once it would be impossible for the human to respond fast enough to avoid this point of no return. As a result, breaking or the maneuver is often relatively violent. Designers often provide warnings prior to this action to alert the vehicle operator and give them the option to take the appropriate action prior to system activation. These systems require accurate identification of obstacles, and recent automotive testing has shown that some automobiles are unable to detect important obstacles, such as pedestrians. Terrain maps used in aircraft sometime contain spurious data points that cause the system to activate even when cruising at altitude. Also, humans can over rely on these systems, using them to stop at the last moment or simply not bothering to break.

5.3.3 Human Error Reduction

While accident-avoidance systems save lives, it is desirable to not enter the dangerous situations where we deploy these systems. Accidents often happen when other tasks distract humans or when humans experience cognitive decline due to factors such as fatigue. Therefore, systems have been developed to understand human physiologic state (Boeke et al., 2015; Parasuraman et al., 2007). For example, some automobiles observe a driver's eyes, looking for signs of drowsiness and alert the driver that he or she should rest. More aggressive systems might automatically activate lane-keeping AI or provide other assistance in response to this detection. These systems seek to avoid conditions that might require the need for more extreme safety aids, such as automated braking systems. However, designers can couple situation avoidance with error avoidance systems, providing additional information on the ability of the operator to take appropriate actions, which the AI can use to adjust the behavior of other safety systems. For example, if an aircraft ground collision avoidance system was aware that a pilot was unconscious, it might recover the aircraft sooner than if the pilot was alert and focused on achieving their goals.

5.3.4 Directed Attention

One source of human error is that the human becomes focused on particular information and does not sense or perceive other information which is critical to a near-term decision (Rantanen & Goldberg, 1999). Designers can use visual or auditory signals to redirect human attention to other objects. Word processors often use highlighting to indicate words that are misspelled or misused, helping the human identify potential errors but permitting them to use an uncommon spelling, acronym, or technical term the system is not aware of. Using augmented vision systems, we can highlight objects in the environment by imposing graphics near them or even provide graphical depictions of objects that are important to the operator but not yet visible. As a demonstration, proponents of automated driving systems have shown that the use of advanced sensors, such as lidar, could detect items such as a bicycle as it crosses an intersection, even when an intervening truck or other barrier obscures the driver's vision of this object. If the AI were aware of this oncoming obstacle, it could provide a graphical depiction of this object on an augmented reality display, directing the driver's attention to this "invisible" object, helping the driver to respond appropriately. These systems can distract the driver from other important tasks, and thus understanding the relative importance of information in the human's task is necessary for successful implementation.

5.3.5 Understanding Complex Information Spaces

As humans, we live in a three-dimensional environment. Therefore, we are often able to think or visualize the influence of two variables upon a third variable. However, as the number of variables increases, it becomes increasingly difficult for us to understand the relationships among the variables. The statistical techniques employed by most modern AI agents can create models among large numbers of variables. Consequently, given the data, these algorithms can understand extremely complex relationships. As such, these algorithms can model the relationships among tens of variables to understand complex interactions among them. This then provides these systems with the ability to predict data points within complex data sets with high accuracy. Therefore, these tools can observe relationships in large data sets that have challenged humans for decades. Unfortunately, just because values are correlated or have an apparent relationship, it does not mean that changes in one variable reliably result in changes in another variable. A classic example is the observation that both ice cream sales and drowning deaths increase during certain months and decline in other months. Does this mean that consuming ice cream increases your chances of drowning? It is true that everything else being equal, that drowning deaths will increase as ice cream sells increase. The relationship between these two variables is a correlation that exists in the data because we both consume more ice cream and swim more often when the temperature is high. While an algorithm might observe the correlation between these two variables and assume that a relationship exists, as humans understanding the data, we can evaluate whether we can use this relationship to predict certain events or suggest alternate models that might be more robust. Nevertheless, AI can understand complex data sets and provide insights that are not apparent to humans, permitting further exploration and improving human understanding.

5.3.6 Option Generator

We often value creativity. However, in society, we learn to apply rules that aid our ability to evaluate and navigate in the world by reducing the option space. These rules are helpful under most circumstances, but they often constrain our ability to innovate (Bates, 2018). AI is not encumbered by these same rules. Further, if the mathematical representation of the value of each option is available, and we can describe the option space in terms of potential variables that can be changed, AI can theoretically generate and evaluate every practical option in the option space. Thus, AI can create options that we might never consider. An example comes from the Defense Advanced Research Project Agency program called AlphaDogfight, which pitted an AI against a human pilot (Tucker, 2020). The AI bested the human pilot in five of five simulated air battles. When the design team analyzed the wins, they reported that the AI performed activities that the Air Force trained the pilot never to perform. For example, the AI would initiate a nose-to-nose attack, which might result in a nose-to-nose collision. However, because the AI agent performed these activities, the human pilot was unable to use the strategies they trained and practiced, providing the advantage to the AI agent. Under the right constraints, the ability of AI agents to generate and select novel options can be advantageous. However, these options can also have negative consequences as the AI only understands the consequences that the designers include in training criteria. Therefore, a novel option may have negative consequences that the AI will not anticipate.

5.4 MODELING: UPDATING GOAL HIERARCHY AND GOAL–RESPONSIBILITY DIAGRAM—IDENTIFYING PURPOSE OF REDESIGN

5.4.1 Modeling Objective

It is quite possible that most, if not all, of the responsibilities captured in the Goal–Responsibility Diagram could benefit from the inclusion of an AI agent. However, design teams face limited resources to apply to the design of any system. Thus, it is important to narrow the focus and leverage those implementations that would be the most valuable. Identifying the purpose of the redesign enables us to focus on those responsibilities where AI agent(s) implementation(s) will achieve the intended benefits we seek from system redesign.

Note on cost reduction: When considering cost reduction as a goal of automation, including AI agent implementation, it is common to expect that the automation will completely replace a human, thus replacing the entire labor cost. However, the human performs many responsibilities, and while the automation may replace the ability to do individual tasks, or even an entire responsibility, it is highly unlikely that an agent or even multiple agents would be capable of replacing all aspects of the human's contribution to the system and team. Additionally, the introduction of the automation(s) will require additional coordination and handoffs, adding new tasks and new responsibilities. When making a cost comparison, in addition to deducting the payroll cost of the human, the designers should account for additional costs for accomplishing the additional responsibilities. It is also common to only consider the upfront costs of developing and installing AI. This is the equivalent of only considering the hiring and recruitment costs for a human worker. Just as the human worker has an ongoing payroll and training cost, the automation has an ongoing operations and maintenance cost, including the requirement for technology refresh to maintain its currency and compatibility with other systems and processes as they evolve.

The purpose of the redesign may impact the goals or responsibilities of the current system. This impact can originate from four sources.

1. New Goals provided by new or existing Stakeholders;
2. Existing Goal with new or updated Quality Goal threshold;
3. New Responsibilities on an existing Goal to meet quality threshold, for example, to overcome, bottlenecks, productivity issues; and
4. Existing Responsibilities of not meeting the existing quality goal threshold to overcome bottlenecks or productivity issues.

5.4.2 Diagram Components and Composition

The goal and goal–responsibility diagrams for the "To-Be" system follow the same guidelines as the goal and goal–responsibility diagrams discussed in Chapter 3. The goal diagram consists of a hierarchy of functional goals that can be annotated with quality goals. However, at this stage of development, you should avoid assigning agents to any of the goals as this activity will occur later in the design process. Similarly, the goal–responsibility diagram(s) should include the same functional goals as the goal diagram, potentially with quality goals and should associate responsibilities the team needs to fulfill to achieve their goals. Once again, agents should not be assigned to goals or responsibilities at this stage of development.

5.4.3 Considerations

As you create "To-Be" goal and goal–responsibility diagrams, you should revise the "As-Is" diagrams to fulfill your stakeholder's needs and desires for system revision. Typically, these revisions will include new goals, changes in quality goals, or additions and changes to responsibilities. As you create these diagrams, remember that new or existing stakeholders can provide new goals or convert implied or tangential goals into higher priority goals when initiating a system redesign. Changes in goals will typically also require you to add or revise existing responsibilities.

For the Diagnostic Center, an example would be a new goal of *Prioritize High-Risk Patients*. While the *Primary Care Physician* may have already been conducting this activity, as seen in the concept map, it was not previously a goal of our HAT. A change in leadership or leadership priorities could provide our HAT with a new goal of *Prioritizing High-Risk Patients* so that the diagnostic center can ensure an efficient use of their resources in both detections and interventions. Supporting this new goal will require new responsibilities. These new responsibilities might include responsibilities such as: *Identify High-Risk Patients* to give them the prioritization, *Update Electronic Patient Record* to reflect their high-risk status, *Establish High-Risk Patient Procedures* to account for what will be done differently to provide the prioritization, *Implement Identification Criteria into Screening Systems* so that our screening procedures account for the fact that particular patients are high-risk, *Coordinate with PCM to Implement High-Risk Patient Procedures* to include an increased screening for high-risk patients, *Update Risk Assessment* to consider high-risk status, *Update Anomaly Detection Criteria* to consider high-risk status, *Update Intervention Evaluation* to consider high-risk status, and *Update Intervention Implementation* to consider high-risk status.

Alternatively, the stakeholders may initiate a redesign to accomplish an existing goal that has a new or updated quality goal threshold. This may entail simply updating the existing goal with the quality goal or new quality goal value. Changes in quality goals will also likely require the addition of new responsibilities. For example, assume our current Diagnostic Center has a False Alarm rate of 30%. A new quality goal threshold might be to achieve a reduction in the false alarm rate to 10% or less. The design team might achieve the goal by providing additional training for human team members, making a technological improvement to the existing equipment, or improving the diagnostic aid, in which case the design team will not add any new responsibilities for the HAT. Alternatively, the design team could achieve the goal by adding new responsibilities, such as adding a process to *Incorporate State of Art Knowledge into Evaluation* or a process to *Identify Personnel Errors* resulting in False Alarms and then *Provide Error Feedback*. Additionally, you might decide to implement all these system changes to ensure adequate changes in false alarm rate.

When attempting to change a quality goal, it is important to remember that as a team prioritizes one goal or quality goal, they may trade performance on another goal. For instance, changes in false alarm rate might be made by shifting the threshold of detection, such that as false alarm rate decreases, misses will increase or perhaps the efficiency of the team will decrease as they provide further scrutiny when evaluating any anomaly. When making changes to quality goals, it is important to identify such trade-offs and establish criteria for these trade-offs that are acceptable for the stakeholders. If the stakeholders are willing to accept decreases in efficiency or increases in misses, different team designs might be considered than if the stakeholders will not accept these trades. In the latter case, your design should consider ways to monitor and control parameters associated with efficiency and misses when considering responsibilities while you may not need to establish these responsibilities in the former case.

The purpose of the redesign may not affect the goals of the system, but rather might establish new responsibilities for an existing Goal to permit the HAT to meet existing quality thresholds that the HAT is currently unable to meet. The stakeholders may express the need for such a redesign as a desire to reduce or eliminate bottlenecks in the system or to improve system productivity. For example, if the HAT was unable to meet the responsibility to *Obtain Quality Image* satisfactorily, this might be resolved by adding a new responsibility to *Provide Feedback from Radiologist to the Radiology Technician*.

However, if existing responsibilities are not meeting the existing quality goal threshold, the design team may not need to add new responsibilities. Instead, the design team should consider alternative methods for implementing the HAT's existing responsibilities. For example, in our current system, the HAT performs manual tasks to meet the responsibility *Interpret Health History for Risk Level*, with the Radiologist reading through the patient history. The design team could design an alternative method for accomplishing this responsibility, using a more rigorous analytical method. If this alternative method could improve the HAT's ability to identify risk level and improve the HAT's ability to accomplish the *Provide Correct Diagnosis* goal, this redesign could achieve the design goal.

5.4.4 Example Application: Redesign Purpose of "To-Be" System

For the purposes of our modeling example, we selected the incorporation of a new goal, *Prioritize High-Risk Patients*, as the objective of the redesign effort. Figure 5.3 contains the updated Goal Hierarchy, with this new goal added, *Prioritize High-Risk Patients* (shown in dark gray for visibility) as a subgoal to *Maximize Team Performance*.

Figure 5.4 provides the updated Goal–Responsibility Diagram, representing the transition from the "As-Is" system to the "To-Be" system. The diagram shows the new *Prioritize High-Risk Patients* goal, along with its responsibilities: *Identify High-Risk Patients, Update Electronic Patient Record, Establish High-Risk Patient Procedures, Implement Identification Criteria into Screening Systems, Coordinate with PCM to Implement High-Risk Patient Procedures, Update Risk Assessment, Update Anomaly Detection Criteria, Update Intervention Evaluation*, and *Update Intervention Implementation*. Additionally, existing responsibilities that the new *Prioritize High-Risk Patients* goal impacts are highlighted in dark gray. These responsibilities provide us with focus as we proceed in the identification of potential AI agent implementation.

5.5 CHAPTER SUMMARY

To redesign the HAT, it is important to form a sharp vision of the purpose for including AI agents. AI designers typically assume that AI agents will replace humans, reducing the direct monetary cost to the operating organization. Historically, the introduction of AI or AI agents has reduced the required number of human operators in many applications. However, the adoption of AI has introduced significant complexity, increased the required skills and knowledge of the remaining crew, and increased the number and skills of support personnel, as well as reduced the resilience

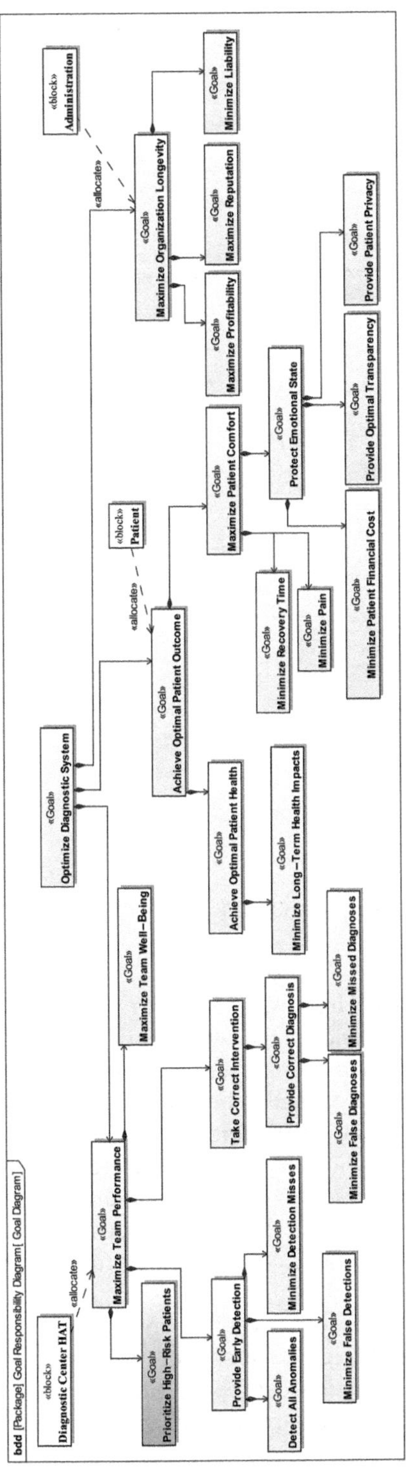

FIGURE 5.3 Updated goal hierarchy of "To-Be" System.

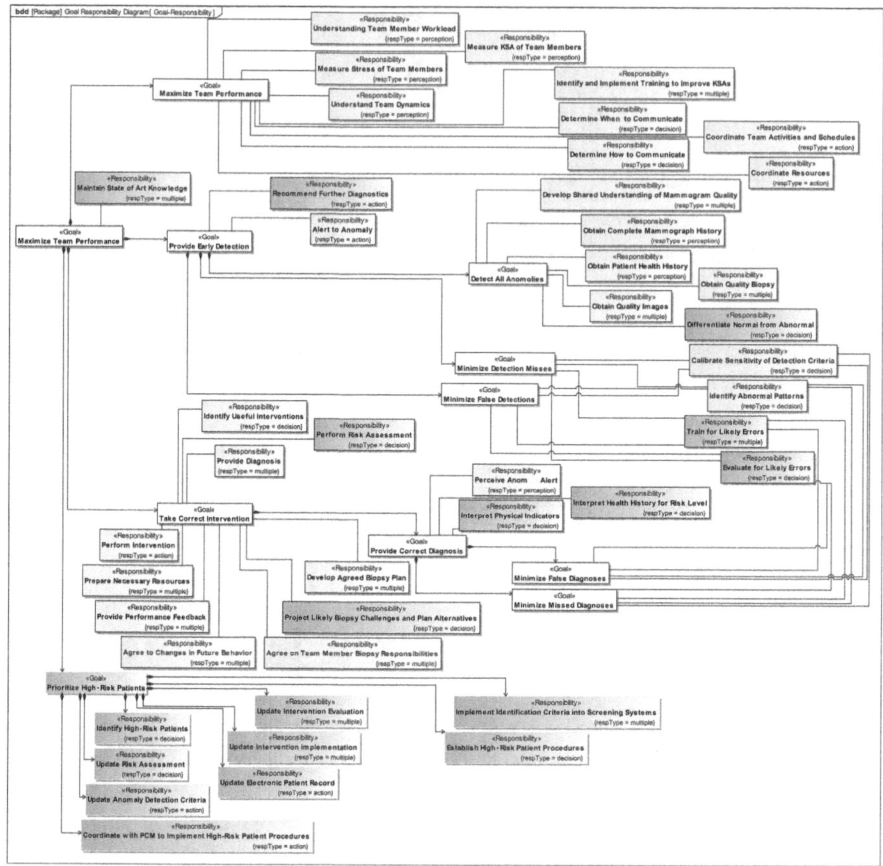

FIGURE 5.4 Updated Goal–Responsibility Diagram of "To-Be" System.

of the system to unexpected circumstances. Both the direct and indirect costs of automation are important to understand when assessing the benefit of introducing AI to enhance or replace human operators, and the designers should not assume that adding AI to the system will provide this benefit.

System designers and operating organizations often place humans in situations where they do not have the mental capacity or knowledge to perform a task, which truly results in human error, resulting in system failures. AI agents, when designed appropriately, can provide backup behaviors to reduce the burden on the human, particularly, when the cognitive load on the human is undesirable, reducing the likelihood of errors and mishaps.

Current artificial intelligence consists of specific or narrow intelligence, in which the AI is highly proficient at a specific task(s) or in limited domain(s). While these specific intelligences can be employed to search parameter spaces in efficient fashion, these systems are only applicable when the parameter space is well known, and the outcome of the system response based upon the modification of the parameter space is well characterized. However, the variability in the environment produces circumstances which the system designers cannot fully anticipate or compensate for during AI system design. The system must be capable of adapting to these situations to maintain safe performance. It is the human's ability to anticipate and innovate, due to our general intelligence, which separates and elevates our cognitive assets as compared to modern AI systems, motivating the need for human–AI teams rather than a complete replacement of humans with AI agents.

Identifying the purpose of the redesign enables us to focus on the AI agent implementation most relevant to achieving the redesign objectives. This redesign purpose will impact the goals or responsibilities of the current system thus resulting in an updated Goal Hierarchy and Goal–Responsibility Diagram for the "To-Be" system.

5.6 QUESTIONS AND DISCUSSION TOPICS

1. Provide an example from your own life in which the incorporation of automation resulted in a reduction in cost, increased safety, or improved decision-making. What are some additional indirect costs or additional new tasks or activities that resulted from the incorporation of this automation?
2. Describe a task that a HAT can accomplish with the use of AI. How does the AI's approach to task accomplishment differ from that of a human intelligence approach? In what ways is AI stronger? In what ways is AI weaker?
3. Select an AI implementation from Table 5.1 and provide a one-paragraph long description of how a designer could use an AI to achieve a goal within the scenario. What additional design considerations do you believe the design team should consider to mitigate listed implementation consequences? Do you foresee any other unintended consequences that the design team should consider?

5.7 PROJECT QUESTIONS AND ASSIGNMENTS

1. Identify at least three potential redesign objectives for your system. Identify how these objectives add or revise existing goals and responsibilities. Be sure to consider:
    1. New goals provided by new or existing stakeholders;
    2. Existing goals with a revised updated quality goal threshold;
    3. New responsibilities for existing goal to meet quality threshold; and
    4. Existing responsibilities not currently meeting existing quality goal threshold.
2. Select which redesign objective(s) you will implement in your "To-Be" system. Update your Goal Hierarchy and Goal–Responsibility Diagram accordingly.

REFERENCES

Bainbridge, L. (1983). Ironies of automation. *Automatica*, *19*(6), 775–779. https://doi.org/10.1016/0005-1098(83)90046-8

Bates, A. (2018). *Augmented mind: AI, humans and the superhuman revolution*. Neocortex Ventures LLC.

Bice, G. W., Skoog, M. A., & Howard, J. D. (1990). *Aircraft ground collision avoidance and autorecovery systems device* (Patent No. 4,924,401). United States Patent 4,924,401.

Billings, C. E. (1996). Human-centered aviation automation: Principles and guidelines. In *NASA technical memorandum* (Issue A-961056). Ames Research Center, National Aeronautics and Space Administration.

Boeke, D. K., Miller, M. E., Rusnock, C. F., & Borghetti, B. J. (2015). Exploring individualized objective workload prediction with feedback for adaptive automation. In S. Cetinkaya & J. K. Ryan (Eds.), *Proceedings of the 2015 industrial and systems engineering research conference* (pp. 1437–1446). Institute of Industrial and Systems Engineers.

Bradshaw, J. M., Hoffman, R. R., Woods, D. D., & Johnson, M. (2013). The seven deadly myths of "autonomous systems". *IEEE Intelligent Systems*, *28*(3), 54–61. https://doi.org/10.1109/MIS.2013.70

Canellas, M., & Haga, R. (2020). Viewpoint unsafe at any level. *Communications of the ACM*, *63*(3), 31–34. https://doi.org/10.1145/3342102

Endsley, M. R. (2015). *Autonomous horizons: System autonomy in the air force—A path to the future: Vols. 1: Human-A*. Office of the Chief Scientist, United States Air Force.

Endsley, M. R. (2023). Ironies of artificial intelligence. *Ergonomics*, *66*(11), 1656–1668. https://doi.org/10.1080/00140139.2023.2243404

Gates, D., & Baker, M. (2019, June 22). The inside story of MCAS: How Boeing's 737 MAX system gained power and lost safeguards. *The Seattle Times*. https://www.seattletimes.com/seattle-news/times-watchdog/the-inside-story-of-mcas-how-boeings-737-max-system-gained-power-and-lost-safeguards/

Gordon, A. (2023). *'Week of Cone': Activist group is protesting driverless cars by disabling them with traffic cones*. Motherboard. https://www.vice.com/en/article/bvjv48/week-of-cone-activist-group-is-protesting-driverless-cars-by-disabling-them-with-traffic-cones

Holbrook, J. (2021). Exploring methods to collect and analyze data on human contributions to aviation safety. *49th International Symposium on Aviation Psychology*, 110–115. https://corescholar.libraries.wright.edu/isap_2021https://corescholar.libraries.wright.edu/isap_2021/19

Hubert, T., Schrittwieser, J., Baker, L., Hui, F., Hassabis, D., Antonoglou, I., Silver, D., Bolton, A., Simonyan, K., Sifre, L., Lillicrap, T., Graepel, T., van den Driessche, G., Chen, Y., Lai, M., Huang, A., & Guez, A. (2017). Mastering the game of go without human knowledge. *Nature*, *550*(7676), 354–359. https://doi.org/10.1038/nature24270

Inagaki, T. (2003). Automation and the cost of authority. *International Journal of Industrial Ergonomics*, *31*(3), 169–174. https://doi.org/10.1016/S0169-8141(02)00193-2

Kahneman, D. (1973). *Attention and effort*. Prentice-Hall. https://doi.org/10.2307/1421603

Klein, G. (1993). A recognition-primed decision (RPD) model of rapid decision making. In G. Klein, J. Orasanu, R. Calderwood, & C. Zsambok (Eds.), *Decision making in action: Models and methods* (pp. 138–147). Ablex Publishing Corporation. https://doi.org/10.1002/bdm.3960080307

Lopes, L. L. (1991). The rhetoric of irrationality. *Theory & Psychology*, *1*(1), 65–82. https://doi.org/10.1177/0959354391011005

Marks, P. (2020). *Driverless cars face laser sabotage*. Communications of the ACM. https://cacm.acm.org/news/243377-driverless-cars-face-laser-sabotage/fulltext

Norris, G. (2016). *Auto-GCAS saves unconscious F-16 pilot—declassified US air force video | Air combat safety content from aviation week*. Aviation Week.

Parasuraman, R., Barnes, M. J., & Cosenzo, K. a. (2007). Adaptive automation for human-robot teaming in future command and control systems. *The International C2 Journal*, *1*(2), 30. http://oai.dtic.mil/oai/oai?verb=getRecord&metadataPrefix=html&identifier=ADA503770

Patterson, R., & Eggleston, R. G. (2017). Intuitive cognition. In *Journal of cognitive engineering and decision making* (Vol. 11, Issue 1, pp. 5–22). SAGE Publications Inc. https://doi.org/10.1177/1555343416686476

Patterson, R. E., Pierce, B. J., Bell, H. H., & Klein, G. (2010). Implicit learning, tacit knowledge, expertise development, and naturalistic decision making. *Journal of Cognitive Engineering and Decision Making*, *4*(4), 289–303. https://doi.org/10.1177/155534341000400403

Rantanen, E. M., & Goldberg, J. H. (1999). The effect of mental workload on the visual field size and shape. *Ergonomics*, *42*(6), 816–834. https://doi.org/10.1080/001401399185315

Schutte, P. C. (2017). How to make the most of your human: Design considerations for human—machine interactions. *Cognition, Technology and Work*, *19*(2–3), 233–249. https://doi.org/10.1007/s10111-017-0418-2

Shively, J. R., Brandt, S. L., Lachter, J., Matessa, M., Sadler, G., & Battiste, H. (2016). Application of human-autonomy teaming (HAT) patterns to reduced crew operations (RCO). In D. Harris (Ed.), *Engineering psychology and cognitive ergonomics: 13th international conference* (pp. 244–255). Springer International Publishing.

Simon, H. A. (1956). Rational choice and the structure of the environment. *Psychological Review*, *63*(2), 129–138. https://doi.org/10.1037%2Fh0042769

Strauch, B. (2017). Ironies of automation: Still unresolved after all these years. *IEEE Transactions on Human-Machine Systems*, 1–15. https://doi.org/10.1109/THMS.2017.2732506

Tucker, P. (2020). An AI just beat a human F-16 pilot in a dogfight—Again. *Defense One*. https://www.defenseone.com/technology/2020/08/ai-just-beat-human-f-16-pilot-dogfight-again/167872/

6 AI Methods and Agent Architectures

6.1 DESIGNING THE "TO-BE" SYSTEM

Having developed an understanding of our "As-Is" system and designed the goals of our "To-Be" system, we are now prepared to begin to design our "To-Be" system. We explain the process steps for designing the "To-Be" system, beginning in this chapter through Chapter 9. Figure 6.1 shows these process steps. This figure shows the overall AOSM process on the left, with the fourth step highlighted. The right panel of Figure 6.1 shows the process steps discussed in Chapters 6 through 9. As shown, this step begins with developing a Responsibility-Capability diagram in this chapter. This step is followed by developing a Capability Assessment Table in Chapter 7. In Chapter 8, we arrange the capabilities into roles, creating a "To-Be" internal block diagram for each role. We then synthesize these IBDs into an agent diagram. We assign Responsibilities and then create Agent Capability Diagrams. In Chapter 9, we develop the information architecture to support decisions in our "To-Be" system. Thus, we complete these steps throughout the four chapters involved with the overall step of Designing the "To-Be" system using AOSM.

6.2 AI: THINKING ABOUT INTELLIGENCE

When discussing intelligence, we consider entities that pursue the action process shown in Figure 3.2 in which they establish a goal, decide upon an action after observing the environment, implement the action, observe the result of the action, compare the result to the original goal, and then decide whether to continue pursuing the same action, to take a different action, or perhaps pursue a different goal. Importantly, we can further decompose decision-making into determining one or more actions, determining which of these actions are likely to have a desired effect, and selecting one or more of the actions based upon their likelihood of producing the desired effect. Therefore, as we consider artificial intelligence, it is important to determine the system's ability to perform these steps.

Additional attributes are likely necessary for intelligence. Endsley has recently described an intelligent system as one that can "recognize situations, adapt to changes and generate solutions to even novel problems, and can act to optimize performance" (Endsley, 2023). She points out that modern AI, including significant innovations such as ChatGPT, can perform feats such as generating text that is often useful but can provide inaccurate and biased answers as it rephrases existing text without the ability to make sense of and reason about the content of the text. Thus, having the ability to complete the perceptual loop for known conditions may not be

DOI: 10.1201/9781003428183-6

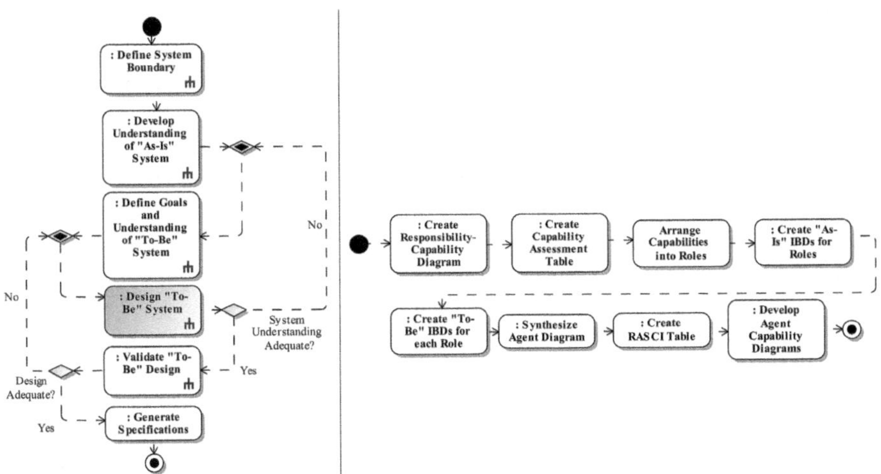

FIGURE 6.1 A depiction of the AOSM process on the left with the current sub-step highlighted. The right panel shows the sub-steps associated with Design the "To-Be" System step.

adequate to claim intelligence. Instead, the ability to form an understanding by creating an abstract understanding of the information, and the ability to reason using this abstract form of the information may be a necessary element of intelligence. Thus, these systems have "artificial" intelligence as they mimic intelligence without the ability to perform sensemaking.

While Minsky proposed the concept of AI over six decades ago (Minsky, 1961), initial systems failed to provide these features; thus, the field has repeatedly garnered interest and investment with novel advancements, followed by loss of interest and investment as the field failed to deliver on its original promises. Therefore, it is common for AI scientists to discuss the fact that AI has undergone three waves. Generally, AI scientists have focused predominantly on constructing algorithms to perceive and integrate perception of the world, reasoning about this information, and solving problems based on this information, using language, and learning (Copeland, 2023).

6.2.1 Wave 1: First Principles AI

The first wave of AI applied the basic strengths of the computer. The ability to execute programmed logic statements, perform computation to determine physical relationships, and to execute search of precise data are features that humans are not capable of doing at the speed of a computer. However, to complete the action cycle, during the time of the first wave, sensors and actuators were quite expensive, relegating AI to either decision support systems that do not require sensors or actuators or to applications where funding the expensive requisite sensors and actuators was possible. As a result, limited applications occurred in military and commercial aviation, as well as manufacturing. Furthermore, as the AI systems needed to be custom-built, this attribute limited application of this technology in manufacturing facilities as the manufacturer often needs to retool these facilities as products evolve. Therefore, the AI designer needed to rebuild the AI with each product evolution. The investment in AI waned, ending the first wave with limited application.

Today, AI systems often involve these same basic technologies. However, AI researchers have developed other first principles of AI techniques since. These researchers augmented technologies with optimization techniques, including linear programming, dynamic programming, and methods that use multi-criteria decision-making. These techniques permit exact computation, the designer can explain these methods to the user, and the decisions these AI systems make are repeatable and can be altered, as necessary.

It is important to recognize two essential elements of technologies in this category. First, the fact that this technology results in deterministic decision-making. That is, when given a specific set of inputs, the equations and logic produce the same output regardless of when or how many times these same inputs are observed by the system. Therefore, it is possible for humans to review the logic that the AI uses to arrive at the decision. The fact the system can display the method and key inputs that the AI used to make decisions allows the end user to understand how the system works and build trust in the performance of these systems. As we will demonstrate later, many of these systems permit the human to provide input on the importance of factors that the AI applies to inform a decision, permitting an active exploration of the decision space by the user.

However, because these technologies are deterministic, some technologists question whether the algorithms are a form of AI. In fact, many individuals might state that systems employing these technologies are a form of automation, rather than AI, as these technologies have been applied to provide automation of human activity for decades. In this text, we view these technologies as a form of AI. Therefore, these technologies can be used to develop AI agents if the technology is able to sense and respond to changes in the environment.

6.2.2 Wave 2: Expert Systems

AI scientists developed expert systems two decades after Minsky's original work. These systems are based on knowledge gained from experts and encoded into databases. This knowledge includes objects or states of objects that are associated through relationships. These systems use reasoning engines to rely upon multiple portions of the knowledge representation to draw conclusions beyond the knowledge the AI designer directly coded within the database. As an example, the knowledge base might include information like "most birds and insects are animals", "birds have feathers", "only birds and insects can fly", "birds are bigger than insects", and "insects do not have feathers". By combining these concepts, it might be reasonable to conclude that larger flying animals have feathers. Expert systems provided languages that permit AI system developers to encode and reason across information. Expert systems made it possible to reuse knowledge across AI systems, making AI more useful in specific manufacturing environments. Once again, this technology garnered significant interest and investment. However, the utility of these systems is dependent on how well the designer encodes information and the completeness of this information within the system. Therefore, these systems are often brittle when the AI system designers do not logically order data or apply incomplete data. As a result, these systems again found limited application, and the interest in these AI techniques waned.

Today, we see continued use of expert systems. One application area is ontology. In this area, concepts are related to one another, and AI designers apply tools, such as Protege, to reason across the concepts. Ontologies also often include specific data associated with the concepts in the ontology. For example, recent work on understanding the anthology of enslaved Americans during the 18th and 19th centuries has been tied to data for individuals who were shipped into the United States, as well as records of sale to provide data that allows one to reason across individuals who were involved in common relationships (Shimizu et al., 2020). Thus, one can reason not only across the terms in the ontology but the underlying data as well. Recently, modular software approaches have been applied to this space to increase the rate of ontology development and the ability to develop robust expert systems (Shimizu et al., 2023).

Similar to the technology in the first wave, technology in this second wave is also deterministic. That is, given a specific input, algorithms in this class will produce the same output. Furthermore, the AI applies rules that a human can review to understand the logic that the AI applies. Therefore, there is a method for the AI to explain its reasoning to the human.

6.2.3 Wave 3: Statistical Machine Learning

Each of the first two waves relied upon humans to understand the underlying relationships in data and to encode knowledge of this relationship into a computer. They permitted the computer to calculate or reason faster than a human could to implement this same approach but did little to extend the knowledge or understanding of the human. However, in the real world, the relationships we understand only explain a portion of the variation we see in most real-world data sets. Scientists refer to this unexplained variation as random variation. This random variation might originate from truly random processes in the world or in our measurement techniques or it might originate from relationships that we simply do not yet understand. We could explain this later source of variation if we understood the underlying source of the variance, and we could develop methods to calculate these effects. However, one can never explain truly random variation. Regardless of its source, the presence of this unexplained variation in data makes it difficult to see many of the effects in a data set, particularly when the phenomenon we are trying to understand is complex, that is, it originates from the presence of many underlying factors, many of which have complex, nonlinear effects on one another.

Over the years, numerous methods have been developed to analyze data to understand relationships in complex data. These techniques permit one to interrogate a dataset to find complex trends within the data set, permitting us to gain insights that we, as humans, might not have been able to determine without these methods. These techniques aim to account for underlying, not yet understood, effects and attempt to separate them from random variability.

We can examine a simple linear regression example to illustrate a simple statistical machine learning technique. Let us assume we have two columns of data, one of which represents Ice Cream Sales and the other which represents Drowning Deaths per Million People. We can perform a linear regression to obtain a figure like Figure 6.2. This approach assumes that a linear relationship exists between these two variables and creates a line that minimizes the sum of squared vertical distances between the line and each point (Wasserman et al., 1996). The resulting line indicates that one can estimate drowning deaths by multiplying millions of dollars in ice cream sales by 0.0389 and subtracting 0.0458 from this value. As shown, the fit between these two variables is not perfect. Instead, this figure shows the random error as the vertical distance between each point and the line in the figure. In this fabricated example, the equation accounts for about 84% of the variance, while the remaining variance is random.

Effects within real-world data may be causal, meaning that changing one variable causes a change in another variable, but other effects may not be causal, meaning that a change in one variable might not cause a change in another variable, but a third variable might cause a change in the first two variables, such that the two variables appear to be related. A common example is the relationship between ice cream sales and the number of drownings where analyzing the total ice cream sales and drownings per month across the United States will show a correlation such that the number of drownings per month increases with increasing ice cream sales. If we were to use this relationship to make decisions, we might decide that banning the sale of ice cream would eliminate, or at least reduce, the number of drownings across the United States.

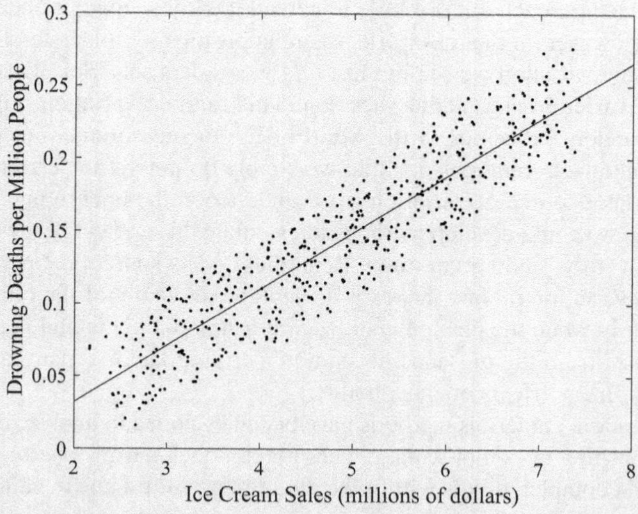

FIGURE 6.2 Example linear regression between two fabricated variables.

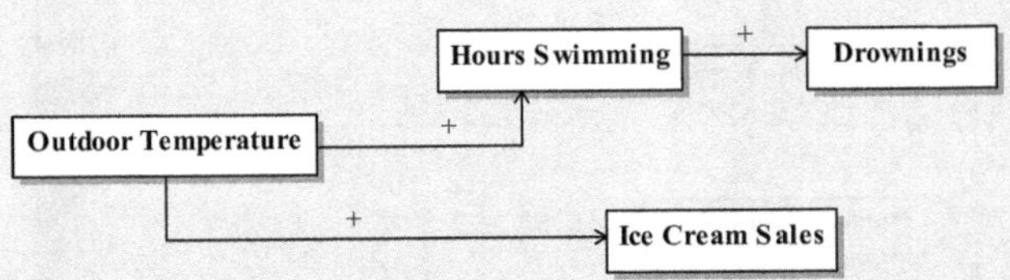

FIGURE 6.3 Diagram illustrating causal effects associated with ice cream sales and drowning.

However, implementing this policy would have a negligible effect on the number of drownings as the underlying effect is that ice cream sales increase with increasing outdoor temperature. Similarly, the number of hours people spend swimming increases as temperature increases, and the number of drownings increase as the number of hours people spend swimming increases. Figure 6.3 shows the causal effects as indicated by the arrows. As outdoor temperature affects ice cream sales and drownings, these variables are merely correlated since there is no direct relationship between these variables.

It is also important that we can use this type of data analysis both to support understanding and to aid decision-making. In the previous example, we illustrated that deciding on the basis of a noncausal effect, that is, eliminating the sale of ice cream to reduce the number of drownings, leads to poor decisions. However, when we are trying to simply predict when the number of drownings will increase, then some might argue that measuring and using ice cream sales for this prediction might be just as good as basing this decision on outdoor temperature. While this may be true, at least over brief time horizons, the fact remains that other factors influence these variables. As shown in Figure 6.4, it might be the case that ice cream sales are dependent upon disposable income. Therefore, during recessions, individuals purchase less ice cream (i.e., as these disposable income and ice cream sales are positively correlated, reducing disposable income reduces ice cream sales), thus this relationship might be influential if we collected the data on which this prediction is based during a time of economic expansion but apply the relationship during a recession. Further, other variables might intervene that we have little experience with. For example, we now know that disposable income and video game time are positively correlated, thus as disposable income increases, the time spent playing video games increases. It is likely that as the number of hours playing video games increases, the hours spent swimming will decrease, likely decreasing drownings. These example relationships, also shown in Figure 6.4, begin to illustrate the complexity of real-world decisions and the complexity that can be present in real-world data.

If we were to collect data for each of these variables over several years and begin to examine it visually, especially without knowledge of the relationships among the variables, it would be difficult for us to determine all relevant relationships from simply looking at and even plotting the data. However, by

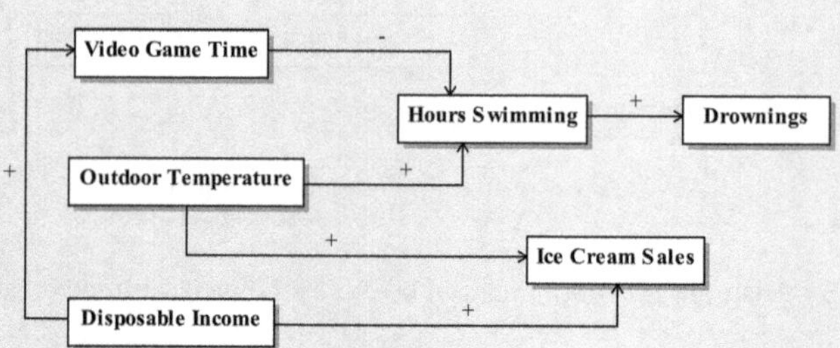

FIGURE 6.4 Theoretical causal diagram, indicating the relationships between variables impacting ice cream sales and drowning deaths.

applying the correct statistical machine learning techniques, these methods could derive the correlational relationships among the variables.

Unfortunately, while statistical machine learning techniques can be used to assess the structure of the data, they have a number of shortcomings: These include: (1) They are unable to differentiate causal from noncausal effects, (2) they are unable to determine if all important relationships exist in the data, (3) most individual techniques have some inherent assumptions about the structure of the underlying data and can produce poor results if these assumptions do not apply, and (4) many of the techniques require that someone identify the variables that are likely to predict versus the variables that are to be predicted (i.e., separate and provide both input or independent variables, as well as response or dependent variables).

6.2.3.1 Feature Space Definition

The previously highlighted example included numbers which are interpretable by a computer and meaningful for the AI developer's evaluation. Assume that we wish to build an algorithm to construct a self-driving vehicle. As an input to the agents in this vehicle, assume we capture the latitude and longitude of all other vehicles in the region around our vehicle. Based on this data, can we determine if or when we might have a collision with each of these vehicles? This data alone tells us little about whether we will collide with another vehicle. After all, if we do not know where our vehicle is located, then we do not know how much distance exists between our vehicle and other vehicles. Is our vehicle or the other vehicles moving? If not, then there is no concern about a collision. If one of the vehicles is moving, are the vehicles moving toward one another or away from one another? Is our vehicle or the other vehicles accelerating or decelerating? Each of these pieces of information is potentially much more useful to anticipating a collision than the location of other vehicles in our environment. This example illustrates that we may represent data in diverse ways. Some representations will make it easier to compare and interpret information than others (Bindewald et al., 2016).

Traditional statistical machine learning techniques to include support vector machines, neural networks, random forest algorithms, classification, regression, learning vector quantization, and other algorithms require that the algorithm designer identify the feature space (i.e., the proper input or independent variables). In these methods, the way that the data is represented in this feature space is extremely important to the utility of the algorithm. Therefore, while statistical machine learning algorithms rely upon data to learn relationships rather than requiring humans to know and encode relationships, the design of the feature space and human understanding of causal relationships are both necessary to ensure the robustness of traditional statistical machine learning techniques.

6.2.3.2 Ground Truth

As indicated earlier, for AI developers to create statistical machine learning algorithms, it is important that the AI developers create multiple inputs which include the proper independent variables, that is, the items represented in the feature space, as well as dependent variables, that is, the items that are to be predicted or interpreted. These algorithms fall into two categories, classification algorithms, which classify items into discrete bins or categories, and prediction algorithms, which attempt to predict a value on a continuum. Depending upon the purpose of the algorithm, the algorithm may require data sets that provide a large number of data samples for the independent variables, associated with data for the dependent variables. To train the algorithms, the AI developer must label the data by annotating each member of the input data, referred to as the training set, with the proper categories or value representing the dependent variable. The exact values for the dependent variable depend on the algorithm. For example, the dependent variable will include data classified into categories, if the algorithm is to provide a classification (e.g., dog or cat) as the output, but it will include a continuous value if the algorithm is to predict a continuous response variable (i.e., the number of drownings per year). Often, the AI developer must assign and check these annotated values, often referred to as *labels*, for correctness when creating the dataset that they will use to train the statistical machine learning algorithm.

In these algorithms, particularly classification algorithms, it is important that the AI developer design the dataset used to train the algorithm to include not only illustrated positive examples of each class but also negative examples which might confuse the algorithm. As an example, in the early days of digital photography, cameras or cellular phones provided a bright flash to illuminate the subject of the photograph when a photographer took a picture. Under certain circumstances, people's eyes in the photograph reflected this light back to the camera, only the blood vessels within people's eyes filtered the reflected light. As a result, people in the photograph would have a bright red pupil rather than the black pupil that we usually see when looking at another person's eyes. Digital camera developers referred to this artifact as red eye. Digital camera producers developed algorithms to find these red eyes in photographs and turn them black. Among the first algorithms the developers created was an algorithm that looked for a small red area, having a particular red color that was typical of this artifact. To train this algorithm, the developers obtained thousands of photographs, and two individuals selected the exact location of all pixels in each image that were part of each red eye. The developer compared the results from the two people. If the results differed between the two person's responses, a third person either chose one of the two images or corrected one of the images. The developer then used these images with hand-labeled red eye pixels as the ground truth, and the developer used the images to train a support vector machine algorithm. In early testing, the algorithm performed exemplary, removing red eyes in photographs. After release in products, the algorithm also appeared to function well. However, after the Christmas holiday, customers began complaining that the red lights on their Christmas trees were all black. The developers traced this artifact to the red eye removal algorithm. Notably, the original training set contained few, if any, images of Christmas trees, and this failure mode of the algorithm went undetected. To repair the algorithm, the designers added features to the feature space, that is, features indicative of a face around the red eyes. The designers then retrained the algorithm with a training set that expanded the original training set by including several images of Christmas trees with red lights. Once the designers trained and evaluated the algorithm, the product provided the desirable function for Christmas trees. Over time, customers complained of additional artifacts, and the designers applied similar procedures to update and correct the algorithm. This example illustrates that a ground truth database must include inputs that contain the items which the algorithm is to classify, that is, the red eyes. This training set must include many examples that illustrate many positive differences in images, for example, different skin tones, images where one or both eyes are present, and images with different sized faces. Further, the training set must contain all possible negative examples where the algorithm is likely to be confused, for example, pictures of Christmas trees with red lights. Finally, the algorithm must contain relevant and necessary features, within its feature set, to differentiate positive from negative examples.

Once again, this class of algorithm became popular throughout the 1990s. However, it was expensive to create the hand-labeled data sets, thus the data sets were often relatively small. Additionally, the algorithms were computationally intensive to train, and training computation increased with increasing numbers of features. Therefore, the designers maintained minimal size feature sets. During this period, digital sensors began to become affordable, and these algorithms began to find application in larger markets, such as consumer photography. However, the algorithms were often not highly reliable, required significant expertise and human labor to create and train, and thus again found limited, but expanding, applications. As the feature sets were small and the algorithms accomplished very narrow tasks, it was possible to understand and explain these algorithms, supporting human interaction. However, as the red eye example illustrates, the result of applying the algorithm is not always predictable.

6.2.4 Deep Learning

One of the algorithms in the statistical machine learning category is neural networks. This class of algorithms were investigated within the neurophysiology and psychology literature prior to the advent of AI (Mcculloch & Pitts, 1990; Rosenblatt, 1958), but other algorithms outperformed early neural network instantiations within the AI literature. These original, neural networks consisted of a few layers of "artificial neurons" which developers designed to mimic the behavior of certain neurons in the human brain. These neural networks originally consisted of around three layers of interconnected neurons, where each connection had a weight assigned to it as shown in Figure 6.5. As shown, this neural network contains three layers, which transform an input to an output. The input flows into the first layer of neurons, which perform some type of mathematical transformation, usually a nonlinear transformation, on the input signal. The algorithm multiplies the output of the first layer by a weight (w) between each node in the first and second layer. The algorithm then sums the inputs to the neurons in the second layer, designated as j1 through j4, and then it applies another mathematical transformation. The algorithm then subjects the result to weights associated with the input to neurons in the third layer, designated as k1 through k3. Once again, the neurons in the third layer apply a transform to the sum of the input values to the neuron. The algorithm then sums the outputs of this final layer. The resulting value is either used as the output if the algorithm outputs a continuous value or thresholds this value if the algorithm is a classification algorithm.

As with all statistical machine learning algorithms during training, the algorithm calculates the number of correct classifications or an error signal for every input in the labeled training set of data. Then the training algorithm adjusts the weights within the algorithm, indicated by the w's in Figure 6.5, and the algorithm calculates the number of correct classifications or error signal again for the entire training set. The algorithm repeats this process of adjusting the weights again and again, using a structured approach to maximize the number of correct classifications or minimize the error signal. Through this process, the algorithm learns the optimal set of weights to minimize the error signal between the learned response and the response, which is indicated in the training set. Once trained, the designer can use the neural network to assess novel inputs to predict outputs. However, the algorithm only has knowledge of the data that was input to it, and so it may well predict improperly for novel conditions that did not exist in the training set.

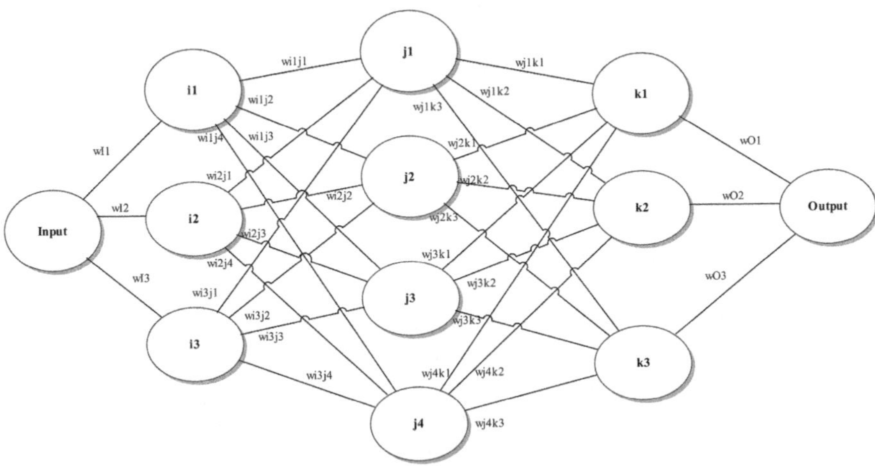

FIGURE 6.5 Example neural network structure.

This class of algorithms underwent momentous change in the early 2000s through three synergistic efforts. First, digital sensors and storage became less expensive, providing an explosion of digital data that was available for processing. Second, corporations labeled large sets of images using processes such as Captcha and online databases of consumer-labeled images to acquire exceptionally large sets of labeled data at an affordable price. Finally, AI researchers discovered that they could use graphics processing units, installed in most video cards designed for gaming, to rapidly train the weights in neural networks, enabling the use of deeper and larger neural networks at an affordable cost. These innovations permitted the training of neural networks with eight or nine layers of neurons where each layer might be as large as 256 by 256 arrays of neurons. Overnight, neural networks were performing tasks such as image classification with accuracy that was rivaling human capacity, at least for images that were typical of the images in the training set. The AI implementations of neural networks only included feedforward pathways, unlike the original research papers, which recognized the importance of "circular" or feedback loops within the networks (Mcculloch & Pitts, 1990; Rosenblatt, 1958).

AI designers now use deep neural networks in numerous AI applications, often outperforming other AI methods. However, they require large sets of labeled data for training. The algorithms can learn from trends that exist in the data; however, like all statistical machine learning algorithms, they assume that the data is stationary, that is, relationships do not change, and they create a statistically optimal explanation of the training data. Thus, if designers apply these algorithms to interpret data that is different than the training data, their response may not be correct, and the algorithms can learn biases from the training data. Further, if there are conflicts in the data set, the algorithms might learn to respond to new stimuli but unlearn behavior they had previously learned, such that they perform less well when analyzing or responding to data they previously performed well on. Additionally, the algorithms are brittle when examples do not exist in the training dataset and do not have the ability to understand the likelihood of an error or even blatantly incorrect responses that humans would rely on common sense to discount. Finally, these algorithms are extremely complex, and it is impossible to understand or explain what they are doing. Thus, they often appear as a black box to a human and at times produce results that are not desired and not explainable, which degrades human trust in the output of these algorithms (Choi, 2021).

6.2.5 Unsupervised Learning

Although statistical machine learning algorithms typically require training data, it is possible to avoid the use of training data. Unsupervised learning finds patterns and trends in unlabeled data and has a variety of uses from clustering to large language models. Consider a game like Chess or Go. In these games, there is an objective method of determining a winner. When a chess player's King is in check and the player has no way of overcoming this check, the other player wins. Thus, it is possible to construct two algorithms or deep learning networks to play chess and permit them to play against one another. If one can win most games over the other, then the statistical weights within its algorithm must be superior to the weights within the other algorithm. Through constructing two algorithmic players and having them play repeatedly against one another, using win–loss ratios, one can tune the weights within the algorithm based upon winning percentage. As a result, the AI designer does not need to develop ground truth data, only the ability to judge correctness or goodness using objective criteria. Using this approach, pairs of algorithms can play thousands of simulated games against one another rapidly within the computer, learning the best set of weights to win the game, providing training without ground truth. While this technique of using so-called "Adversarial" networks can facilitate training without the use of labeled data sets, AI designers can only implement this approach when there is a clear and objective metric for determining success.

Although we discussed unsupervised learning as any algorithm that does not require labeled response data, these algorithms can be subclassified into multiple categories. Traditional unsupervised learning employs an objective mathematical criterion that is computed to provide a response criteria or virtual label for an algorithm. Agents that play one another as exemplified by the game example employ adversarial learning where multiple agents serve as adversaries to one another, and their win and loss ratios provide the objective metrics necessary to support and guide training. Large language Models (LLMs), such as ChatGPT, Claude, Llama, Bard, or others, employ what can be discussed as self-learning where they consume large amounts of assumedly correct text, learn statistical patterns in this text, and then employ these statistical relationships to generate text that follows these same statistical relationships. While early language models treated each word as a unique entry, LLMs learn phrases, as well as individual words, which are used in similar ways and thus likely have similar meaning (Amazon Web Services, 2023). These phrases are then assumed to be like one another, providing multiple methods to represent concepts. These algorithms consist of an encoder, which is used to analyze the existing text to create the deep learning model and decoders that are used to produce unique text. Based on examples within the training set, these algorithms can perform tasks such as generating novel text or software code, classifying text based upon similarity, and answering general knowledge questions based on the input text. These algorithms do not understand the meaning of the text they consume or create, only the statistical relationships within the input data. However, because language is constrained to have a limited number of symbols (words or phrases) and methods for forming meaningful sentences from these symbols (grammar and sentence structure), reasonable, although not always correct or meaningful, text can be generated by these algorithms (Raiaan et al., 2024). Consequently, these algorithms sometimes "hallucinate", producing text with nonsensical meaning or "facts" that are incorrect. Nevertheless, technologies such as LLMs and future versions could be useful in creating natural language interfaces between AI agents and humans, increasing the naturalness or intuitiveness of these interfaces in future human–AI agent teams.

6.2.6 AI Techniques and Agents

Thus far, this chapter has discussed AI algorithms. As shown, there are distinct types of algorithms that have different capabilities and limitations. These are not agents, simply algorithms. Importantly, an agent might use one or several AI algorithms to facilitate the interpretation of sensed information, decision-making, and actuator control. Further, while AI scientists created different algorithms within each wave of AI development, each technique provided additional tools rather than necessarily replacing the tools developed in the earlier waves. An AI agent might include a deep learning algorithm that accepts images as input from a camera and segments the image into objects. Another deep learning algorithm might classify the object in each segment to provide meaning for decision-making. An algorithm based on a combination of first principles algorithms and expert systems might then decide upon actions to take, based upon information sensed from the camera. Finally, another deep learning algorithm might provide control inputs to motors on robot limbs to cause the robot to move in a way that permits it to implement the decision. This is an example of an AI agent which is based upon a series of AI algorithms of distinct types (Weiss, 2013).

6.3 AGENTS AND ARCHITECTURES

Although we have provided a general definition of an agent, it is important to recognize that designers develop AI agents to have distinct characteristics and functionality. Before proceeding further, it is useful to understand some of these basic differences.

6.3.1 AI Agent Definitions

We could classify AI agents as either embodied or disembodied agents. An *embodied agent* is a single physical entity, for example, a robot, capable of meeting the requirements of an agent. In this instance, the sensors necessary to effect perception, the processor necessary to support decision-making, and actuators necessary to exhibit action are all present in a single physical entity. However, this is not necessary. For example, an AI agent might be comprising sensors located along a waterway, which are capable of sensing water levels, a processor capable of interpreting changes in water level consistent with flooding, and an alarm system which the processor could activate to issue warnings to residents along the waterway. Note that in this example, the AI agent does not exist in a specific location, but its function is distributed across a large geographic region. However, this *disembodied agent* can exhibit all the attributes required of an AI agent. According to the current definition, we do not require the AI agent to issue a command, but it could simply issue an alert to a human who is then responsible for the final decision to issue the alert. That is, an AI agent, according to our definition, can either act on its own or function as a decision aid to another agent.

Sterling and Taveter define other classifications of agents (Sterling & Taveter, 2009). An *intelligent agent* is an agent which can be both reactive and proactive, as well as social. Each of these characteristics are worthy of further definition. A *reactive agent* can perceive (i.e., sense and interpret) its environment and respond to environmental changes in a timely fashion. In an automotive example, we might see an animal enter our lane, and we react by breaking to avoid a collision. A *proactive agent* does not simply respond to changes in the environment but is able to pursue goal-directed behavior in an opportunistic fashion. As such, a proactive agent might anticipate a change within its environment and act to change or avoid a future event. If we use a driving example, we might see an animal standing on the edge of the highway and begin to decelerate in anticipation that the animal might suddenly approach and enter the highway. In this case, the anticipatory behavior is trying to avoid a future circumstance that might require more violent breaking.

Many of the AI agents of interest in this text will be *interactive agents*. These AI agents will be capable of interacting directly with other agents. Ideally, these interactive agents will be capable of interacting with human agents, as well as other AI agents within the system under design. The concept of a social agent extends the requirements for an interactive agent. A *social agent* can interact, when appropriate, with other agents to complete their own activities and, when able, help other agents complete the activities for which they are responsible. To be effective, social agents when interacting with humans must not only have the "know how" to perform activities within their environment but also have the knowledge that Pacaux-Lemoine and Debernard have referred to as the "know how to cooperate" so that they communicate appropriately with team members (Pacaux-Lemoine & Debernard, 2002), which may include many of the concepts we discussed in Chapter 4.

6.3.2 AI Agent Architectures

Wooldridge describes four architectures for artificial agents, including logic-based, reactive, belief-desire-intention, and layered architectures (Wooldridge, 2013). As we define these architectures, it is important to note that some of the architectures are inspired by biologic systems.

6.3.2.1 Logic-Based Architectures

In logic-based architectures, the AI agent designer symbolically codes properties of the world in a database within the AI agent where combinations of particular property values are associated with a particular action the agent might perform. For example, if an AI agent has a sensor that can determine that material (e.g., water) is deposited on a windshield and it has another sensor that can determine that the windshield wipers are off, the AI may use this information to explore the symbolic information in its database, find the condition where each of these are true, and retrieve the prescribed action, for example, activate the windshield wipers at a slow speed.

Although straightforward to implement, this architecture has potential pitfalls. First, the agent must appropriately interpret the sensor data to provide the correct symbolic code. If the sensor is faulty, and the interpretation of the sensor information is not correct, the AI agent can fail. For example, suppose the material on the windshield was ice droplets rather than liquid water, but the sensor indicated the presence of water. In this case, the logic would activate the windshield wipers, and the ice might damage the wipers. Therefore, the sensors must provide the correct symbolic information, and the symbolic information provided by the sensor must be present in the database for all important conditions. Further, as the number of logic statements increases, the time to process the information increases. Therefore, the likelihood that the environment will change before the agent takes an action increases. The AI agent might then take action based on outdated information. Further, representing even simple procedural knowledge (i.e., what to do) can be time-consuming, and for complex systems it is difficult to anticipate all conditions prior to system operation. On the other hand, the information in the database is known, and the user can review the information for accuracy, making system behavior predictable, assuming that the sensors accurately translate the sensed environmental information into appropriate symbolic states.

6.3.2.2 Reactive Architectures

These AI agents assume that intelligent behavior is an emergent property of simple behaviors that occur as one interacts with the environment. A particular reactive architecture, referred to as the subsumption architecture, can be thought of as consisting of a hierarchical set of goals where lower-level goals are given priority over higher-level goals (Brooks, 1986). The agent then executes the behavioral rules associated with these goals upon environmental feedback. For example, a robot might have the goal of avoiding collisions and a secondary goal of navigating to a destination. When no obstacles are present in the robot's path, the robot pursues the higher-level goal, executing actions to move toward the destination. However, when an obstacle is present in the robot's path, the agent triggers the lower-level goal of avoiding a collision, and the robot pursues actions consistent with this goal. As such, the robot reacts to information that is present in the environment to select specific actions.

This architecture has many advantages, including that it is simple and economical to implement, as well as relatively robust. It does have disadvantages—for example, the agent must sense the appropriate information accurately, and local information must be present to support appropriate decision-making. Further, because it reacts to local information only, the decisions may not be optimal. While the designer can build effective AI agents with a few layers of goals, it is much more difficult for the designer to construct effective AI agents as the numbers of layers increase. Finally, the agents do not decide behavior prior to implementation, but the behavior emerges as the system interacts with the environment. Therefore, it is difficult to predict behavior under all circumstances.

6.3.2.3 Belief-Desire-Intention (BDI) Architectures

This architecture combines deliberation with means-ends reasoning. During deliberation, an agent decides what they want to achieve, that is, selects or forms the goal they wish to pursue. Next, the agent applies ends-means reasoning to decide how to pursue the goal. During the reasoning process, the agent decides the available options, often called alternatives. The agent then chooses between the alternatives and commits to one or more of these alternatives. The chosen alternatives become intentions, and these intentions feed into and constrain future reasoning processes.

These intentions persist for a period, drive means-end reasoning, constrain future deliberation, and drive beliefs. Means-ends reasoning can occur at multiple levels of abstraction where higher levels of abstraction involve means that are not actionable but constrain the decision space, whereas means at lower levels of abstraction are actionable, providing actions the agent can take.

Ideally, as the agent reasons and selects actions based upon means-ends reasoning, the agent will ideally drive the system toward the intended goal. However, it is possible that this will not occur or will not occur within the available time limits. Therefore, it can be useful to reconsider intentions. Agents which frequently reconsider their intentions are cautious, while agents which charge ahead without reconsidering their intentions are bold.

These AI agents follow a process in which they acquire sensor input which shapes the AI agent's beliefs about the current state and potential future states of the world. As the agent forms beliefs based on the sensor input, the agent revises current beliefs. The AI agent then generates options based on its current beliefs about its environment and its current intentions. The available options form the AI agent's desires. The agent then filters the resulting list of current options through deliberation to form the AI agent's intentions. The agent then selects actions which are consistent with the AI agent's intentions. This architecture is intuitive to understand but can be difficult to implement.

Returning to the robot example, a robot that is navigating space may encounter an obstacle. In a BDI architecture, the robot might then deliberate on a course of action. Does it go over the obstacle, go around the obstacle, or reverse course to find another path? It will then select one of these options based on current beliefs. For example, the robot might believe that going over the obstacle is not possible, and the shortest path is to go around the obstacle, leading to the selection of the option to go around the obstacle. Once the agent selects one of these options, it becomes the intent of the robot, that is, the end it desires to achieve. It then considers the means to achieve this intent. For instance, if it decided to go around the object, it then might deliberate on whether to go to the left or the right. Once this agent selects this option, it becomes the agent's intent and drives further behavior. Note, however, once an agent chooses an intent, the agent no longer considers methods to achieve the options it did not select, and it no longer considers these options unless the agent decides to reconsider their earlier selected intentions. This then reduces the agent's decision space.

6.3.2.4 Layered Architectures

It is worth noting that the BDI agent results in an agent which is generally proactive. The agent is setting goals and working toward these goals as opposed to reactive agents. When implementing layered architectures, the agent combines attributes of these architectures by implementing processes in layers which it can execute simultaneously. For example, the agent might implement reactive components in one layer, while it may implement proactive components in another layer. The agent can execute layers together with the agent reasoning to form its response. Designers can implement various layered architectures, where layers perform distinct functions that are important to the agent's behavior.

6.4 MODELING: RESPONSIBILITY-CAPABILITY DIAGRAM

6.4.1 Modeling Objective

Now that we are familiar with AI techniques, AI agents, and AI architectures, we can start to envision the potential capabilities that the agents in our "To-Be" system could possess and, thus, the responsibilities these AI agents might be able to fulfill.

The purpose of the AOSM Responsibility-Capability Diagram is to capture the full trade-space for agent allocation by documenting all of the relevant and significant capabilities that will be required to meet the Responsibilities associated with our system's Goals (Miller et al., 2020).

6.4.2 Diagram Components and Composition

The AOSM Responsibility-Capability Diagram is a custom SysML Block Definition Diagram, which displays responsibilities and capabilities and depicts the connection between responsibilities and the capabilities, which must be present to permit an agent to fulfill the responsibility. We stereotype activities as Responsibility and Capability blocks. We introduced Responsibilities in Chapter 3 in the Goal–Responsibility diagram and then iteratively updated the Responsibilities for our system in Chapters 4 and 5. The *Capabilities* are attributes of the agents, consisting of the knowledge, skills, and abilities that the agents possess and that are utilized to fulfill the Responsibilities.

Figure 6.6 depicts a generic Responsibility-Capability Diagram. This diagram captures Responsibilities from the Goal–Responsibility diagram in a column along the left-hand side. During construction, the designer can retrieve these responsibilities from the earlier diagrams and place them onto a new SysML Block Definition Diagram. The diagram then displays the Capabilities required to fulfill those Responsibilities in a column on the right-hand side of the diagram. The designers create the capabilities by examining each responsibility and determining the capabilities required to fulfill the responsibility. The designer then links the Capabilities and Responsibilities, usually SysML associations, which provide documentation for the origin of the capabilities. In most cases, the diagram will contain many-to-many relationships, where the designer links each Responsibility to multiple Capabilities, and each Capability to one or more Responsibilities. It is common for the designer to link groups of Responsibilities to groups of Capabilities. The designer should consider allocating these groups to a single Role, and these groups, however, will typically be linked to one or more Agents.

6.4.3 Considerations

To differentiate Responsibilities from the Capabilities required to perform them, recall that responsibilities are the Perception, Interpretation, Decision, and Action activities that the HAT must accomplish to fulfill a goal. That is, they are tasks. Capabilities are the knowledge, skills, and abilities needed to accomplish a Responsibilities. That is, they are qualities, or attributes, of an agent. For each

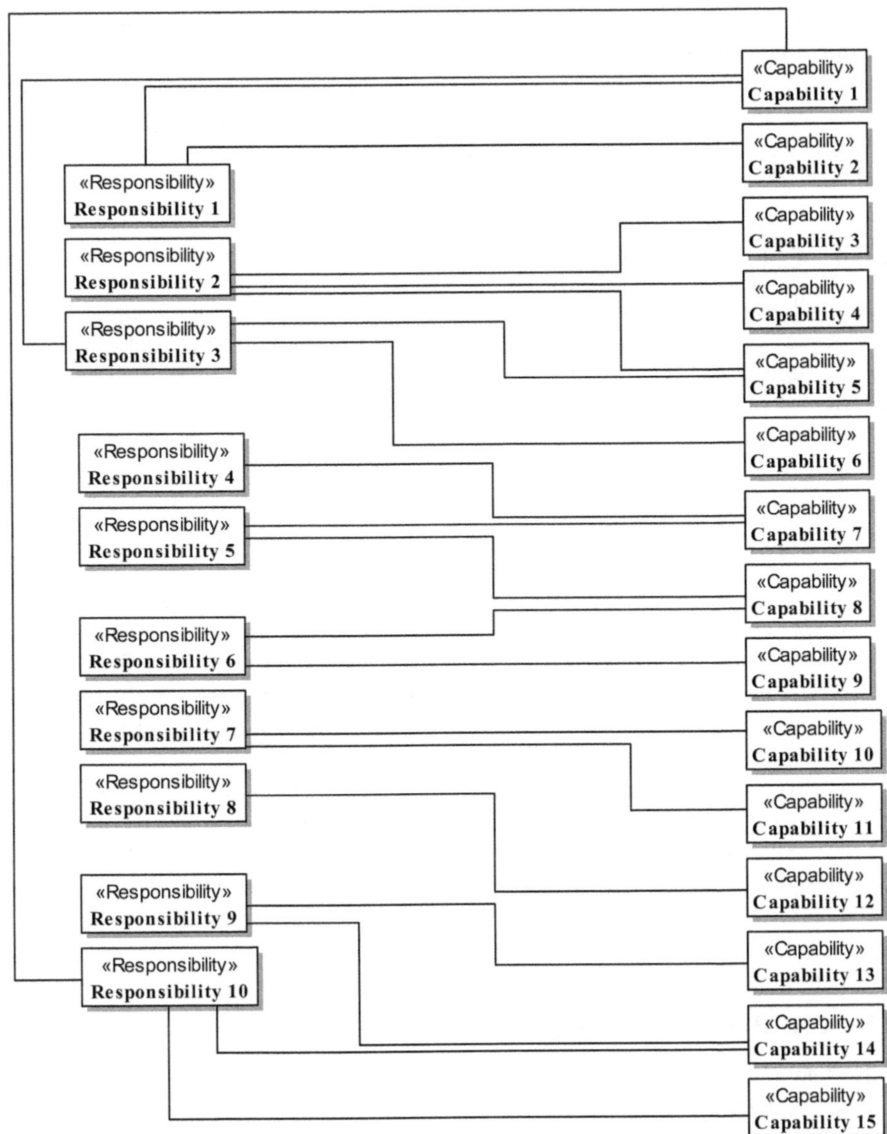

FIGURE 6.6 Generic Responsibility-Capability Diagram containing generic elements.

Responsibility in the Goal–Responsibility diagram, ask about the knowledge, skills, and abilities to accomplish this task successfully. Knowledge consists of information that the agent must know or have access to. Skills are behaviors that the agent must learn and perform to do the task. Abilities are innate capacity to perform skilled behaviors. Many responsibilities require both knowledge and skills or abilities and may even require multiple knowledge or skill capabilities.

You might find it helpful to specifically label the capabilities as either knowledge or ability. It is unnecessary to differentiate between skills and abilities at this stage of analysis, as the "innateness" of the capacity is not relevant in an environment where agents can acquire a skill through training. With training, practice, and experience, the agent can increase their capacity to perform behavioral activities, increasing their ability.

6.4.4 Example Application

Figure 6.7 provides an example Responsibility-Capability Diagram for our Diagnostic Center HAT. For readability, the diagram displays only select responsibilities, including the highlighted responsibilities shown in Figure 5.4. Since the purpose of our redesign is to *Prioritize High-Risk Patients*, this example of the "To-Be" system focuses on those Goals and Responsibilities most connected to this purpose. Thus, we tailored the Goals and Responsibilities associated with *Maximize Team Well-Being* and all those related to performing the biopsy, out of the example, going forward. Figure 6.2 displays the remaining Responsibilities from Figure 5.3 associated with the goals and subgoals of *Provide Early Detection*, *Take Correct Intervention*, and *Prioritize High-Risk Patients*. It also includes the *Maintain State of the Art Knowledge* responsibility that is linked to the *Maximize Team Performance* goal. We linked each responsibility in Figure 6.7 to the required capabilities. For example, we linked the *Recommend Further Diagnostics* responsibility to the capabilities: *Ability to Interpret Risk Level*, *Ability to Communicate with PCM*, and *Recognizing Need for Additional Data.*

It is also noteworthy that the responsibility *Interpret Health History for Risk Level* appeared as a highlighted item in Figure 5.3 but does not appear in this diagram as we decided that this human responsibility could not be fully satisfied. Therefore, we redesignated it as a new high-level team goal in the redesign, removing it as an individual responsibility.

6.5 CHAPTER SUMMARY

This chapter began with an overview of the history of artificial intelligence. The first wave attempted to mimic knowledge by capturing symbolic representations and associated logical constructs. The second wave expanded this by using expert systems to reason across large data sets. The third wave matured statistical machine learning techniques, beginning with tools such as simple linear regression and advancing these techniques to deep learning that involved supervised as well as unsupervised learning. As modern advances during this third wave required large data sets to train the algorithm, it was supported by the availability of large data sets, the use of crowdsourcing to label the data sets, and high-speed processing using general processing units (GPUs) found in video gaming cards.

We then discussed the classification and architectures of AI agents. Importantly, AI agents can be reactive, proactive, or even social, where we might expect proactive,

FIGURE 6.7 Example Responsibility-Capability Diagram for related to prioritizing high risk patients.

social AI agents to behave more like teammates than AI agents. Further, we discussed the fact that designers can construct AI agents from a collage of AI algorithms, where some of these algorithms might rely on first principles' calculations, expert systems, and machine learning techniques.

Progressing in the design of our "To-Be" system, we developed the AOSM Responsibility-Capability Diagram. The Responsibility-Capability Diagram documents the capabilities that the HAT must perform to meet the Responsibilities associated with our system's Goals. This step provides insights into the capabilities that any agent would require to contribute to the associated responsibility and sets the stage to establish the full trade-space for the allocations in our "To-Be" system design.

6.6 QUESTIONS AND DISCUSSION TOPICS

1. For each Wave of AI, provide an example, from everyday life of an AI algorithm or agent, which is representative of these points in AI history. Are your examples proactive agents and/or social agents? What characteristics do they possess that would lead you to classify them as proactive or social?
2. Why do you believe the AOSM method constructs Responsibility-Capability Diagrams rather than immediately assigning responsibilities to agents as the next step?
3. Name one characteristic of the algorithms from each wave of AI, which makes this class of algorithm useful, as well as one characteristic that might hinder their use within HATs.

6.7 PROJECT QUESTIONS AND ASSIGNMENTS

1. For your system, select a potential activity that could involve an AI agent. In separate paragraphs, describe two different implementations using two different architectures: Logic-based, reactive, belief-desire-intention, and layered.
2. For your "To-Be" system, develop a Responsibility-Capability Diagram. Answer the following questions about this diagram once you have created it.
    1. Which responsibilities required multiple capabilities?
    2. Which capabilities are associated with many responsibilities?
    3. Does your diagram have significant interconnectedness or are there "islands" of isolated groupings?

REFERENCES

Amazon Web Services. (2023). *What are large language models*. https://aws.amazon.com/what-is/large-language-model/#:~:text=The encoder and decoder extract,that transformers perform self-learning.

Bindewald, J. M., Peterson, G. L., & Miller, M. E. (2016). Clustering-based online player modeling. In T. Cazenave, S. Edelkamp, & M. Winands (Eds.), *International joint conference on artificial intelligence (IJCAI)—Computer games workshop*. Springer International Publishing.

Brooks, R. A. (1986). A robust layered control system for a mobile robot. *IEEE Journal on Robotics and Automation*, *2*(1), 14–23. https://doi.org/10.1109/JRA.1986.1087032

Choi, C. Q. (2021). 7 revealing ways AIs fail: Neural networks can be disastrously brittle, forgetful, and surprisingly bad at math. *IEEE Spectrum*. https://spectrum.ieee.org/ai-failures

Copeland, B. J. (2023). *Artificial intelligence*. Britannica. https://www.britannica.com/technology/artificial-intelligence

Endsley, M. R. (2023). Ironies of artificial intelligence. *Ergonomics*, *66*(11), 1656–1668. https://doi.org/10.1080/00140139.2023.2243404

Mcculloch, W. S., & Pitts, W. (1990). A logical calculus nervous activity. *Bulletin of Mathematical Biology*, *52*(l), 99–115.

Miller, M. E., McGuirl, J. M., Schneider, M. F., & Ford, T. C. (2020). Systems modeling language extension to support modeling of human-agent teams. *Systems Engineering*, *23*(5), 519–533. https://onlinelibrary.wiley.com/doi/abs/10.1002/sys.21546

Minsky, M. (1961). Steps toward artificial intelligence. *Proceedings of the IRE*, *49*(1), 8–30. https://doi.org/10.1109/JRPROC.1961.287775

Pacaux-Lemoine, M.-P., & Debernard, S. (2002). Common work space for human—machine cooperation in air traffic control. *Control Engineering Practice*, *10*, 571–576. https://doi.org/10.1016/S0967-0661(01)00060-0

Raiaan, M. A. K., Mukta, S. H., Fatema, K., Fahad, N. M., Sakib, S., Mim, M. M. J., Ahmad, J., Ali, M. E., & Azam, S. (2024). A review on large language models: Architectures, applications, taxonomies, open issues and challenges. *IEEE Access*, *12*(February): 26839–26874. https://doi.org/10.1109/ACCESS.2024.3365742

Rosenblatt, F. (1958). The perceptron: A probabilistic model for information storage and organization in the brain. *Psychological Review*, *65*(6), 386–408. https://doi.org/10.1037/h0042519

Shimizu, C., Hammar, K., & Hitzler, P. (2023). Modular ontology modeling. *Semantic Web*, *14*(3), 459–489. https://doi.org/10.3233/SW-222886

Shimizu, C., Hitzler, P., Hirt, Q., Rehberger, D., Estrecha, S. G., Foley, C., Sheill, A. M., Hawthorne, W., Mixter, J., Watrall, E., Carty, R., & Tarr, D. (2020). The enslaved ontology: Peoples of the historic slave trade. *Journal of Web Semantics*, *63*. https://doi.org/10.1016/j.websem.2020.100567

Sterling, L. S., & Taveter, K. (2009). *The art of agent-oriented modeling*. MIT Press.

Wasserman, W. F., Neter, J., & Wasserman, W. (1996). *Applied linear statistical models*. Irwin/McGraw-Hill.

Weiss, G. (2013). *Multiagent systems*. MIT Press.

Wooldridge, M. (2013). Intelligent agents. In G. Weiss (Ed.), *Multiagent systems* (2nd ed., pp. 3–51). The MIT Press.

7 Allocation

7.1 ALLOCATION DEFINED

In Chapter 5, we briefly mentioned the term *allocation*. The *Merriam Webster* dictionary defines allocate as "to divide and give out for a special reason or to particular people or things". In the systems engineering literature, allocation is a step of the engineering process in which system designers assign functions to physical resources. This process is a crucial step in creating the physical architecture or implementation of any system as the physical resource to which the function is allocated is instantiated in the world and has capabilities and constraints which dictate how fast, accurate, and costly the function will be performed.

When discussing automation, Hollnagel and Woods list three different allocation strategies, including the leftover principle, the compensatory principle, and the complementarity principle (Hollnagel & Woods, 2005).

7.2 LEFTOVER PRINCIPLE

Given that a common goal when engineering systems which employ automation is to replace the human, engineers often approach the problem of designing systems with the aim of automating every function which can be automated, with any remaining "leftover" functions being allocated to the human. In most complex systems, it is not practical to automate all functions performed by the human. Therefore, when applying this principle, the designer does not make the allocation decision deliberately and judiciously or with concern for the human. Instead, the designer assumes that if automation can relieve the human from some of the activities they perform, the automation must aid human performance. Unfortunately, this assumption has associated flaws which have been well documented.

One of the flaws with this assumption is that replacing any task will reduce the effort required of the human and improve system performance. As we discussed in Chapter 4, humans are cybernetic organisms, which pursue goals by acting in their environment and using feedback to determine whether their action is consistent with pursuit of the goal. Therefore, we operate in a closed loop process, that is, the perceptual loop (Neisser, 2014). When designing automation, it is possible for the designer to automate a task within the perceptual loop, such as detecting threats. However, the human needs information obtained from this step, including not only the presence of the threat but also its location, movement, and type to perform higher-level cognitive processing to complete the perceptual loop. Therefore, the automation of threat detection is only helpful if the system provides information to the human in a fashion which enables them to perform the higher-level processing steps which follow threat detection. Although this fact seems obvious, system designers often overlook this fact. Further, as the automation is often fallible, by overlooking this fact, the designer

DOI: 10.1201/9781003428183-7

assumes the human will verify the automation's performance or accept the risk of error. Verification implies that the humans must perform threat detection in parallel with the automation and then perform the added task of cross checking their detection results with the automation's results. Therefore, automating this task does not necessarily reduce the load placed on the human but can increase the task work the human needs to perform.

Another important, but often overlooked, fact when considering automation is that human decisions often vary in complexity. Thus, it is incorrectly assumed that if a human decides to pursue an action which results in a catastrophic outcome, the decision was flawed, and the accident could have been avoided by a professionally designed artificial system. In fact, in real-world circumstances, adequate information is not always present to make an accurate decision, or random events affect the outcome in ways that designers cannot anticipate. As such, decisions made using appropriate processes can produce undesirable outcomes. Chapter 9 discusses decision-making and the effect of missing, incomplete, or inaccurate information in more detail. However, it is important to recognize that in real-world situations, an entity can follow a robust and logical decision process that results in catastrophic outcomes. Thus, the presence of these outcomes does not imply that another entity would have made a decision that was less likely to result in a similar outcome.

Designers using the leftover principle should be cognizant of whether the entire decision-making process is being automated. The implementation of the leftover principle without sufficient analysis could add increased complexity and uncertainty to the functions allocated to the human.

One last flaw of the leftover principle is that this approach often places little emphasis on redesigning the human's role and interface in the system to enable the user to understand how they should interact with the automation. Yet, modifying the system in this fashion will change the role of the human within the system and how they could best interact with the system (Klein et al., 2004). As humans are still responsible for system operation, that is, the legal system holds humans accountable for violations, the human has no recourse other than to do their best within the confines of the given automated system. For many operators, this may mean disabling the automation, or choosing not to activate the automation (as is often the case with vehicle cruise control), to avoid surprises, thus defeating any advances the automation could have potentially provided. As a result, any potential return on investment and the promises of benefits from automation development are unrealized. Therefore, employing the leftover principle, while potentially being less expensive from a design and development point of view, is often not an efficient or effective design strategy.

7.3 COMPENSATORY PRINCIPLE

In brief, the compensatory principle assumes that the automation designer should design automation to compensate for deficiencies in human performance. That is, if we understand the functions that the machine or automation might perform better than the human, we should assign these functions to the automation rather than the human.

In the Human Factors Engineering literature, the foundational literature regarding allocating functions to people or portions of the system is a technical report written in 1951 by then US Air Force Captain Paul Fitts, summarizing the outcome of a workshop on the future of Air Traffic Control systems. This report included a pair of figures, which have become known as Fitts' List (Fitts, 1951). In these figures, Fitts enumerates the 11 generic functions shown in Table 7.1. In this workshop, the committee agreed upon the categorization of these functions as functions that are performed "better" by either humans or machines.

Fitts' List has persisted, as the allocation literature consistently cites this report 70 years after it was written. In fact, the capability of any given agent to perform a required function is an important consideration, although not the only consideration that a skilled designer should apply.

Given the age of Fitts' List an important question is: "Does it apply in today's world?" Technology has improved over the intervening years, with a 2015 survey reevaluating the 11 statements of Fitts' List, where individuals perceived that humans continue to surpass machines in their ability to improvise and use flexible procedures, ability to reason inductively, and the ability to exercise judgment (de Winter & Hancock, 2015). However, they perceived that machines are better at detecting tiny amounts of visual or acoustic energy as well as storing and recalling vast amounts of information. There was a split in the number of individuals indicating whether humans or machines are better at perceiving patterns of light or sound.

It is not surprising that individuals would state that machines are better at storing and recalling mass amounts of information as we have all become reliant on internet searches to inform everyday activities, and recent developments, such as ChatGPT,

TABLE 7.1
Items Performed Better by Humans or Machines (Fitts, 1951).

Humans Appear to Surpass Present-Day Machines in Respect to the Following:	Present-Day Machines Appear to Surpass Humans in Respect to the Following:
1. Ability to detect a small amount of visual or acoustic energy	7. Ability to respond quickly to control signals and to apply great force smoothly and precisely
2. Ability to perceive patterns of light or sound	8. Ability to perform repetitive, routine tasks
3. Ability to improvise and use flexible procedures	9. Ability to store information briefly and then to erase it completely
4. Ability to store exceptionally copious amounts of information for long periods and to recall relevant facts at the appropriate time	10. Ability to reason deductively, including computational ability
5. Ability to reason inductively	11. Ability to manage highly complex operations, that is, to do many different things at once
6. Ability to exercise judgment	

are likely to further increase the use of AI for information retrieval. The fact that individuals perceive that machines are better at detecting tiny amounts of energy indicate the general improvement in sensors, especially for sensing information beyond the limits of the human sensory system. The ability to sense information in the infrared portion of the electromagnetic spectrum to apply radar and lidar to sense information beyond line of sight also increases the ability of automation to gather information the human cannot. Further, the ability to synthesize information from multiple sensors, perhaps in neighboring vehicles or mounted at various locations in the environment, further extends the ability of the machine to sense information that humans cannot.

The disagreement about perceiving patterns arises from the fact that in 2015, machine learning algorithms could find patterns in logical data structures that humans could not but severely underperformed in visual pattern recognition. As machine learning with respect to pattern recognition continues to progress, this category is likely to continue to shift toward the machine's favor. However, humans are likely to continue to outperform machines in making sense of unfamiliar patterns, permitting them to excel in novel situations.

Although Fitts' original report did not support using this list alone when allocating functions between humans and machines, individuals often assume or interpret this list in this fashion. As a result, researchers often criticize Fitts' list for promoting at least three fallacies. These include:

1) "Best" Agent Fallacy: Functions or activities should be allocated to humans or machines based only upon the decision of which entity can best perform the function;
2) Single Agent Fallacy: Functions or activities should be allocated to the human or machine during design, and only that entity can perform the function or activity; and
3) Single Performer Fallacy: Functions or activities should be allocated completely to either the human or the machine.

These fallacies are important to understand as they all support assumptions which lead to suboptimal designs.

7.3.1 "Best" Agent Fallacy

The first of these fallacies implies that one entity or the other will always be better at performing a function than the other. Under any given set of circumstances, this statement is false as many functions do not comprise solely one of Fitts' 11 items. Further, human, and often machine, capacities vary depending upon circumstances; that is, while one entity may be better at providing a capability under one set of circumstances, the reverse may be true under a different set of circumstances.

Figure 7.1 shows a notional depiction of a two-dimensional space which represents functions (T1–T6) with respect to human and machine performance. This figure illustrates that it is possible for some functions to fall within the black region near the origin where neither the human nor the machine can satisfactorily perform the function. If functions lie within this region, the designer should redesign the system, or the system

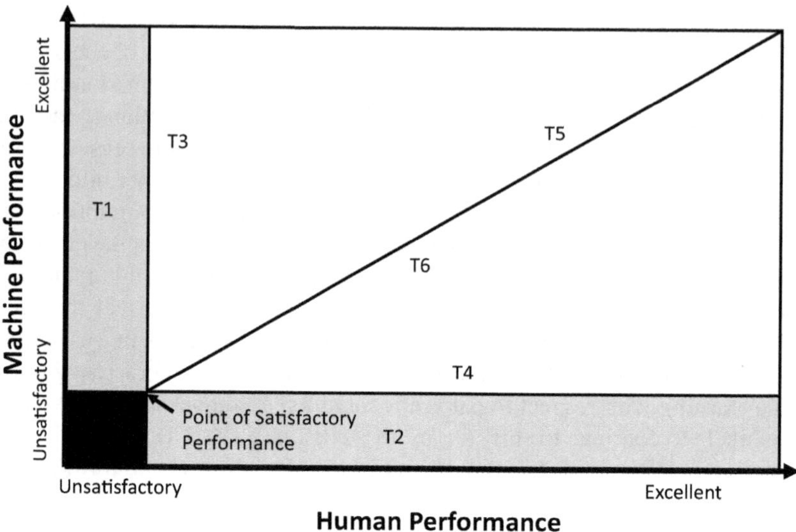

FIGURE 7.1 A depiction of functions differentiated by human and machine performance.

Source: adapted from Price (1985)

will fail. Also shown are functions where either the machine (T1) or the human (T2) can perform a function satisfactorily, while the other entity is not capable of successful completion. Functions within these regions are the functions where the entity's capability should drive allocation. That is, these functions should be allocated to the entity which is capable of satisfactory performance. However, either entity can perform the remaining Functions T3 through T6; thus, these functions can be reasonably allocated to either entity. Either entity can perform functions near the 45-degree line (i.e., functions T5 and T6) at the same level; thus, the designer cannot rely upon the capability of the entity to inform the allocation decision. As this discussion illustrates, there are a number of functions where human and machine performance will not be sufficiently different to support an allocation decision based only on the "Best" Agent criteria.

It is also known that human, and at times machine, performance is not stationary but changes depending upon the circumstances. Fitts points out that humans "are poor monitors" eluding to the occurrence of decreased human performance under low levels of "activation" or stimulation, as well as the occurrence of decreased performance of both humans and machines when either of these entities is "overloaded". This change in human performance as a function of human arousal or workload, often referred to as the Hebb–Yerkes–Dodson Law (Teigen, 1994), is well known in the human performance literature. Thus, it is inappropriate to assume that one entity will be able to perform better than the other under all circumstances. The knowledge of the variability in human performance led to an adaptation of the compensatory principle in the 1990s which was referred to as "adaptive automation" in which the automation's activation adapts to the situational needs of the human (Byrne & Parasuraman, 1996; Scerbo, 1996).

7.3.2 Single Agent Fallacy

In the previous section, we discussed assigning a function to a single entity at design time. It is important to recognize, however, that this assumption is not only unnecessary but often detrimental to team performance. Proponents of adaptive automation recognized the fallacy of this assumption. *Adaptive automation* attempts to compensate for the change in human performance as a function of factors, such as human workload. The concept requires the system to monitor either the environment or the human's performance and to alter the allocation based upon these variables. For example, a vehicle that monitored the environment would implement automatic braking, if the human does not brake within a set distance of an object in front of the vehicle. With human monitoring, an adaptive vehicle could monitor the human for symptoms of drowsiness and then trigger automated steering, acceleration, and braking when the human falls below a threshold level of alertness.

Before we discuss this in detail, however, it is important to further discuss some terminology. In the previous section, we discussed function allocation. However, the literature sometimes uses the term "function" and, at other times, the term "task". Although one might assume these terms are synonyms, they have very different meanings and implications for design within the human factor's literature. A *function* is a process which accepts an input and generates an output. Functions transform an input to an output, and a designer can apply a function in any situation that requires this transformation. For example, a function might receive an image as an input and provide the locations of potential targets in the image as an output. Thus, the function performs "target detection". This is a generic process which can be instantiated or implemented in software or performed by a human. Therefore, it is reasonable to discuss allocating the function to a human or machine during design. However, if we assume that we will allocate the function of target detection to a machine, the implication is that during operation, the machine will perform this function anytime that target detection is required.

A ***task*** is an activity which an entity performs in real time during operation. Therefore, during design, the function of "target detection" might be allocated to either the human or the machine. In this case, only the human or machine to which the function was allocated can perform the task of target detection. Alternately, during design, the designer could allocate the function of "target detection" to both the human and machine. During operation, it is then acceptable for either the human or the machine to perform any individual task of "target detection". Therefore, the human may perform the task of "target detection" for a first image the system acquires, but the machine may perform the task of "target detection" for a second image the system acquires.

For example, Figure 7.2 shows an alternate rendition of the previous figure. In this figure, functions T5 and T6 are in the gray region which overlaps the 45-degree line, indicating that the performance of the human or the machine would, under "normal" circumstances, be nearly equivalent. However, in adaptive automation, when human workload is less than the human's capacity, the tasks associated with functions T5 and T6 are allocated to the human to increase the human's level of activation, thus increasing their overall performance. As human workload increases beyond human

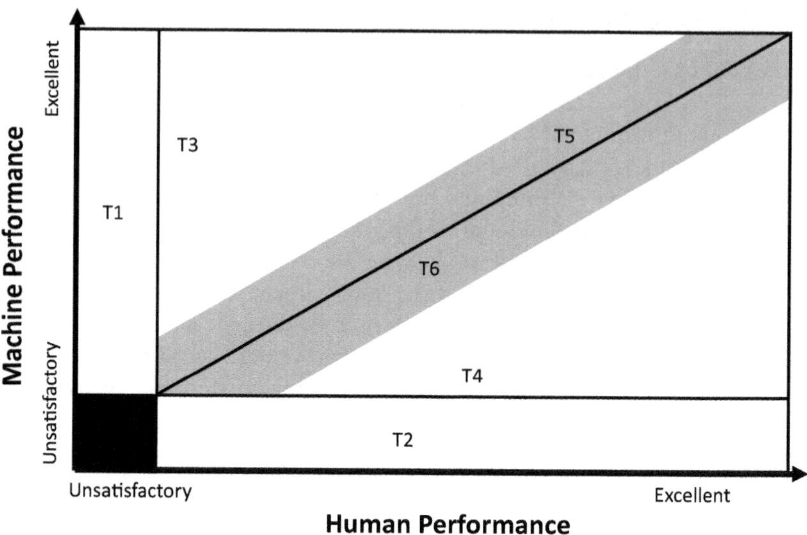

FIGURE 7.2 A depiction of functions differentiated by human and machine performance and illustrating functions in the gray band, which might be adaptively allocated between the human and machine during system operation.

Source: adapted from Price (1985)

limits, the tasks associated with these functions are allocated from the human to the machine to relieve the human of some of their duties. Thus, the system allocates more tasks to the human when doing so might improve the human's performance and allocates more tasks to the machine when the human's performance on their remaining items (e.g., T2 and T4) is high to avoid overload conditions. This concept recognizes that human performance changes based upon the situation and attempts to design the system to compensate for this factor, illustrating an incorrect assumption of the second fallacy.

In an adaptive automation, the system allocates at least a subset of the functions (e.g., T5 and T6) to both the human and the AI. Either the human or machine can then perform the corresponding task during system operation. However, many functions (e.g., the functions corresponding to T1, T2, T3, and T4) are wholly allocated to either the human or machine. Based on this discussion, it should now be clear that functions allocated to the human or machine during design can be allocated to either or both types of entities. Further, if functions are allocated to more than one entity during the design phase, any of the entities to which this function was allocated might perform the associated task during system operation.

When functions are allocated to more than one entity, the designers must develop the software and hardware associated with the functions allocated to the machine. Similarly, designers must develop human training, procedures, and interfaces for the functions allocated to the human. Therefore, allocating functions to more than one entity increases the development cost of the system but, ideally, provides more flexible, and more robust, operation of the system. Superior system performance has been demonstrated for systems which employ adaptive automation as compared to systems which employ fixed automation (Kaber et al., 2005; Parasuraman et al., 2009). However, confusion can occur as the human and machine must both be aware of any changes in task allocation as they occur (Kaber et al., 2001). This design often requires the human to engage in cognitive processing to maintain awareness of the tasks they must perform.

7.3.3 Single Performer Fallacy

In the last section, we illustrated that it is possible, and at times desirable, to allocate functions to more than one entity, for example, the human and the machine. In that discussion, we saw that the baseline adaptive automation literature assumes that all the tasks associated with one task would be allocated to either the human or the machine at any given time during system operation. However, this assumption is also not necessary and at times less than desirable. There are at least two options for real-time allocation of tasks to both a human and an AI agent during operation.

The first option for allocating tasks to more than one entity during operation is to require both entities to perform each task. Returning to our "target detection" example, this task might be allocated to both the human and the machine during design. In this option, both the human and machine perform the task of "target detection" during system operation, the results are compared, and a final decision is made. In systems where the target detection task is not sufficiently reliable, but the comparison and decision-making tasks are reliable, this arrangement can increase the overall reliability of the system. It does, however, create new tasks, for example, comparison and final decision-making, which may reduce the efficiency of the overall task. The value of this trade-off depends on the overall goals of the system.

The second option for allocating tasks to more than one entity during operation is to permit each agent to perform any task, depending upon availability of the entities. Adaptive automation researchers originally discussed this concept as a method to encourage human engagement with a system, while permitting the machine to assume increasing amount of work as the human's ability to perform the tasks declines. A concept for accomplishing this is to place tasks into a queue of uncompleted tasks, for example, shown as a list within the user interface. The human draws tasks from the queue as quickly as they can complete tasks. However, as the human falls behind, the machine begins processing tasks and assumes the duty for more of the tasks as the task list becomes longer (Rouse, 1977).

Other mechanisms, besides queues, may also be useful in permitting task accomplishment where either the human or machine can perform each task. For example, Goodman and colleagues employed a graphical interface to a game called Space Navigator (Goodman et al., 2017). In this game, the user's goal was to draw routes from spaceships, which appeared near the edge of a display, to planets which were in the interior of the display. The human was to draw these routes to avoid collisions among the spaceships, avoid no-fly zones that appeared in random locations, and to pick up bonuses which also appeared in random locations. The human could draw a route or redraw an existing route using their finger on a touchpad. An agent was available to aid the human. If the human did not assign a spaceship a route at a specified time after it appeared, the agent would simply draw a straight line from the ship to its destination planet. The human interface depicted routes drawn by the agent with a color that was distinct from the color of routes drawn by the human, permitting the human to understand the work performed by the agent rather than the human. Ships without routes provided an indication that routes needed to be drawn, providing an analogy to a cue. In his work, Goodman demonstrated that even the rudimentary agent, that is, straight-line-drawing agent, improved team performance, especially under conditions of increased workload.

Canzonetta expanded this research by demonstrating a mixed-initiative adaptive allocation of tasks based on workload. Canzonetta utilized the advantages of having human and AI agents within this same game adapt the AI agent delay time, dependent upon the number of ships without routes (Canzonetta & Miller, 2019). Specifically, as more ships were present without routes, the AI agent drew routes more rapidly relieving the human of workload, but as the backlog decreased, the AI agent slowed its response time to increase human activity and, thus, human engagement. The results of this work demonstrated that human engagement and performance improved over a broad range of conditions when the adaptive AI agent timing paradigm was employed.

The designer can divide work between humans and agents or machines by permitting the human and the agents to understand the work to be accomplished and providing a real-time division of this work, depending upon factors, such as the human's available time and their level of engagement.

7.3.4 A Quick Review of the Compensatory Principle

To quickly review this section, the compensatory principle assumes that automation should be employed to overcome human limitations—for example, to perform work when the human is experiencing overload. The compensatory principle has its basis in Fitts' List, which researchers have criticized, justifiably or not, for promoting the "Best Agent", "Single Agent", and "Single Performer" fallacies. Most compensatory methods make one or more assumptions consistent with one of these three fallacies. The reader should note, however, that the imposition of these assumptions provides useful simplification of the design of the system. Therefore, these assumptions have been both promoted and disputed with advancements in the science associated with human–agent teaming. On the other hand, technology to overcome these fallacies, such as adaptive automation, requires a complex design with the agent perceiving the real-time situation and applying knowledge of the performance or mental state of the human to understand their workload, or a related measure, for the system to alter function or task allocation in real time during system operation.

7.4 COMPLEMENTARITY PRINCIPLE

The complementarity principle focuses on longer-term changes or adaptation of the interface between the human and the machine. This principle is based on the understanding that humans learn and evolve over longer periods of time, such that their skills and knowledge change over time. Individuals who are new to a job may need much more support than individuals who have performed the relevant tasks over months or decades. As such, the complementarity principle acknowledges that the type of automation support which would be most beneficial to individuals may also change over time. Further, this principle acknowledges that humans must be relied upon to perform work to gain an appreciation for the nuances and decisions that the work requires. Therefore, automating or providing aids, which would be useful to the least-skilled user, and maintaining these aids in place constantly may not permit the user to gain an in-depth knowledge of the work domain. Thus, it may be necessary to adapt the system to permit the user to gain new skills while providing the necessary support structure to avoid unacceptable failures. To employ automation according to this principle requires designers develop an in-depth understanding of the work and the paths to human skill development to properly implement automation in a HAT during system design.

7.5 ALLOCATION DESIGN AND HUMAN–AI AGENT TEAMING

The three principles we have reviewed in this chapter represent philosophies for allocating functions and tasks between the human and AI agents. However, they also influence the design of the interface between a human and an AI agent, which then influences the degree to which these systems support teaming. Most traditional systems we are familiar with apply function allocation where the system supports certain functions, and we expect humans to rely upon the system to perform these functions while they perform other functions.

Many of the autonomous systems we use today were designed to retain the human as the responsible agent by designers who designed according to the leftover principle, while attempting human replacement. Systems in this domain, including many modern vehicles capable of "autonomous driving" and commercial aviation interfaces, utilize interfaces that provide "Supervisory Control", where the human watches over and ideally can recover the system when the automation fails. We have discussed these interfaces and their challenges in previous chapters. In these systems, the designer allocates most driving functions to the AI agent(s) in the system but provides the ability for the human to assume control. Therefore, most driving functions are allocated to both the AI agents and the humans, but it is assumed that the tasks associated with driving will be allocated to the AI agents unless the human assumes control and then the human will be responsible for performing all necessary tasks. Neither systems in this "autonomous" category nor the previous functional allocation category demonstrates behavior that behavioral scientists would classify as teaming behaviors.

Modern driving assistance systems allocate many of the driving functions both to AI agents and to humans. However, in these systems, the designer assumes that the human will perform all the necessary tasks associated with driving unless they request an AI agent to perform select tasks (i.e., maintaining speed, controlling headlight dimming, controlling windshield wipers). In the literature, these systems are referred to as adaptable automation systems, as the human can adapt or select the specific functions to be automated, offloading specific tasks to automation (Chavaillaz & Sauer, 2017). These interfaces again have little resemblance to human teaming. However, they permit the human to hand off tasks to the AI agents, which can reduce the human's workload and permit them to focus some of their attention on other tasks.

As discussed in Table 5.1, automotive systems employ Human Error Reduction implementations, such as driver drowsiness detection, function by monitoring the driver, determining degraded driver states, and then adapting the automation in response. These systems are examples of adaptive automation systems where an AI agent senses a characteristic of the driver and causes a change in the behavior of itself or another AI agent in the system to provide backup behaviors to the human operator. This behavior is a teaming behavior where the AI is providing support to the driver. Earlier discussions also mentioned mixed-initiative examples where either the human or an AI agent can assume responsibility for a task, providing backup and sharing of taskwork.

Overall, most systems employing AI agents today perform task work, while few systems attempt to monitor, respond to, and interact robustly with human agents in the environment. The discussion of allocation within this chapter focuses on how designers design and share task work between team members. As we begin to consider the design of human–AI agent teams, it is important to consider the allocation of functions, the allocation and real-time distribution of tasks among team members, as well as the teamwork that is necessary to create a reliable interaction among members of the human–AI agent team.

As part of these teaming behaviors, it can be useful to enable a desirable level of implicit communication and coordination, particularly for teams performing in time-critical environments (e.g., action or production teams). Importantly, adopting implicit coordination does not mean that we need to abandon communication. As noted earlier, teams develop specialized vocabulary that conveys significant information about coordination behaviors, permitting details of a plan to be communicated implicitly during the action phase of teamwork by relying upon information that was communicated explicitly during transition or interpersonal processes that are conducted outside the action phase (Marks et al., 2001). Examples include systems which employ coordination strategies, such as play calling (Funk et al., 2005; Miller, 2017). In these systems, developers design automated agents to seek certain goals when a human provides a high-level command, activating a series of general behaviors that are adapted on the basis of agreed-upon situations. This paradigm is like a play in a sport, where members of a team learn certain goals and behaviors prior to a game and associate it with a name, that is, the name of the play. Each member's plan differs based upon their role on the team, but the team members each understand their objectives, the cues which will cause selection of certain behaviors or plans and their interactions with other members of the team for a specific "play". Operators define and learn these plays prior to the game, mission, or action phase, but the machine executes their portion of the play when the operator calls the play during the action phase. Each team member is responsible for understanding the cues which trigger specific plans for attaining a goal and executing the appropriate plan in response to the cues which trigger a specific plan for attaining the goal. These systems also exhibit certain teaming behaviors in that the members of the team share work and leverage a degree of shared knowledge to facilitate coordination. However, they do not necessarily adapt their behavior based on humans in the system other than through the plays they call.

7.6 MODELING: AGENT CAPABILITY ASSESSMENT TABLE

7.6.1 Modeling Objective

The purpose of the AOSM Agent Capability Assessment Table is to capture the full trade-space for agent allocation by documenting which agent types (human, AI, or both) can perform the Responsibilities identified in the Responsibility-Capability Diagram. We are not allocating specific agents to specific responsibilities; thus, this table should just capture whether it would be practical or feasibly possible for a human or AI agent to perform at least part of a responsibility through providing a specific capability.

7.6.2 Diagram Components and Composition

Figure 7.3 shows a generic Agent Capability Assessment Table. The rows of this table present the Capabilities identified in the Responsibility-Capability Diagram and then Human, AI agent, or Either, for the column labels. For each responsibility, we place a relationship at the intersection of the row and column to indicate which class of agent(s) could acquire or possess the capability, in the context of the system. For the human agent, one should consider potential training (or recruitment) that could result in the capability. For the AI agents, one should consider current technological developments. In the case of the "Either" column, this includes when either one could possess the capability or, when combined, both could possess the capability.

7.6.3 Considerations

Remember that we are not currently allocating responsibilities; therefore, we should be as open minded as possible to the potential for an AI agent or a human agent to perform a particular responsibility. It is likely that many of the responsibilities can be performed by either or both classes of agents. Additionally, it is likely, and desirable, that more than one agent will jointly perform multiple responsibilities.

The capability assessment should be useful; thus, it is necessary to be practical when assessing the feasibility of an entity to perform a responsibility. For example, while humans may be technically able to update thousands of records, it would not be practical for our system to have a human responsible for regularly manually updating thousands of records. Similarly, there are likely to be some decisions that, while they could be made solely by an AI agent, the safety, risk, or criticality will require the human involvement for acceptance in our system. As we are not performing allocation, it is not essential that this assessment be correct. This initial assessment is

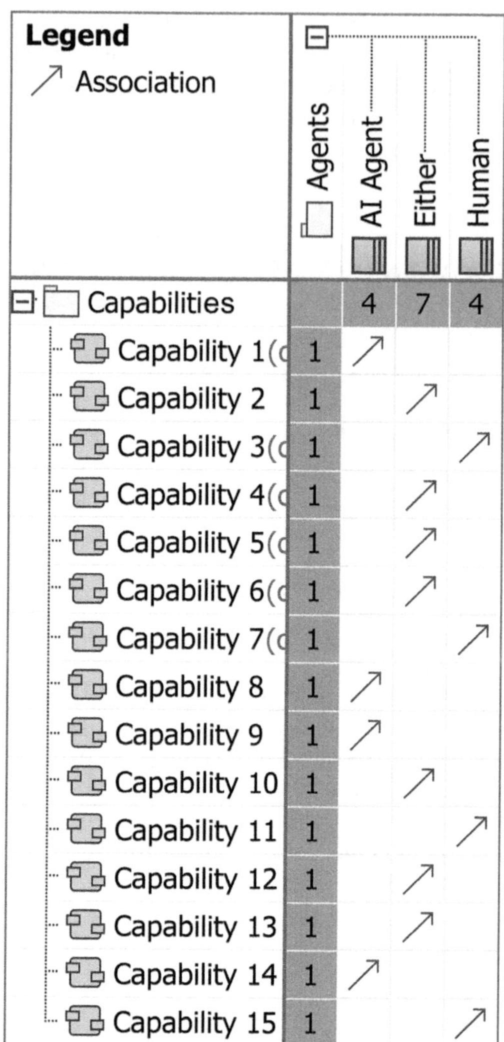

FIGURE 7.3 Generic Agent Capability Assessment Table containing generic elements.

useful as we begin to further consider the architecture (i.e., included agents and their interfaces), as well as potential allocations.

7.6.4 Example Application

Figure 7.4 provides the example Agent Capability Table corresponding to the Responsibility-Capability Diagram in Figure 6.8. The capabilities identified as specifically being possessed only by humans require physical operation within the workspace (*Ability to Determine Placement for Image Composition*, *Ability to*

Legend
↗ Association

	Responsibility	AI Agent	Either	Human
Goal Responsibility Diagram		5	20	8
Ability to Assign Risk based on Factors (context Risk Assessment Agent)	1	↗		
Ability to Annotate Source of Record Update	1	↗		
Ability to Communicate with PCM	1	↗		
Ability to Determine Placement for Image Composition (context Radiol	1			↗
Ability to Determine Quality of Image	1		↗	
Ability to Disseminate New Information	1		↗	
Ability to Feedback on Image Quality (context Radiologist)	1		↗	
Ability to Identify how Medical Research Affects Current Protocols (co	1			↗
Ability to Identify Risk Factors in Patient History	1	↗		
Ability to Identify Unique High-Risk Patient Procedures	1		↗	
Ability to Implement New Information	1			↗
Ability to Interpret Risk Level (context Radiologist)	1			↗
Ability to Manipulate Patient Position for Composition (context Radiolo	1			↗
Ability to Populate Patient Record	1		↗	
Ability to Provide Feedback on Errors	1		↗	
Ability to Select Screening Workflows (context Radiologist)	1			↗
Ability to Synthesize Indicators	1		↗	
Ability to Update Screening Workflows (context Radiologist)	1			↗
Access to Current Medical Research	1		↗	
Accessing Patient Record	1		↗	
Associate PCMs with Patients	1	↗		
Determine Correct Exposure (context Radiology Technician)	1		↗	
Interpret Biopsy Results	1		↗	
Interpret Images	1		↗	
Knowledge of DIfficult Cases	1		↗	
Knowledge of Financial Cost	1		↗	
Knowledge of Intervention Options	1		↗	
Knowledge of Invasiveness	1		↗	
Knowledge of Outcomes of Interventions	1		↗	
Knowledge of Previous Errors	1		↗	
Knowledge of Unintended Consequences of Interventions	1		↗	
Recognizing Need for Additional Data (context Radiologist)	1			↗
Select Correct Exposure (context Radiology Technician)	1		↗	

FIGURE 7.4 Example Agent Capability Assessment Table.

Manipulate Patient Position for Composition), adaptation of unwritten workflows and processes (*Ability to Update Screening Workflows, Ability to Select Screening Workflows, Ability to Identify How Medical Research Affects Current Protocols*), or requires human in-the-loop judgment or authority (*Ability to Interpret Risk Level, Ability to Implement New Information, Recognizing Need for Additional Data*). Several capabilities would be too burdensome to be performed by a human, given the number of patients involved, and thus would need to be performed by an AI agent (*Associate PCMs with Patients, Ability to Communicate with PCMs, Ability to Assign Risk based on Factors, Ability to Identify Risk Factors in Patient History, Ability to Annotate Source of Record Update*). When creating the example, the authors assumed that either, or in many cases both, the human and AI agents could perform the remaining capabilities.

7.7 CHAPTER SUMMARY

This chapter discussed both function and task allocation. Designers perform function allocation during design. Operators perform task allocation during system operation in systems where designers provide the operator the flexibility to perform the function or automate the function. This chapter discussed three different philosophies or principles for the allocation of functions and tasks in human–AI teams. It is important that teaming requires the ability for agents to perform backup behaviors. To enable teaming, designers must design the system such that operators and agents can collaborate and dynamically allocate tasks among human and AI agents in real time during operation. While this is a necessary requirement for teaming systems and is only present for current designs which employ certain embodiments of the compensatory or complementary principle, it is not sufficient to achieve teaming. Teaming additionally requires the design of interfaces between the agents which are responsible for systems that permit them to coordinate their activities. As illustrated, most current human–AI agent teaming systems achieve this by providing methods for the humans to command AI agents to perform tasks, either explicitly or through high-level communication protocols. Some examples exist in which an AI agent can proactively support the human, assuming further behaviors that are common in human teams. Thus, as we consider the design of future human–AI teaming systems, we must consider the allocation of functions and tasks to support teaming, but we must also consider the design of the interaction between agents to support teamwork behaviors, including different forms of interdependence, which is discussed in more detail in subsequent chapters. We begin the allocation process by creating an Agent Capability Assessment Table.

7.8 QUESTIONS AND DISCUSSION TOPICS

1. Using Fitts' list of 11 items, create your own "personal Fitts' List" division of this list between humans and machines, based on items you believe are "better" at performing.

2. Imagine a robotic teammate, designed to assist with gardening. List two to four gardening functions that you would allocate to the machine and two to four that you would allocate to the human, based on your "personal Fitts' List" using the assumptions of the "best" agent fallacy.
3. For your gardening HAT, describe how one of the functions could be allocated such that there is not a single performer.
4. For your gardening HAT, describe how an existing function, or a new function that you add, could be allocated using adaptive automation.
5. For your gardening HAT, describe an example of how the AI agent in your gardening HAT could proactively provide a backup behavior by assuming a responsibility that you would normally perform. In your response, include a list of capabilities the AI agent would need to be capable of to enable it to initiate this backup behavior.

7.9 PROJECT QUESTIONS AND ASSIGNMENTS

1. For your "To-Be" system, develop an Agent Capability Assessment Table. Which tasks would you label as human-only tasks, and why? Which tasks, if any, would you label as AI-only tasks, and why?
2. Are there any responsibilities in the Capability Assessment Table, which you believe should be assigned "Either" and that you had not previously envisioned a Human or AI agent performing prior to reading this chapter? Provide at least one example and explain your original and your changed rationale.
3. This table shows only potential allocations, that is, a human, an AI agent, or both could be endowed with these capabilities and thus contribute to the associated responsibilities and goals. However, select a capability that you have assigned as "either" and a responsibility that it is associated with.
   1. Provide a description of how the implementation of this responsibility and capability might change if you were to assume best agent, single agent, and single performer fallacies.
   2. How could you reallocate this capability and associated responsibility to enable functional allocation to both a human and the AI agent?
   3. What are the benefits and limitations of this allocation?
   4. How could you incorporate adaptive automation?

REFERENCES

Byrne, E. A., & Parasuraman, R. (1996). Psychophysiology and adaptive automation. *Biological Psychology*, *42*(3), 249–268. https://doi.org/10.1016/0301-0511(95)05161-9

Canzonetta, D. J., & Miller, M. E. (2019). Identifying a possible function for artificial agent adaptation in variable task rate environments. *20th International Symposium on Aviation Psychology*, 375–380.

Chavaillaz, A., & Sauer, J. (2017). Operator adaptation to changes in system reliability under adaptable automation. *Ergonomics*, *60*(9), 1261–1272. https://doi.org/10.1080/00140139.2016.1261187

de Winter, J. C. F., & Hancock, P. A. (2015). Reflections on the 1951 Fitts list: Do humans believe now that machines surpass them? *Procedia Manufacturing*, *3*(Ahfe), 5334–5341. https://doi.org/10.1016/j.promfg.2015.07.641

Fitts, P. (Ed.). (1951). *Human engineering for an effective air navigation and traffic control system*. National Research Council, Committee on Aviation Psychology.

Funk, H., Goldman, R., Miller, C., Meisner, J., & Wu, P. (2005). *A playbook TM for real-time, closed-loop control* (pp. 1–7). Defense Technical Information Center. https://apps.dtic.mil/sti/tr/pdf/ADA439281.pdf

Goodman, T. J., Miller, M. E., Rusnock, C. F., & Bindewald, J. M. (2017). Effects of agent timing on the human-agent team. *Cognitive Systems Research*, *46*, 40–51. https://doi.org/10.1016/j.cogsys.2017.02.007

Hollnagel, E., & Woods, D. D. (2005). *Joint cognitive systems: Foundations of cognitive systems engineering*. CRC Press.

Kaber, D. B., Riley, J. M., Tan, K.-W., & Endsley, M. R. (2001). On the design of adaptive automation for complex systems. *International Journal of Cognitive Ergonomics*, *5*(1), 37–57. https://doi.org/10.1207/S15327566IJCE0501_3

Kaber, D. B., Wright, M. C., Prinzel, L. J., & Clamann, M. P. (2005). Adaptive automation of human-machine system information-processing functions. *Human Factors*, *47*(4), 730–741. https://doi.org/10.1518/001872005775570989

Klein, G., Woods, D. D., Bradshaw, J. M., Hoffman, R. R., & Feltovich, P. J. (2004). Ten challenges for making automation a "team player" in joint human-agent activity. *IEEE Intelligent Systems*, *19*(6), 91–95. https://doi.org/10.1109/MIS.2004.74

Marks, M. A., Mathieu, J. E., & Zaccaro, S. J. (2001). Remember me a temporally based framework and taxonomy of team processes. *Academy of Management Review*, *26*(3), 356–376.

Miller, C. A. (2017). The firefox fallacy: Why intent should be an explicit part of the external world in human automation interaction. In *Cognitive systems engineering: The future for a changing world* (pp. 269–294). https://doi.org/10.1201/9781315572529

Neisser, U. (2014). Cognitive psychology: Classic edition. In *Cognitive psychology* (Classic). Psychology Press.

Parasuraman, R., Cosenzo, K. A., & De Visser, E. (2009). Adaptive automation for human supervision of multiple uninhabited vehicles: Effects on change detection, situation awareness, and mental workload. *Military Psychology*, *21*(2), 270–297. https://doi.org/10.1080/08995600902768800

Price, H. E. (1985). Allocation of functions in systems. *Human Factors*, *27*(1), 33–45. https://doi.org/10.1177/001872088502700104

Rouse, W. B. (1977). Human-computer interaction in multitask situations. *IEEE Transactions on Systems, Man & Cybernetics*, *7*(5), 384–392. https://doi.org/10.1109/TSMC.1977.4309727

Scerbo, M. W. (1996). Theoretical perspectives on adaptive automation. In R. Parasuramn & M. Mouloua (Eds.), *Automation and human performance: Theory and applications* (pp. 37–63). Lawrence Erlbaum Associates.

Teigen, K. H. (1994). Yerkes-Dodson: A law for all seasons. *Theory and Psychology*, *4*(4), 525–547.

8 Human–AI Agent Team Architectural Patterns

8.1 HUMAN–AI AGENT TEAM PATTERNS

Patterns can provide inspiration during design. As Larman indicates, a good design pattern has at least four characteristics (Larman, 2004). To support easy communication, the pattern includes a name. Second, the pattern describes a well-known problem and solution pair a designer can use in other contexts. Third, the pattern provides advice on how to apply the pattern in novel situations. Finally, the pattern acknowledges trade-offs, as well as alternate implementations and variations. Lesser-known, unproven, arrangements which might become patterns once proven can be referred to as "prototype patterns" (Neerincx et al., 2019).

As we consider the design of human–AI teams, it is important to explore existing patterns and methods for describing these patterns, as each can provide insight into the design of future systems. Although the literature contains documented patterns for dyads that include a human and an AI agent, at the current time, the literature contains limited discussion of larger teams of human and AI agents. In the current chapter, we will explore existing prototype patterns from the literature for human–AI agent dyads and discuss prototype patterns for larger human–AI teams to attempt to motivate the types of human–AI teams that we promote in this text. As suggested by Neerincx and colleagues, it is useful to discuss these patterns using a standardized language (Neerincx et al., 2019).

Another important function of patterns is they permit us to gather observations about these teams and to characterize their differences. Schulte and colleagues separate work patterns from human–AI teaming patterns (Schulte et al., 2016). Work processes have objectives, influences (i.e., environment, inputs), and outputs. Additionally, the objectives in the work process can be arranged in hierarchical or network structures. Van Diggelen and Johnson suggest that other important attributes of the work are whether it is physical or cognitive and whether it is task work (which they refer to as direct work) or teamwork (which they refer to as indirect work) (Van Diggelen & Johnson, 2019).

Patterns are used during architectural design of buildings, mechanical systems, and software to capture and convey best practices. These patterns not only illustrate examples that have worked in the past but also provide a structure for the reuse of design components and inspire designers to examine their problems

DOI: 10.1201/9781003428183-8

through the lens of various past designs. As patterns permit designers to reuse existing knowledge, they can reduce the risk of developing designs that do not work as expected. Human–AI teams reshape work, often human cognitive work, through the inclusion of novel technology. Therefore, the pattern must consider the structure of the work environment as well as each agent's contribution to this work environment. Further common environmental limitations, such as communication infrastructure or time constraints, can significantly affect the work of the team. Thus, HAT patterns must permit the designer to explore the impact of these factors on potential system designs.

When describing human–AI teams, Schulte and Donath point out that humans rely on their cognitive resources to control a machine (e.g., teleoperate a remote vehicle) (Schulte et al., 2016). Relying on this as an analogy, Schulte and colleagues propose that human–AI teams can either embed an AI agent in the machine, providing so called "autonomous" or "robotic" systems, or directly support the human's decision-making, as a decision support, assistant, or "associate" system. They suggest that control of the work can be either hierarchical, where one entity (i.e., human or AI agent) supervises others, or heterarchical in which any one of the entities can command or govern other entities at any moment. Van Diggelen and Johnson also suggest that important attributes of the human–AI team include the dispersion of the team, which they enumerate as colocated, connected, or disconnected, and the communication modalities that the team employs (Van Diggelen & Johnson, 2019).

Although not explicitly captured by either of these researchers, the time requirements for the control of any machine in the system are also an important attribute of the work that imposes additional constraints, as discussed by Sheridan and Verplank (Sheridan & Verplank, 1978). For example, Sheridan and Verplank when discussing undersea welding operations note that communication delays due to the dispersion of the team between the sea floor and the surface require that the real-time decision-making required to weld a seam must be allocated to the remote AI agent on the welding machine, while less time-sensitive decisions, such as selecting the next seam to weld, can be allocated to the remotely located human. These time delays become larger if communications loss may disconnect a remote agent from the human.

8.1.1 Patterns for Human–AI Team Dyads

8.1.1.1 Colocated Supervisory Control

This pattern has been discussed by Van Diggelen & Johnson (2019) and is present in "autonomous" vehicles, such as Tesla's autopilot. The team comprises a human and an AI agent colocated with a machine as shown in Figure 8.1. The human can provide commands to the AI agent when it controls the machine, or the human can assume control of the machine. As shown in the IBD in the right panel of Figure 8.1, communication between the human and AI agent includes communication to support the handover of control from the human to the machine. A crucial difference between

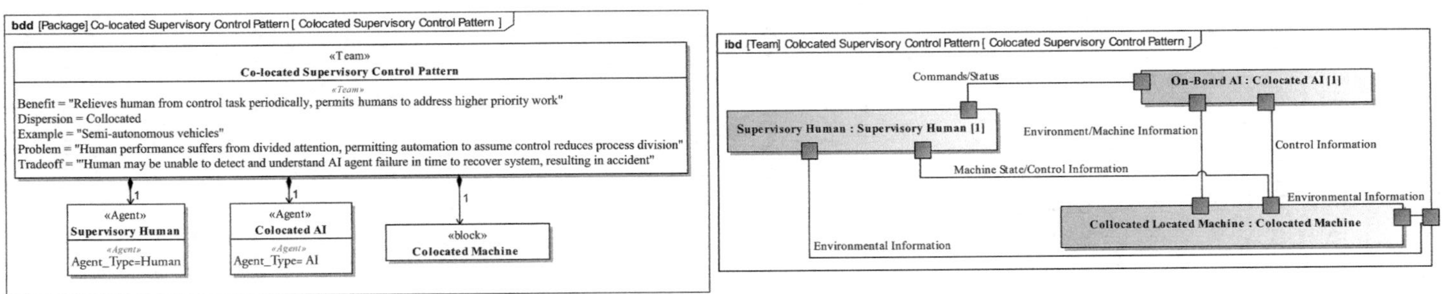

FIGURE 8.1 Block Definition Diagram indicating the components of the colocated supervisory control pattern in the left pane and an Internal Block Diagram illustrating the interfaces and information flows between team components in the right pane.

systems employing this pattern and the Connected Supervisory Control pattern is the ability of the human to rely upon direct observation of the environment, without mediation through sensors on the machine. However, it also acknowledged that information from machine sensors can augment the human's understanding of the environment.

The primary advantage of HATs employing this pattern is to relieve the human, permitting the human to rest or concentrate on more pressing tasks. Systems employing this pattern require the AI agent to be capable of controlling the machine without human intervention or to seamlessly transition control to the human in a timely fashion if the AI agent is incapable of safe and effective operation. In time-sensitive environments, there is a risk that the AI agent will fail to provide adequate control and the human will be incapable of recovering the machine from a hazardous condition. This later condition is more likely to occur than designers anticipate, as the human cannot maintain vigilance in situations where they are not actively involved and therefore will not understand the current situation when the automation fails, requiring the human to assume control.

8.1.1.2 Connected Supervisory Control

Numerous authors have described the supervisory control pattern (Schulte et al., 2016; Sheridan & Verplank, 1978; Van Diggelen & Johnson, 2019). This pattern assumes that the team comprises a human and an AI agent as shown in Figure 8.2. This figure shows the structure of the team in a BDD in the left panel and an Internal Block Diagram (IBD) in the right panel. The IBD shows primary interfaces among elements of the system and the environment. The human and AI agent can be connected or, at times, disconnected. The literature most often discusses the connected supervisory control pattern, where the AI agent is colocated with a machine it controls, and a telecommunications channel connects a remotely located human to the machine and AI agent. Thus, the connected supervisory control pattern differs from the colocated supervisory control pattern because the machine is located remotely, and communication requires the use of a communications infrastructure that is often external to the team. The human can control the machine during teleoperation where sensors on the machine transmit environmental information to the human and receive commands from the human. Alternately, the human can provide commands to the AI agent to control the machine. The machine will sense the environment and provide this sensed information and its status information to both agents. The machine receives control information from one of the two agents. As mentioned, the telecommunications infrastructure is typically outside the design of the system; thus, the IBD shows the communication between the human and either the AI agent or Machine passing in and out of the system boundary. This is important as it recognizes the inability of the designer to truly control this interface and assure interface reliability.

HATs employing this pattern can remove humans from dangerous or dirty environments and often rely upon the AI agent to provide real-time control over the machine to overcome communication delays during teleoperation. Thus, systems employing this pattern extend human capabilities, permitting the AI agent to perform tasks the human could not perform from their remote environment. Unmanned aircraft employing this pattern have been discussed in the literature (Johnson et al.,

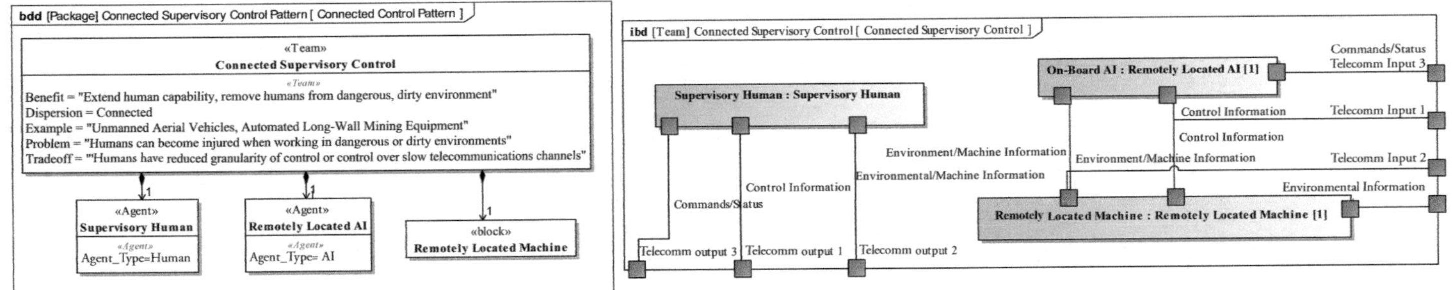

FIGURE 8.2 Block Definition Diagram indicating the components of the connected supervisory control pattern in the left pane and an Internal Block Diagram illustrating the interfaces and information flows between team components in the right pane.

2020). This pattern often reduces the granularity of human control over the machine when commanding the AI agent as the human cannot control all aspects of the machine in real time.

8.1.1.3 Disconnected Supervisory Control

Although typically not intended, the connected supervisory control pattern may lose communication, resulting in disconnected supervisory control. This is a recognized mode of control in many unmanned aerial vehicles (UAVs). This pattern assumes that the team comprises a human and an AI agent as shown in Figure 8.3. This pattern is similar to the pattern shown in Figure 8.1 with the exception that there is no real-time communication between the human and either the machine or AI agent; thus, the IBD in the right panel of Figure 8.3 shows a gray line of connection from the On-Board AI and the supervisory Human, designating that this connection is not always available, but that communication is available prior to the loss of telecommunications. To overcome this situation, it is necessary that the human provide plans and authority to the AI agent to execute these plans before the telecommunication channel is disabled, disconnecting the AI agent and machine from the human. This communication might occur prior to mission execution, or the human might update plans during mission execution. Also note that the human operator could provide similar plans and authority in any other supervisory pattern, but the ability to provide plans and authority for future activity is essential to the success of systems employing this pattern.

The primary advantage of systems employing this pattern is to extend human capabilities, permitting the AI agent to perform tasks the human could not perform from their remote environment. Designers design unmanned aircraft to support this pattern. This pattern necessarily reduces human control but is necessary for mission completion or retrieval of the vehicle from the environment. One could argue that this is not a pattern on its own but a necessary variant of the connected supervisory control pattern. While this is a valid argument, separating these patterns provides a method to explicitly communicate the need for the disconnected variant and its implications.

8.1.1.4 Accident Prevention Pattern

In another pattern involving a Dyad, the human is in control of the machine, but an agent observes the machine's and potentially the human's behavior—intervening by controlling the machine to avoid disaster, thus improving system safety (Schulte et al., 2016). Examples include automatic braking systems in automobiles, which use cameras or other sensors to determine obstacles (Thamarai & Karthik, 2014), and automatic ground collision avoidance systems in aircraft where aircraft location is compared to terrain maps (Bice et al., 1990), each of which are activated only when collisions are imminent. The pattern is similar to the colocated supervisory control pattern. However, it is assumed that the human is controlling the machine unless the AI agent detects an impending collision, at which time the AI agent determines the final point at which action may be taken to avoid a collision, alerts the human who can override activation, and, at the point at which action is required, assumes control of the machine if its action has not been overridden by the human.

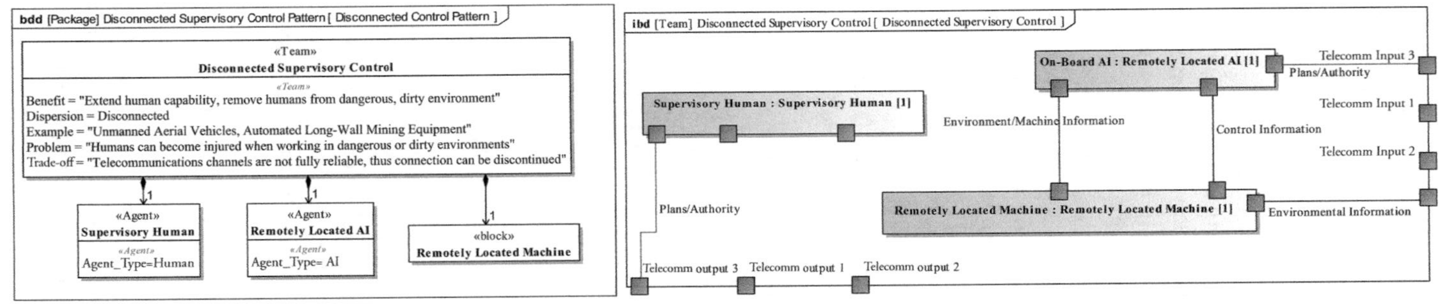

FIGURE 8.3 Block Definition Diagram indicating the components of the disconnected supervisory control pattern in the left panel and an Internal Block Diagram illustrating the interfaces and information flows between team components in the right panel.

The primary advantage of a system employing this pattern is improved system safety. The trade-off involves the necessity to implement user interface elements to warn the human prior to automation activation and the fact that humans may over rely on these systems even when they are not completely reliable. As a result, accidents may occur when the AI agent fails to detect an impeding object, while the human expects the AI agent to avoid the object. Research has also shown that operators can view these systems as a nuisance and deactivate the AI agent if they engage the safety measure too early or are activated in response to the false detection of an obstacle. Although these systems are activated on the basis of comparisons of vehicle location and environmental obstacles, it is acknowledged that many accidents occur when the human is incapacitated or are improperly allocating their attention, such that they do not perceive the impending collision. Therefore, it is reasonable that these systems may also benefit from additional sensors which provide insight into the human's state. However, the systems utilizing human state information have not been fielded, and therefore the pattern shown in Figure 8.4 does not indicate these information flows.

8.1.1.5 Decision Support Pattern

Although each of the systems discussed thus far includes control of a machine, a machine is not a necessary component in a HAT. For example, in command-and-control or aircraft routing systems, AI agents may support human decision-making even though there is no physical machine for the human or agent to control (Layton et al., 1994; Neerincx et al., 2019). In these systems, the AI agent structures information to aid one or more steps in the decision-making process, including understanding the current circumstances, generating options for future actions, understanding trades or selection criteria among these options, and selecting a specific action. Information flows might occur as shown in Figure 8.5 with the human and the AI agent both monitoring the information in the relevant environment. In these systems, the AI agent can help the human make faster decisions or provide a greater understanding of the trade-space such that the human considers more options and ideally selects a better option than they would without the AI agent. Without this type of support, especially in time-constrained environments, the human may not consider a broad array of options but may default to options they have selected under similar circumstances in the past. The trade-off is that decisions offered by the AI agent can bias human decision-making in favor of a potential solution and slow human decision-making if the solution provided by the AI agent is unsatisfactory. We will discuss human decision-making in more detail in the subsequent chapter.

8.1.2 Patterns for Larger Human–AI Teams

Each of the patterns we have discussed thus far include one human and one AI agent. However, as we discussed in Chapter 1, we are interested in HATs that include larger numbers of agents. There are fewer of these patterns available in the literature, which forces us to examine prototype patterns for inspiration.

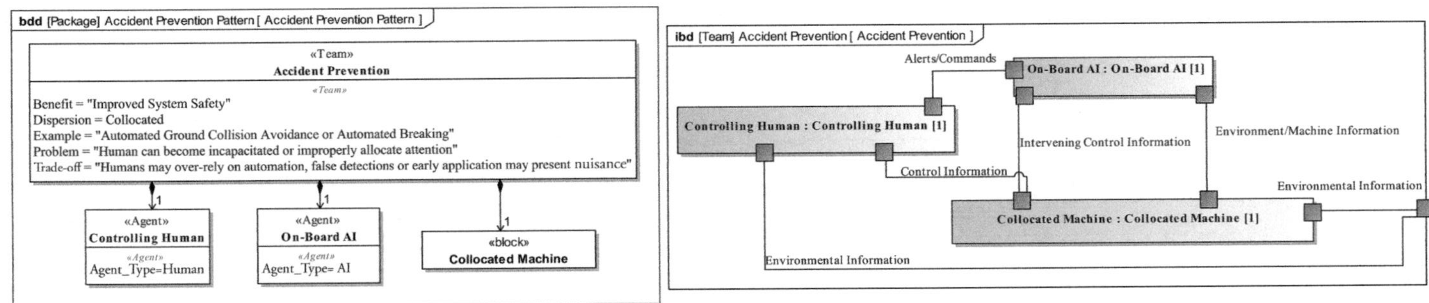

FIGURE 8.4 Block Definition Diagram indicating the components of the connected accident prevention pattern in the left pane and an Internal Block Diagram illustrating the interfaces and information flows between team components in the right pane.

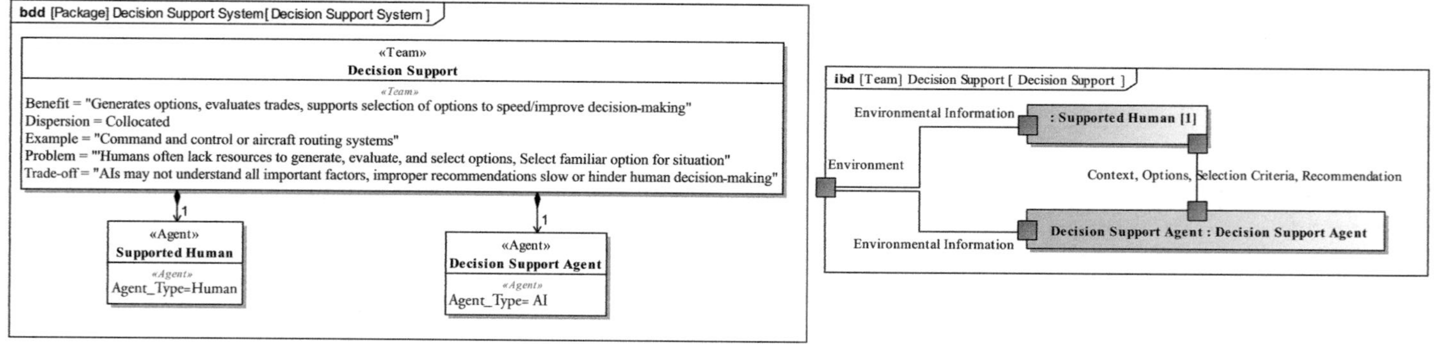

FIGURE 8.5 Block Definition Diagram indicating the components of the decision support pattern in the left pane and an Internal Block Diagram illustrating the interfaces and information flows between team components in the right pane.

8.1.2.1 Dual-Mode Cognitive HATs

The Dual-Mode Cognitive HAT pattern assumes that there is a need for a human to coordinate the activities of multiple machines within an environment. However, we should recognize that a single human does not have the cognitive bandwidth to control multiple machines simultaneously, let alone coordinate their activities. In this architecture, the designer embeds AI agents in each machine to control the behavior of the respective machine, where each human–AI agent dyad follows the Connected Supervisory Control pattern. However, this alone does not provide adequate cognitive support for the human. Therefore, an additional AI agent is colocated with the human to coordinate the activities of the AI agents and ensure that the human is providing timely input to the individual AI agents (LaMonica et al., 2022; Strenzke et al., 2011). Figure 8.6 illustrates this pattern. As shown, the human interacts with the decision support AI agent. The decision support AI agent presents a common operating picture to the human and coordinates activities. The decision support AI agent then converts the coordinated activities to commands and provides the commands to embedded control agents. The embedded control agents then use this information with environmental and system information from the machine to control the machine. Additionally, the embedded controller provides system status and environmental information to the decision support AI agent which fuses this information to create the common operating picture. This pattern includes more than one machine, each having an embedded controller.

This class of systems have gained significant interest in the human–AI teaming literature in recent years (Calhoun et al., 2018; Zacharias, 2019). Its primary advantage is to permit a human to control multiple machines within an environment. This is particularly useful when it is necessary to coordinate the activities of multiple machines where the human director can integrate the effects of the individual machines to provide an integrated response. It has been recognized, however, that additional humans may be necessary to provide a precise control over individual vehicles when the embedded AI agents cannot adequately control the machine(s); thus, the pattern requires the ability to transition to alternate patterns which include more than one human where the additional humans provide commands or provide near real-time control to one or more of the machines.

The decision support agent in this pattern is colocated with the human. The discussion of potential human–AI teams in Chapter 6 included AI agents that monitor human behavior, physiology, or performance to adjust the behavior of the agent and the interaction of the AI agents with the human. While this type of system has received significant research, application of this technology in real-world circumstances has thus far been rare with its application primarily limited to drowsiness detection in modern vehicles. For this, and for space considerations, the current text does not illustrate patterns where the AI agents monitor, assess, and respond to human behavior or state. However, it is anticipated that systems involving this technology will increase in prominence in the future to improve human–AI agent teaming. This technology could be integrated into the decision support agent in this pattern or in other colocated agents within this text.

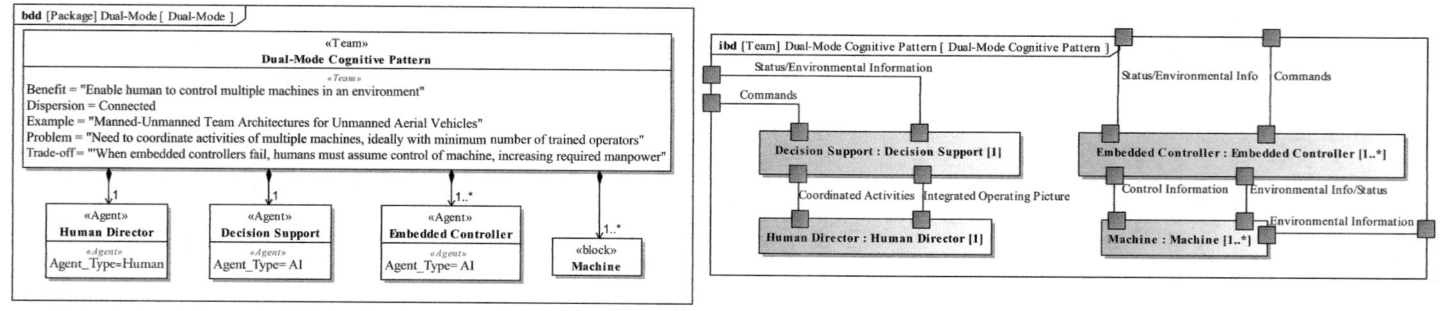

FIGURE 8.6 Block Definition Diagram indicating the components of the dual-mode pattern in the left pane and an Internal Block Diagram illustrating the interfaces and information flows between team components in the right pane.

8.1.2.2 Subsumption Pattern

A common structure that includes multiple AI agents in the robotics literature is based on Brook's subsumption architecture (Brooks, 1986). This architecture provided numerous AI agents where initial components performed specific functions that the designer enclosed in or subsumed in additional AI agents that provided a higher level, more abstract function. Johnson and colleagues pointed out that by giving the human the ability to select which of these AI agents to interact with, the human can decide to enact a precise control of a system through an AI agent at the lowest level of abstraction or relinquish low-level control to shed cognitive work and reduce workload by interacting through an AI agent at a higher level of abstraction (Johnson et al., 2020). For example, the human might directly control surfaces or engine power of the machine, or the human might provide augmented control to an AI agent at the lowest level of abstraction. In this later configuration, the human provides an abstract command, such as move up-down or left-right to the AI agent. The AI agent then converts the human inputs into signals that modify the control surfaces on the machine and provides these converted signals to the machine. Alternately interacting with an AI agent at a higher level of abstraction, the human might provide parametric control information. For example, the human might only indicate the destination, requiring the AI agents to perform all lower-level control to reach that destination. Figure 8.7 shows an internal block diagram representing this pattern.

8.1.2.3 Overwatch Pattern

Certain domains require, or at least benefit, from the faster decisions that an AI agent can provide, as compared to a human. In these domains, for example, stock trading or cyber security, it can be beneficial for an adequate AI agent to respond rapidly to external influences, for example, a sudden change in trend of a stock price. However, experience has demonstrated that these trends can result in reinforcing decision loops, which lead to runaway system behavior (i.e., the faster a stock price declines, the faster the agent sells shares, leading to faster stock price decline). This is especially true when many actors in an environment are applying similar technology. Application of this technology has led to trading halts in major stock indices. Human oversight of such a system, while not being fast enough to make the initial decision, can help mitigate the chances of irrational, "runaway" system behavior when designed appropriately (Bar-Yam, 2017). Therefore, this system includes a Decision-Maker AI agent to make decisions in near real time, while a second AI agent, the Integrator, supports human decision-making by integrating the information, displaying trends, and providing the human with contextual information with which to interpret these trends. The resulting integrated operating picture permits the human to isolate and diagnose potential issues, overriding or redirecting the near real-time decision-maker AI agent when necessary. Figure 8.8 shows a depiction of this pattern.

8.1.2.4 Learner Prototype Pattern

The learner prototype pattern attempts to balance two competing demands. First is the need for faster analysis of input data—for example, analysis of vast amounts of

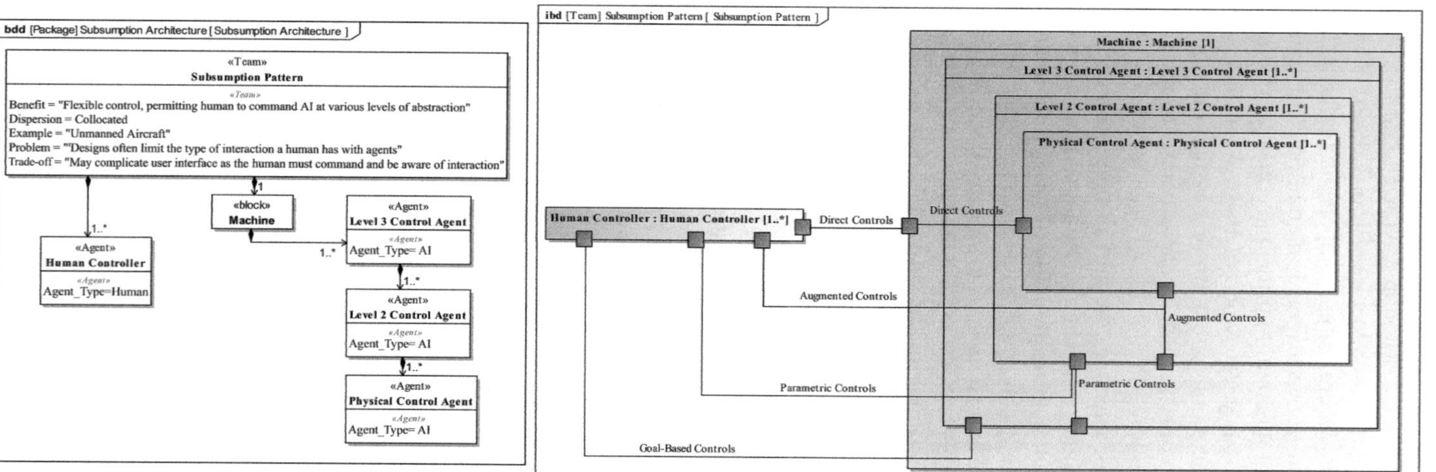

FIGURE 8.7 Block Definition Diagram indicating the components of the subsumption pattern in the left pane and an Internal Block Diagram illustrating the interfaces and information flows between team components in the right pane.

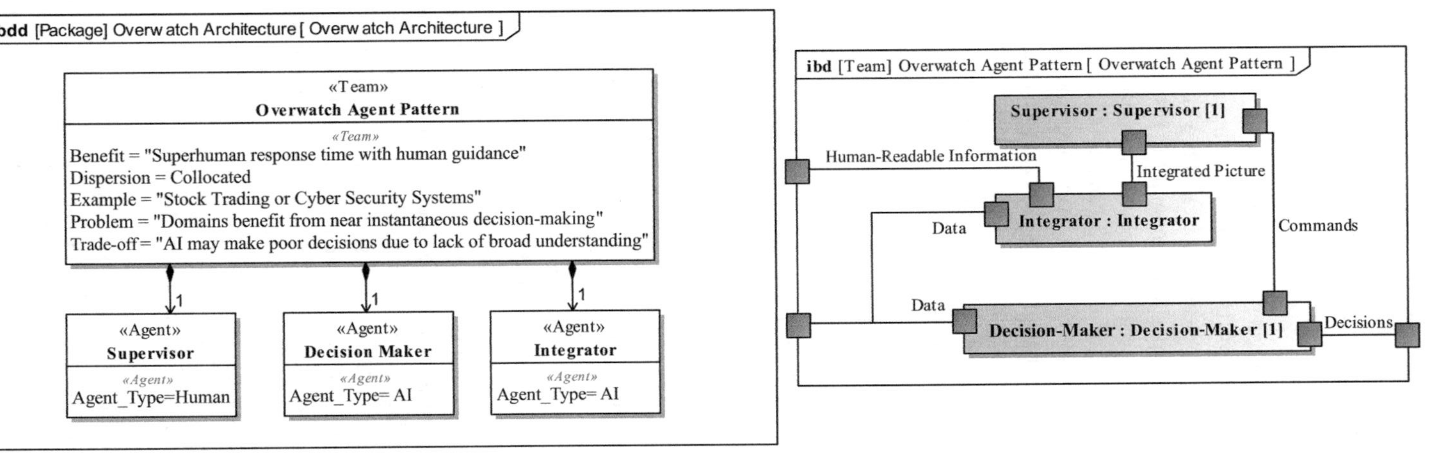

FIGURE 8.8 Block Definition Diagram indicating the components of the overwatch pattern in the left pane and an Internal Block Diagram illustrating the interfaces and information flows between team components in the right pane.

imagery to examine disguised human activity where the disguise is likely to change over time in an unpredictable fashion. The second is the maintenance of a cadre of trained human analysts who are capable of detecting changes in the method of disguise. This prototype pattern includes an AI agent, the Learner agent, which performs image analysis. For example, this agent might determine certain types of vehicles in satellite or aerial imagery. A second AI agent selects a subset of this imagery for analysis by a group of trained but apprentice human analysts. A third AI agent then compares the results from the Learner AI agent with the results from the apprentice human analysts. Expert human analysts resolve significantly different results. Feedback based upon the expert response supports further training for the human analysts or the Learner AI agent, permitting continuous improvement to both the AI and human image analysts. Like the overwatch pattern, this pattern is primarily providing a decision aid that provides faster, more efficient decision-making for common decisions, but it recognizes the need to build human expertise and AI knowledge by training both agents to respond to unforeseen circumstances. Thus, this pattern takes advantage of the human's ability to innovate to determine new modes of camouflage or other deceptive practices, while supporting the adaptation of the AI with time to provide resilience to adaptation by an intelligent adversary. Figure 8.9 depicts such a pattern.

8.1.2.5 Service Negotiator Prototype Pattern

The negotiator pattern is primarily useful when one or more humans are seeking the ability to procure a service from one or more providers. A common everyday event might be working with a group of friends to identify a restaurant for dinner. In this environment, we might consider that each provider has one or more services which are available at a known location and time. Multiple consumers (called Initiator and Participants) wish to coordinate activities to select a relevant service, with a known time and location range. Further, each of the consumers have a set of preferences or desires which they seek to have fulfilled but do not necessarily wish to express these preferences to one another. This pattern assumes that one of the consumers, termed the Initiator, initiates the activity and controls the final list of invitees and selection. Figure 8.10 depicts the flow of information and the agents within the service negotiator pattern. Note that in the language of the current text, the initiator and participants are assuming distinct roles and thus have different responsibilities.

As shown in this pattern, an AI agent, referred to as the Supplier Agent, supports the Supplier and negotiates terms of service with a Negotiator Agent. The Initiator controls the Negotiator Agent. The Initiator and each Participant have a preference agent which works to understand their preferences associated with the service (types of food, mealtime, potential locations, spending, etc.). The Negotiator Agent gathers these preferences from each Participant and Initiator Preference Agent to inform the negotiation. The Negotiator provides options to the Initiator and eventually to each Participant based on the available options that it arrives at through negotiation with multiple Supplier Agents. In the end, this system is a group decision support system which co-optimizes the value functions provided by the preference agents. As is common in decision support systems, the agents make options visible to the Initiator and Participants which would not have been visible otherwise.

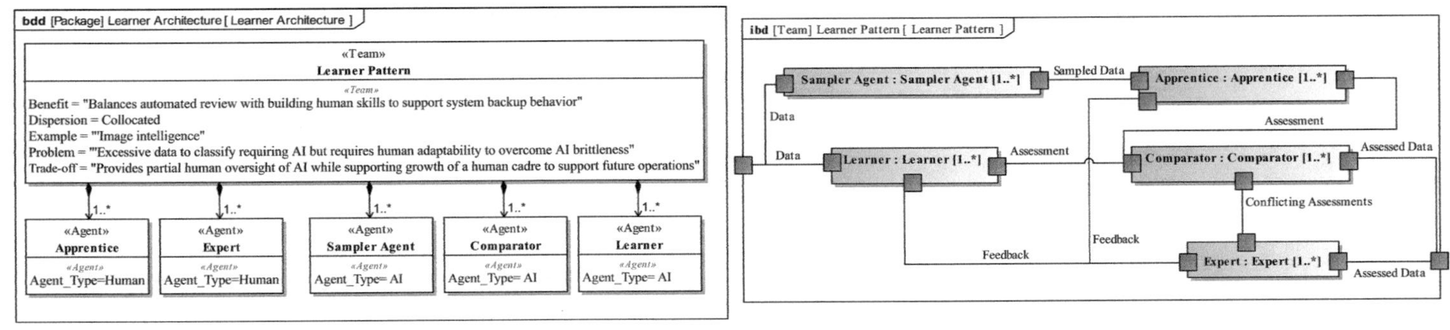

FIGURE 8.9 Block Definition Diagram indicating the components of the learner pattern in the left pane and an Internal Block Diagram illustrating the interfaces and information flows between team components in the right pane.

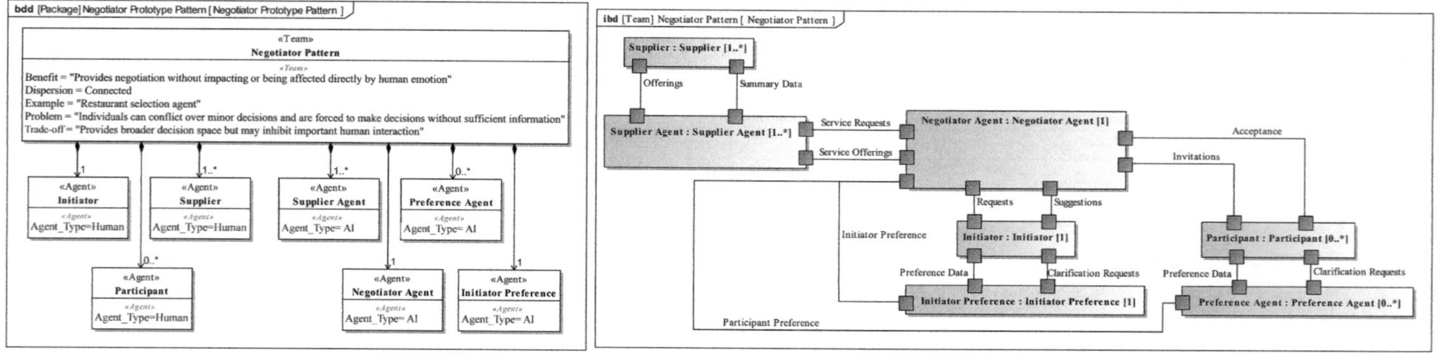

FIGURE 8.10 Block Definition Diagram indicating the components of the service negotiator pattern in the left pane and an Internal Block Diagram illustrating the interfaces and information flows between team components in the right pane.

8.2 DEVELOPING YOUR ARCHITECTURE

As discussed earlier, the patterns provide examples that a designer can adopt in whole or in part within innovative designs. Having analyzed our "As-Is" system, developed an understanding at least some of the uses for humans and AI agents in systems, and the desired improvements to your "As-Is" system, it is time to consider the design of the "To-Be" system. Before illustrating this process and AOSM diagrams to support this process, it is important to realize that this creation step is a creative process that we guide using available knowledge. However, like all creative processes, this creation step can benefit from including multiple individuals with diverse backgrounds. Therefore, this next step should include individuals who participated in understanding the "As-Is" system, individuals with knowledge of system design processes, software engineers, and perhaps some lead users. Because this process is a creative process, it is useful to begin with exercises that generate ideas with little-to-no critique of these ideas, followed by a consideration of advantages and disadvantages of ideas, and finally refinement of initial ideas.

As you work through this process, it is useful to ground the work in the capabilities and with knowledge of the responsibilities and activities that humans currently perform in the environment. Applying the process described in the following section has illustrated a few important concepts.

First, even when told not to, designers, engineers, and software development professionals often begin this process by assuming that a designer can insert AI into the HAT to replace certain human actors or complete activities. This is natural, because it is how educators teach us to design and build throughout our education and careers. Additionally, design goals often include the desire to replace the human to save on labor costs. However, as the design team begins to consider other processes the humans are performing that are difficult, tedious, or nonintuitive, it sometimes becomes obvious that the insertion of technology or simple AI agents to aid these processes can significantly improve human productivity and success. Therefore, even if you think you know the answer, follow the process. You may learn something surprising.

Second, experience has also shown that some AIs can provide significant support to less-skilled humans in the environment. It is useful to consider that knowledge, skills, and abilities of the humans will vary, and these human traits will vary as the scenario changes within the target environment. Adding AI agents to aid the less-skilled operators is a valid consideration as providing support to these individuals enhances overall system performance. The Learner prototype pattern applies design thinking to support the less-experienced operators. You will likely find it valuable to structure part of the brainstorming session to explicitly consider how AI agents can be included to support humans as they build expertise.

Third, it is important not to overly constrain this process by imposing strict limits based on the system boundary we discussed earlier. We can illustrate the potential need for providing a redefinition of the system boundary through a discussion of the service negotiator prototype pattern. The project that produced the original instantiation of this prototype pattern placed the Supplier outside the system boundary. However, during the design process, the designer discovered that considering the supplier inside the system boundary appeared necessary to achieve the project goals. Of course, expanding the project boundary in this way increases expense and risk associated with any real-world project, and a designer should not expand the boundary without deliberation. However, similar expansions of the system boundary often have the potential to provide significant advantages, and therefore designers should not discard concepts that include expansion of the system boundary without considering the trade-space.

Finally, while the examples provided below only include one conceptual redesign of the HAT for each workflow, in practice it may be important to consider many possible redesigns and to compose multiple possible "To-Be" HATs. We only illustrate one of each of these in the example for expediency. The following process should be creative, so assemble your team, relax, and have fun creating concepts for your HAT redesign.

8.3 MODELING: AGENT DIAGRAM

Based on the AI agent types and architectures described in Chapters 5, 6, and 7 and the Architectural Patterns described earlier, you are starting to envision many potential AI agent implementations for your "To-Be" system. Using the Agent Diagram, you will explore potential agent implementations and the corresponding teaming architectures. Based on this exploration, you will then down select the AI agents that you will wish to explore further for the implementation into your "to be" system design.

8.3.1 Modeling Objective

The purpose of the AOSM Agent Diagram is to display human agents, AI agents, and other actors, as needed, along the information flows that exist between them as they conduct their work processes. The Agent Diagram allows us to visualize the teaming structure created by our redesign, along with the handoffs and interactions that will occur between agents. These diagrams are implemented using the internal Block Definition Diagrams that we used to discuss patterns earlier in the chapter.

8.3.2 Diagram Components and Composition

The AOSM Agent Diagram is a SysML Internal Block Diagram, which displays agents and other actors as blocks, with information flows displayed as SysML associations. Figure 8.11 provides a Generic Agent Diagram developed from brainstorming. First, we add each Capability in the Responsibility-Capability Diagram (or Agent Capability Assessment Table) to the diagram, then we arrange the Capabilities in logical groupings that constitute workflows, activities, or processes. As we form these collections, it is useful to reflect on each grouping and whether the capabilities support the entire action process from Chapter 3. If not, then perhaps capabilities are missing, and you need to add other capabilities so that each grouping of capabilities, which comprise a role, can support the entire action process. Alternatively, the groupings may not represent the entire role, and you will need to regroup the responsibilities to include elements from each stage in the action process. Next, you should draft at least two potential architectures for each workflow, considering the responsibilities that the HAT will need to fulfill. The first architecture should be the "As-Is" workflow. If there is no "As-Is" workflow, then we recommend using the

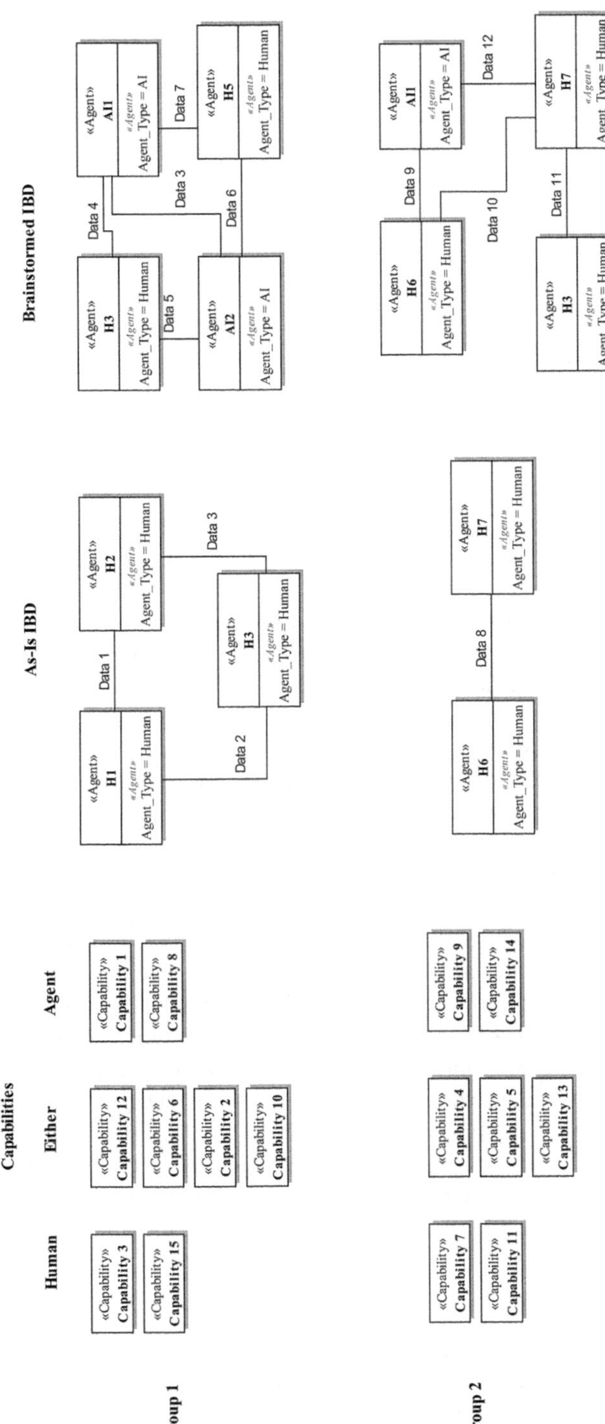

FIGURE 8.11 Generic Agent Diagram brainstorm containing generic elements.

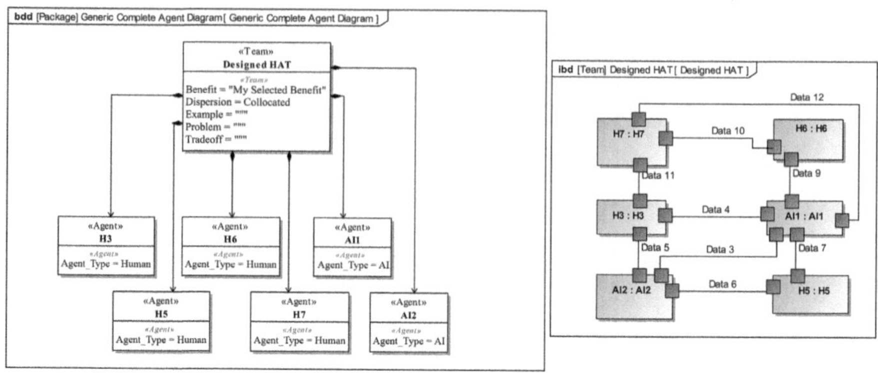

FIGURE 8.12 Generic complete Agent Diagram containing generic elements. The diagram shows a BDD in the left panel and an IBD in the right panel.

"human-only" architecture. This allows us to establish a baseline with the teaming information flows that exist in a human-teaming scenario. The other architecture(s) would be teaming arrangements for the same workflow to capture the incorporation of AI agent(s). We are not committing to an allocation at this time, so feel free to explore numerous architectural patterns and human–AI agent team configurations, especially for the most critical system workflows. In addition to the agents of the team, the designers may need to include external agents or system actors to fully capture the information flows associated with the workflow.

To create the first draft of our complete Agent Diagram, you will need to select an architecture for each of the workflows in the brainstorm. The designer, when selecting architectures, should consider the feasibility of technical implementation, redesign budget, complexity of the teaming architecture, and relevance to the redesign purpose. Upon selection, the designer should remove the referenced Capabilities and rejected architectures. The designer combines the remaining architectures into a single diagram, with each agent represented as a single block, as presented in Figure 8.12. As shown for the patterns, it may be helpful to first form a BDD indicating elements of the HAT and then develop workflows in the IBD. This will require identifying the common agents in your selected brainstormed diagrams, if their interactions with other agents in the system remain consistent. For example, in the generic agent diagram in Figure 8.11, we might assume that we will adopt the *Brainstormed IBDs* for each group of capabilities. In this example, we see that agent H3 is common between these diagrams, and since these diagrams represent nonoverlapping capabilities, we can combine these two diagrams to maintain the flows, where the flows share agent H3.

8.3.3 Considerations

As you identify potential AI agent implementations, consider AI agents across the full Action Process discussed in Chapter 3: Perception, Interpretation, Decision, and Action. Table 5.1 may be useful in imagining distinct types of AI agent implementations, and the Architectures discussed herein may be useful in identifying teaming arrangements. However, you should not consider either the "As-Is" and your initial

"To-Be" architectures to be an exhaustive set. Through brainstorming, you will likely identify novel implementation and teaming architectures that suit your context. Also, beware of the substitution myth; for each workflow option, the version(s) that add AI agent(s) should not be a one-to-one replacement of a human with an AI agent. In the unlikely event that this occurs, at a minimum, the information flows should be different. Additionally, it is quite likely that the designer could better represent the replacement AI agent as multiple AI agents performing distinct aspects of the Action Process.

8.3.4 Example Application

An example brainstorm Agent Diagram for our Diagnostic Center HAT is divided into Figures 8.13 and 8.14. The portion of the diagram shown in Figure 8.13 groups

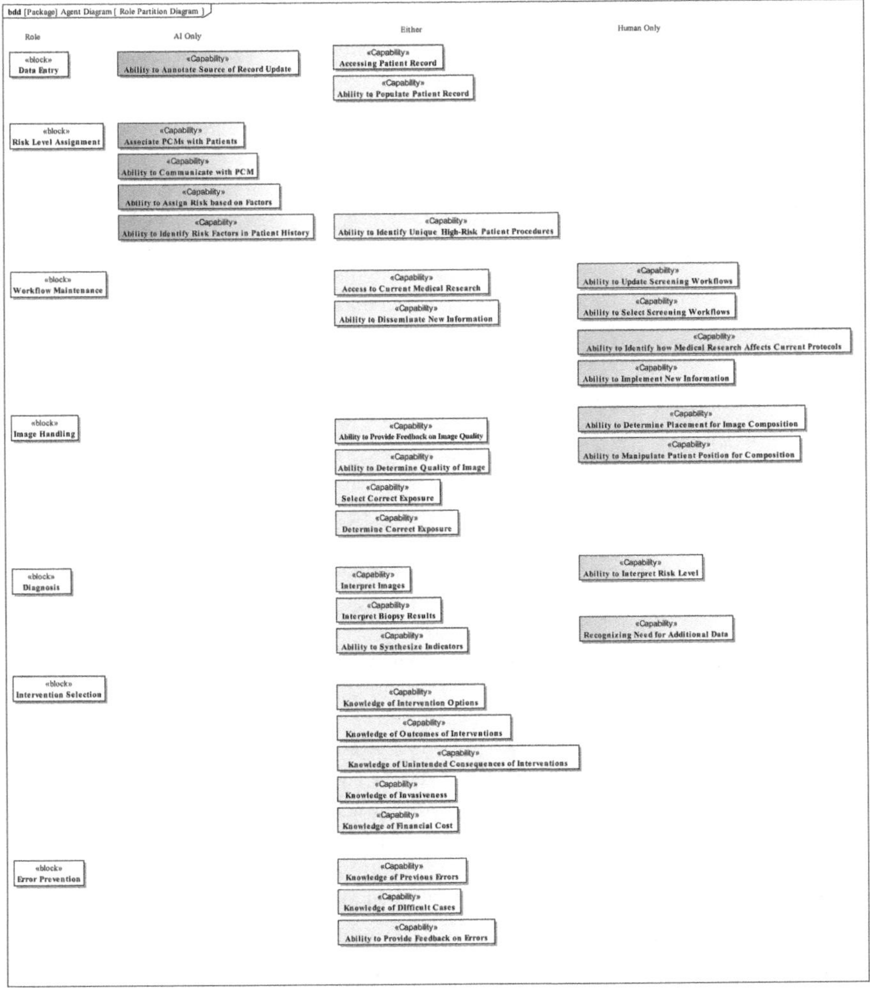

FIGURE 8.13 Role definitions and associated capabilities for the Diagnostic Center HAT.

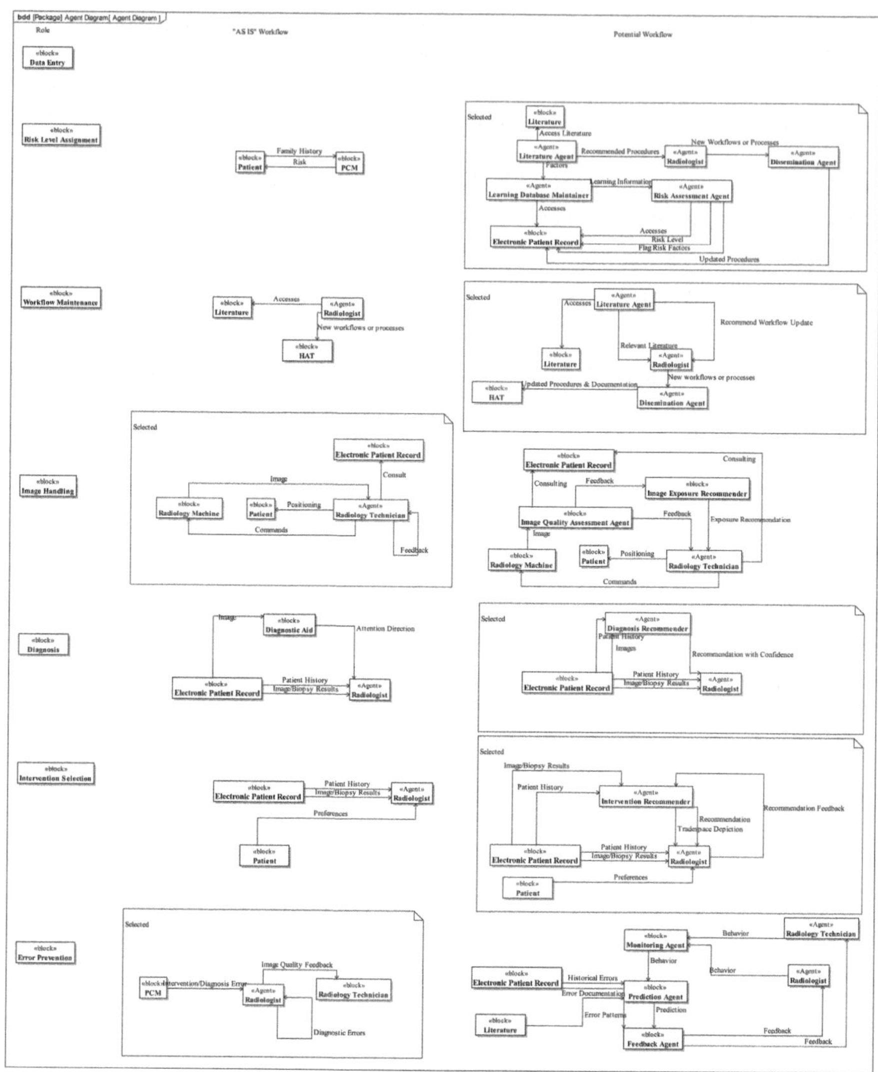

FIGURE 8.14 Example brainstormed Agent Diagrams for each role within the Diagnostic Center HAT.

the Capabilities from the Responsibility-Capability Diagram into Roles: Data Entry, Risk Level Assignment, Workflow Maintenance, Image Handling, Diagnosis, Intervention Selection, and Error Prevention. These roles represent groups of capabilities that require common knowledge and skills. As shown in this figure, the capabilities for each role are shown and divided on the basis of agent type. This grouping of capabilities by both role and agent type provides insight into the types of agents that can be used to perform the work associated with each role.

Once the roles are defined on the basis of capabilities and agent types are identified, this information can be used to develop a human–AI agent team to fulfill each

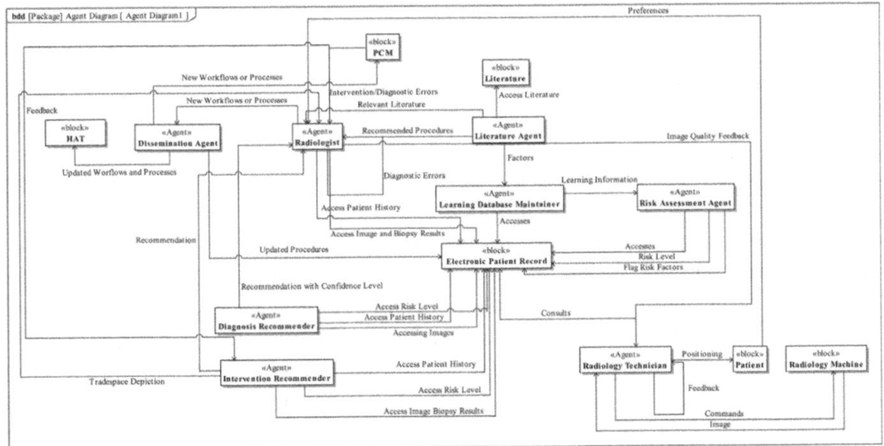

FIGURE 8.15 Complete Agent Diagram for Diagnostic Center HAT.

role. Figure 8.14 repeats the roles but then shows various implementations by diagraming workflows involving different agents and agent types to complete each role. Figure 8.14 presents two workflows for each role with the exception of the data entry role. The data entry role is not a standalone role but is a role that each of the agents must be capable of performing. Therefore, the capabilities associated with this role were assumed to be necessary for each of the other agents. Figure 8.14 shows both the "As-Is" or "human-only" workflow, as well as a potential AI agent workflow. Ideally, one might create many potential AI agent workflows to evaluate, but only one is shown in this diagram due to space limitations.

We then evaluated the workflows for implementation. Based on the difficulty to implement and relevance to the Prioritize High-Risk Patients purpose of redesign, we selected the "As-Is" workflows for the Image Handling and Error Prevention to be preserved in the "To-Be" system. As an aside, we suspect that the effective implementation of the remaining AI agents will have a positive impact on reducing errors, even without the direct implementation of the Error Prevention AI agents depicted in that workflow. We selected the remaining AI agent workflows to go forward with the redesign, and we captured these workflows in Figure 8.15, which shows the complete Agent Diagram. As we created the complete diagram, we detected the need for and added additional information flows, such as *Updates* to *Electronic Patient Record* from multiple agents, and the need for the *Dissemination Agent* to communicate *Updated Workflows and Processes* to the *PCM*.

8.4 MODELING: ALLOCATED AGENT CAPABILITY TABLE

8.4.1 Modeling Objective

At this stage, the complete Agent Diagram is a first draft of the teaming architecture for our "To-Be" system. Over the course of the next several chapters, we will continue to iterate on this design, evaluating and adjusting for effective teaming.

However, it is important to clearly document the capabilities that we are assuming each agent will implement based upon the allocation in our Agent Diagram. To achieve this step, we return to the Agent Capability Assessment table, which documented the potential allocation of each capability to each type of agent. Based on the teaming architecture in our Agent diagram, we can now update the Agent Capability Assessment from Chapter 6 from the Allocated Agent Capability Table to reflect the specific agents within our "To-Be" system. As we perform these steps, we can compare the allocation in our design to our original assessment. If there are areas where the two tables differ, then we may wish to revisit our Agent Diagram to address any inconsistencies. As an example of inconsistency, suppose the original Agent Capability Assessment Table assumed that a human–AI agent was incapable of fulfilling a particular capability and our Agent Diagram assumes that the human must fulfill a responsibility that requires this capability without an AI agent assistant, then this would indicate an inconsistency in our tables. We would then consider methods to remove or justify this inconsistency. Therefore, our objective is not only to create the Allocated Agent Capability Table but also to compare it to the Agent Capability Assessment Table from Chapter 7 to identify inconsistencies.

8.4.2 Diagram Components and Composition

This diagram is similar to the Agent Capability Table described in Chapter 7. As in this table, the diagram is formatted as a table where the rows of the table include each of the capabilities that are assigned to each agent. You can include either the agent types of human, AI, or both in the columns. You can refer to Figure 8.4 to visualize this diagram, replacing the column "either" with "both", indicating that both classes of agents must have this same capability. If you include human, AI, or both, then consider which class of agents your agent diagram assumes will be required to have each capability.

It is desirable to include each agent from your Agent diagram in the columns. If you include each agent, then you will indicate the agents that will require each capability within the cell at the intersection of each row and column. A generic form of this diagram is shown in Figure 8.16. As shown, some capabilities are allocated to only an AI or human agent while other capabilities are allocated to both AI and human agents. The designer can allocate any capability to an indeterminate number of AI and human agents; however, these agents will need to be imbued with this capability as they, their training, and interfaces are developed.

8.4.2.1 Considerations

You should be able to trace the allocation of capabilities between the Human, AI agent, and "Either" columns of the role-based Agent Capability Assessment Table from Chapter 7 and the Allocated Agent Capability Table you are creating. However, a detailed review based on the planned teaming architecture may identify any capabilities that we originally attributed to one agent class, that is, human or AI, which our design assumes should be allocated to a different agent class. Additionally, to support teaming, your design should allocate multiple agents to many of the capabilities

Legend
↗ Allocate

Capabilities	Agents	AI1	AI2	H1	H2	H3	H4	H5	H6	H7
Capabilities		5	5	2	2	2	1	3	2	2
Capability 1	1	↗								
Capability 2	2		↗	↗						
Capability 3	1				↗					
Capability 4	1						↗			
Capability 5	1					↗				
Capability 6	1							↗		
Capability 7	1								↗	
Capability 8	2		↗		↗					
Capability 9	2		↗							↗
Capability 10	2	↗							↗	
Capability 11	2	↗				↗				
Capability 12	2	↗						↗		
Capability 13	2		↗							↗
Capability 14	2		↗					↗		
Capability 15	2	↗		↗						

FIGURE 8.16 Generic Allocated Agent Capability Table.

to effectively execute the workflows with support and interdependence with other agents. In fact, often, you should allocate many of these capabilities to separate agents with at least one human and one AI agent. You will also understand that some of these capabilities may be executed at multiple places in the assumed workflows and that different agents might be assumed to require a capability to execute these different steps in the workflow.

8.4.3 Example Application

Figure 8.17 provides an example of the Allocated Agent Capability Table corresponding to the Agent Diagram in Figure 8.15. As you can see, we require several capabilities for many agents within our system, such as the *Ability to Annotate Source Record Update, Accessing Patient Record,* and *Knowledge of Previous Errors.* The Radiologist has the vast preponderance of capabilities, which is not surprising, since even if the AI agents are performing significant portions of the workload associated with the Diagnostic Center workflows, the Radiologist maintains ultimate responsibility and accountability for the outputs of the team. The *Diagnosis Recommender* and the *Intervention Recommender* also have numerous required capabilities. The shared capabilities indicate potential handoffs, interactions, or at least a need for shared mental models.

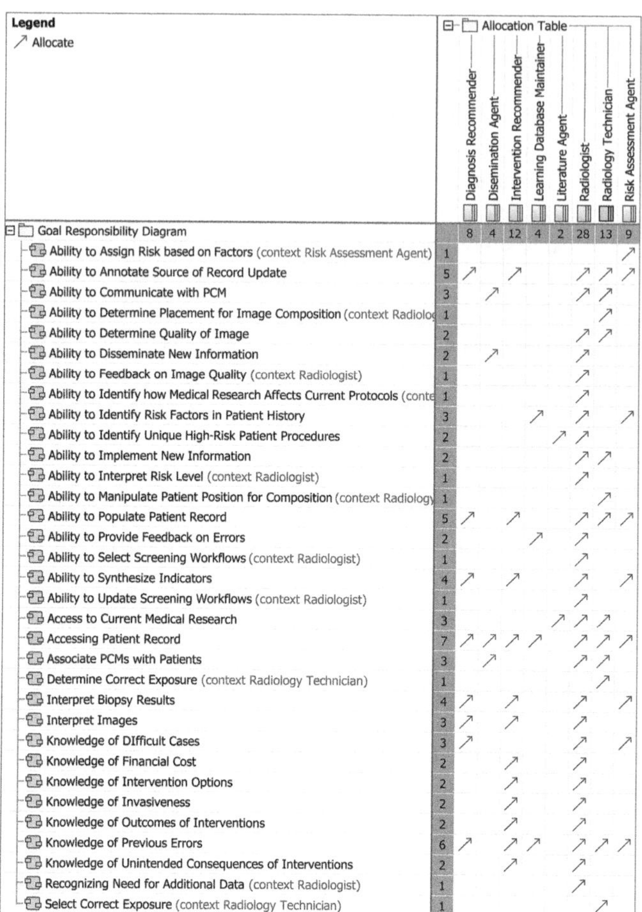

Legend: ↗ Allocate

Goal Responsibility Diagram		Diagnosis Recommender	Disemination Agent	Intervention Recommender	Learning Database Maintainer	Literature Agent	Radiologist	Radiology Technician	Risk Assessment Agent
		8	4	12	4	2	28	13	9
Ability to Assign Risk based on Factors (context Risk Assessment Agent)	1								↗
Ability to Annotate Source of Record Update	5	↗		↗			↗	↗	↗
Ability to Communicate with PCM	3		↗				↗	↗	
Ability to Determine Placement for Image Composition (context Radiolo	1							↗	
Ability to Determine Quality of Image	2						↗	↗	
Ability to Disseminate New Information	2		↗				↗		
Ability to Feedback on Image Quality (context Radiologist)	1						↗		
Ability to Identify how Medical Research Affects Current Protocols (conte	1						↗		
Ability to Identify Risk Factors in Patient History	3				↗		↗		↗
Ability to Identify Unique High-Risk Patient Procedures	2					↗	↗		
Ability to Implement New Information	2						↗	↗	
Ability to Interpret Risk Level (context Radiologist)	1						↗		
Ability to Manipulate Patient Position for Composition (context Radiology	1							↗	
Ability to Populate Patient Record	5	↗		↗			↗	↗	↗
Ability to Provide Feedback on Errors	2				↗		↗		
Ability to Select Screening Workflows (context Radiologist)	1						↗		
Ability to Synthesize Indicators	4	↗		↗			↗		↗
Ability to Update Screening Workflows (context Radiologist)	1						↗		
Access to Current Medical Research	3					↗	↗	↗	
Accessing Patient Record	7	↗	↗	↗	↗		↗	↗	↗
Associate PCMs with Patients	3		↗				↗	↗	
Determine Correct Exposure (context Radiology Technician)	1							↗	
Interpret Biopsy Results	4	↗		↗			↗		↗
Interpret Images	3	↗		↗			↗		
Knowledge of DIfficult Cases	3	↗					↗		↗
Knowledge of Financial Cost	2			↗			↗		
Knowledge of Intervention Options	2			↗			↗		
Knowledge of Invasiveness	2			↗			↗		
Knowledge of Outcomes of Interventions	2			↗			↗		
Knowledge of Previous Errors	6	↗		↗	↗		↗	↗	↗
Knowledge of Unintended Consequences of Interventions	2			↗			↗		
Recognizing Need for Additional Data (context Radiologist)	1						↗		
Select Correct Exposure (context Radiology Technician)	1							↗	

FIGURE 8.17 Revised Agent Capability Assessment Table.

8.5 MODELING: RASCI ALLOCATION TABLE

8.5.1 Modeling Objective

The AOSM RASCI (responsible, accountable, supporting, consulted, and informed) Allocation Table is a variation of the RACI (responsible, accountable, consulted, and informed) responsibility assignment matrix frequently used in project management. The RACI chart is used to assign individual resources to activities according to the corresponding level of roles and expectations for that individual (Project Management Institute, 2013). We pronounce both RACI and RASCI as "ray-see". The purpose of the RASCI Allocation Table is to clearly capture the level of involvement implied by the agent allocations depicted in the selected teaming architectures of the Agent Diagram. This enables us to see which Responsibilities team members share, and which ones are solely reliant upon a single agent. It also reveals which agent(s) will be the "heavy lifters" in our current system design, carrying a significant responsibility for the execution of the team's workflows.

8.5.2 Diagram Components and Composition

Figure 8.18 shows a generic RASCI Allocation Table. The rows of this table present the Responsibilities identified in the Responsibility-Capability Diagram, and the table lists each human and AI agent separately for the column labels. For each responsibility, we place a relationship at the intersection of the row and column to indicate the level of involvement that each agent(s) is expected to contribute to that responsibility. We list the AOSM RASCI definitions below in decreasing order of involvement/responsibility. Variations may be necessary to capture the specific relationships within your system of interest; thus, we recommend that you document the definitions you choose to use.

- ***Accountable*:** The agent that must bear the risk and consequences of the outcome. We expect this agent to be capable of justifying decisions made and is often the approval authority.
- ***Responsible*:** The agent that executes the activities required. They may seek assistance or delegate portions of the work, but they maintain the obligation to ensure task completion.
- ***Supporting*:** The agent that aids or can aid in accomplishing task activities.
- ***Consulted*:** The agent that provides opinions, guidance, feedback, or information as assistance in accomplishing task activities. This might include AI agents that recommend specific actions.
- ***Informed*:** The agent that is apprised of task progress, decisions, and outcomes.

When constructing this table, one determines the highest level of involvement or responsibility each agent is likely to have at any moment during system operation. We only entered this highest level into the table.

	H3	H5	H6	H7	AI1	AI2
Responsibility 1	A		C	I		
Responsibility 2	A		C			R
Responsibility 3	A	I				R
Responsibility 4	A		I			R
Responsibility 5	A				R	I
Responsibility 6	A	S			S	
Responsibility 7	S		A			
Responsibility 8	S		A			
Responsibility 9	A	I	I	I		S
Responsibility 10			A		R	I

FIGURE 8.18 Generic Allocation Table containing generic elements.

8.5.3 Considerations

When constructing the RASCI Allocation Table, we recommend that only one agent is Accountable for the responsibility. Scenarios in which the designer divides accountability prevent clear lines of authority, conflicting authorities, and lack of accountability. In most systems, each responsibility will have one agent that is accountable, and all the other levels of involvement (responsibility, supporting, consulting, and informed) are optional. An assignment of Accountable, when no corresponding Responsible assignment is present, implies that the Accountable entity is also Responsible for task accomplishment.

Three tips to aid with differentiation are as follows:

- Responsible and Supporting both do task activities. Responsible has the obligation, Supporting is just assisting or, using earlier terminology, is providing soft interdependence.
- Supporting and Consulted both aid in task accomplishment. Supporting aids in task execution, Consulted aids by providing information, options, or analysis to support decisions.
- Consulted and Informed both deal with flows of information. Consulted is a two-way dialogue that provides input or recommendations to the accountable or responsible agent. Informed depicts a one-way receipt of a status update, usually from an agent that performs one task to another agent that will need information from the first agent to synchronize or perform another task.

As you populate the RASCI Allocation Table, keep in mind that while the involvement of the Responsible, Supporting, and Consulted agents occur during the task accomplishment, informed often occurs after and Accountable may occur before, during, or after task accomplishment. In the case of Informed, be on the lookout for agents that are not involved but use the outputs of the task execution as inputs to their own task execution. Also remember that AI agents will be narrow in function. As such, AI agents will only take part in some of the responsibilities within your allocation table. If you find that AI agents are taking part in a very large number or all of your responsibilities, then you should consider further decomposing your AI agents into narrower functions as the AI agent you are indicating in this table will likely be implemented as an ensemble of AI agents that your design process is not explicitly capturing. Thus, you are not applying the AOSM process appropriately.

8.5.4 Example Application

Figure 8.19 provides an example RASCI Allocation Table for our Diagnostic Center HAT. One aspect that might draw your attention is the considerable number of Responsibilities that the Radiologist is Accountable for. This is partly due to the Radiologist being the leader of the team, but this aspect is also due to the assumption that while an AI agent within our system can have the responsibility for performing a task, ultimately a human holds the accountability for ensuring the task is done correctly and will need to answer for the outcomes of our system. Also of note is that

	Diagnosis Recommender	Dissemination Agent	Intervention Recommender	Learning Database Maintainer	Literature Agent	Radio-logist	Radiology Technician	Risk Assessment Agent
Alert to Anomaly	S			I		A	S	S
Calibrate Sensitivity of Detection Criteria	R		I			A	S	
Coordinate with PCM to Implement High-Risk Patient Procedures		R				A		S
Develop Shared Understanding of Mammogram Quality						A	R	
Differentiate Normal from Abnormal	R					A		S
Establish High-Risk Patient Procedures		S			S	A		
Evaluate for Likely Errors	S					A	R	
Identify Abnormal Patterns	S					A	R	
Identify High-Risk Patients	I	I	I	I		A		R
Identify Useful Interventions			R		S	A		
Implement Identification Criteria into Screening Systems						A	S	
Interpret Health History for Risk Level				S	S	A		R
Interpret Physical Indicators	R					A	R	
Maintain State of Art Knowledge	I		I	I	R	A	S	I
Obtain Complete Mammograph History	R		R	S		R	R	R
Obtain Patient Health History	R		R	S		R	I	R
Obtain Quality Biopsy	I		I	I		A		
Obtain Quality Images	I		I	I		C	A	
Perceive Anomaly Alert	R		R	I		A		
Perform Intervention						A	R	
Perform Risk Assessment	I		I	S	S	A		R
Provide Diagnosis	R		I	I		A		
Recommend Further Diagnostics	R		R			A		
Train for Likely Errors						A	R	
Update Anomaly Detection Criteria	I			R	S	A		I
Update Electronic Patient Record	R		R			R	R	R
Update Intervention Evaluation			R		S	A		
Update Intervention Implementation						A	S	
Update Risk Assessment	I		I	S	S	A		R

FIGURE 8.19 Example RASCI Allocation Table for Diagnostic Center HAT.

there are three responsibilities, Obtain Complete Mammography History, Obtain Patient Health History, and Update Electronic Patient Record, in which there is no agent listed as Accountable. This is because the Accountability for these items rests with the PCM, who we did not show in Figure 8.19, since the PCM is outside of our system. This is acceptable as the team can identify the accountable agent, but this assumption should be documented and communicated to stakeholders.

8.6 CHAPTER SUMMARY

This chapter has reviewed HAT patterns to provide an overview of designs that have been instantiated in previous systems and to provide further discussion of the interaction among agents in these systems as well as the advantages and disadvantages of these interactions. As shown, the well-documented patterns consist of single humans interacting with single AI agents. However, we also showed patterns where arrangements of one or more humans interact with one or more specialized agents.

Further, the chapter discussed arranging similar capabilities to form Roles. We then brainstormed individual, elemental HATs to fulfill each role. We selected individual embodiments of these HATs to meet the system design goals and aggregated these individual HATs to comprise a larger system level HAT. We mapped the capabilities and responsibilities to agents in this HAT. Finally, we evaluated each responsibility relationship using the Responsible, Accountable, Supporting, Consulted, Informed (RASCI) framework to gain more insight into the contribution of each agent to the HAT responsibilities.

8.7 QUESTIONS AND DISCUSSION TOPICS

1. Describe a dyad pattern from the text that you have witnessed in the real world, including personal observations on the utility or drawbacks of the pattern.
2. Examine one of the multi-human or AI agent patterns that might be useful in your system and describe how and why you believe it is applicable.
3. Describe and differentiate Responsible and Accountable in the RASCI framework.
4. Describe and differentiate Supporting and Consulting in the RASCI framework.

8.8 PROJECT QUESTIONS AND ASSIGNMENTS

1. For your project, arrange your capabilities into roles.
2. For each role, document the “As-Is” HAT arrangement using an Agent diagram.
3. Provide your rationale, select brainstormed agent diagrams for each role, and use these agent diagrams to compose an agent diagram for your system.

4. Develop an Agent Capability Assessment Table, associating your capabilities with your agents in your final agent diagram.
5. Develop a RACSI Allocation Table, associating your responsibilities with the Agents in your system.

REFERENCES

Bar-Yam, Y. (2017). *Artificial and human intelligence*. New England Complex Systems Institute.

Bice, G. W., Skoog, M. A., & Howard, J. D. (1990). *Aircraft ground collision avoidance and autorecovery systems device* (Patent No. 4,924,401). United States Patent 4,924,401.

Brooks, R. A. (1986). A robust layered control system for a mobile robot. *IEEE Journal on Robotics and Automation*, *2*(1), 14–23. https://doi.org/10.1109/JRA.1986.1087032

Calhoun, G. L., Ruff, H. A., Behymer, K. J., & Frost, E. M. (2018). Human-autonomy teaming interface design considerations for multi-unmanned vehicle control. *Theoretical Issues in Ergonomics Science*, *19*(3), 321–352. https://doi.org/10.1080/1463922X.2017.1315751

Johnson, C., Miller, M. E., Rusnock, C. F., & Jacques, D. R. (2020). Applying control abstraction to the design of human—agent teams. *Systems*, *8*(2), 10. https://www.mdpi.com/2079-8954/8/2/10

LaMonica, D. A., Drnec, K., & Miller, M. E. (2022). Employing MBSE to assess and evaluate human teaming in military aviation command and control. *2022 IEEE 3rd International Conference on Human-Machine Systems (ICHMS)*, 1–7. https://doi.org/10.1109/ICHMS56717.2022.9980801

Larman, C. (2004). *Applying UML and patterns: An introduction to object-oriented analysis and design and iterative development* (3rd ed.). Prentice Hall.

Layton, C., Smith, P. J., & McCoy, E. (1994). Design of a cooperative problem-solving system for en-route flight planning: An empirical evaluation. *Human Factors*, *36*(1), 94–119.

Neerincx, M. A., Diggelen, J. Van, & Stolk, M. L. (2019). *Design patterns and ontologies for situated human-agent collaboration at organization, team and interaction level* (pp. 1–10). North Atlantic Treaty Organization STO-MP-HFM-300.

Project Management Institute. (2013). *A guide to the project management body of knowledge (PMBOK® Guide)* (5th ed.). Project Management Institute, Inc. https://www.pmi.org/pmbok-guide-standards/foundational/pmbok

Schulte, A., Donath, D., & Lange, D. S. (2016). Design patterns for human-cognitive agent teaming. *Lecture Notes in Computer Science (Including Subseries Lecture Notes in Artificial Intelligence and Lecture Notes in Bioinformatics)*, *9736*, 231–243. https://doi.org/10.1007/978-3-319-40030-3_24

Sheridan, T. B., & Verplank, W. L. (1978). Human and computer control of undersea teleoperators. *DTIC*. http://oai.dtic.mil/oai/oai?verb=getRecord&metadataPrefix=html&identifier=ADA057655

Strenzke, R., Uhrmann, J., Benzler, A., Maiwald, F., Rauschert, A., & Schulte, A. (2011). Managing cockpit crew excess task load in military manned-unmanned teaming missions by dual-mode cognitive automation approaches. *AIAA Guidance, Navigation, and Control Conference 2011*, 1–24. https://doi.org/10.2514/6.2011-6237

Thamarai, P., & Karthik, B. (2014). Automatic braking and evasive steering for active pedestrian safety. *Middle—East Journal of Scientific Research*, *20*(10), 1271–1276. https://doi.org/10.5829/idosi.mejsr.2014.20.10.114135

Van Diggelen, J., & Johnson, M. (2019). Team design patterns. *HAI 2019—Proceedings of the 7th International Conference on Human-Agent Interaction*, 118–126. https://doi.org/10.1145/3349537.3351892

Zacharias, G. (2019). *Autonomous horizons: The way forward*. Air University Press.

9 Decision-Making and Decision Support Systems

9.1 RETURNING TO HUMAN–AI AGENT TEAM PATTERNS

The previous chapter discussed two patterns that included a discussion of *D*ecision *S*upport *S*ystems (DSS). The first pattern included an example of an aircraft routing or navigation system. In this application, good decision-making processes include gathering data pertaining to many different dimensions, synthesizing this data into information that supports trades among important attributes pertaining to each dimension, and then applying judgment to arrive at a final decision. For example, during navigation for a passenger aircraft, two attributes that might be important to the decision which results in the selection of the route are fuel consumption for the flight and passenger comfort. In the case where a thunderstorm passes through the aircraft's route, these two elements often conflict with one another. The pilot might choose routes that go around the weather, increasing the number of miles flown, which consumes additional fuel, or flying through or close to the storm, which will result in turbulence and passenger discomfort. While the pilot trades between these two attributes, data does not exist for either of these attributes. Flight distance, altitude, and aircraft speed are all data elements that affect fuel consumption. By applying first principles physics equations, one can estimate fuel consumption if the flight path, altitude, and aircraft speed are known. However, this is not a calculation that the humans can perform exactly and rapidly in their head, although they might be able to estimate this value based on experience. There is no metric for passenger comfort, but given proximity to the storm or an estimate of turbulence, an experienced pilot can estimate this quantity. Weather models can estimate storm speed of movement, location, size, and intensity, but these are only estimates that the pilot might use to estimate turbulence at various locations at specific times. This example illustrates a set of conditions under which a DSS can be useful. Some attributes, for example, fuel consumption, can be determined much more reliably by AI, but other variables, for example, passenger comfort, are difficult for existing AI to assess, and the desired trade between these two attributes might require expert judgment. For instance, an experienced pilot might prefer to fly around the backside of the storm rather than the frontside to avoid flying through the storm if the pilot anticipates that the storm might move faster than expected, which the designer of an AI algorithm might not consider.

A second pattern involving a DSS in the previous chapter was the Dual-Mode Cognitive HAT. In this case, an operator is controlling multiple machines, and the

DOI: 10.1201/9781003428183-9

DSS is present to aid the operator's awareness of the need to decide as the pilot is spreading their attention across controlling multiple vehicles, as well as to increase the speed of decision-making. In this application, if the human was controlling a single vehicle, they might have the capacity to monitor and control this vehicle, but when they are responsible for monitoring and, at times, controlling each vehicle, the human simply does not have the capacity to monitor, understand when decisions are required, and make all necessary decisions in a timely manner. Here, the DSS additionally plays the role of reducing the cognitive load on the operator, permitting them to make the necessary decisions, given their limited capacity. Thus, we can see that the presence of time pressure and system complexity can increase the need for use of a DSS.

Overall, DSSs are most useful in complex, complicated, and uncertain environments or when the operator must make decisions under time pressure. *Complicated environments* include a large number of options with a large number of potential outcomes. *Complex environments* include numerous known inputs which interact, often in a nonlinear fashion, to affect the outcome. *Uncertain* environments, which are common in the real world, include factors and future events that are unknown or unknowable.

9.2 DECISION-MAKING AND THE IMPORTANCE OF JUDGMENT

Decision-making is an essential element of the action process. Once an agent obtains cues or information about the environment, the agent can retrieve or develop alternative actions that could permit them to achieve their goal. Therefore, a simplistic description of a generic decision-making process includes interpreting the information from the environment, generating alternative possible actions or sequences of actions, evaluating each alternative's likelihood of achieving the desired goal, and selecting an alternative to be acted upon (Clemen, 1996).

An agent can make decisions under three different conditions: certainty, risk, and uncertainty. When making decisions under certainty, information and models are available that allow an agent to precisely predict the future, all alternatives are known exactly, and an agent can evaluate the result of selecting each alternative, predicting the future states that will occur as the agent selects each alternative. For example, suppose that all bank deposits are safe, interest rates are stable, and an interest-bearing bank account, for example a certificate of deposit, is desired. If the variables are all known, an agent could simply identify the interest rates offered for each different certificate of deposit for each different bank and select the bank account that would provide the greatest return on the deposit over the life of the certificate. This decision is easy as it involves direct comparison of known numbers along a single dimension—interest rate.

The second condition of decision-making is decisions made under risk. In this condition, an agent can determine all possible alternatives, that is, the agent knows all the banks, the certificates the bank offers, and their interest rates. The agent also knows all future outcomes, for example, each bank might either stay in business or fail, because federal insurance only preserves the investment but does not guarantee the return on its investment in a failed bank. Finally, the agent knows the likelihood of all potential future outcomes. That is, the agent knows the likelihood that each bank will fail. Under these conditions, there is not necessarily a correct decision. However, the agent can establish objective criteria to support the decision, such as the highest expected value, knowing that there is a probability that the agent will receive either its full-interest payment or no-interest payment, depending upon a random process that is based on the likelihood values. Under this condition, the agent can maximize the expected value for opening this account in each bank. This would be a rational decision. However, because the agent only knows the likelihood of failure, it is possible the bank the agent chooses will fail and the agent will lose its return while another bank might not fail and would have been a better alternative. When making decisions under risk, an agent cannot determine the future, making this category of decision-making more difficult than decisions under certainty. This discussion also permits us to understand an important attribute of real-world decisions. That is, we cannot judge decision quality by the outcome. If we used a process that was more likely to provide a return on our investment than others, then following this process is justifiable, regardless of the unpredictable, real-world result.

Most real-world decisions fall into the third category. Therefore, an agent acting in the real world must make decisions under the condition of uncertainty. In this condition, the agent can identify some, but not all, alternatives. The agent might be able to project some future outcomes, but not all future outcomes, and the agent is unable to precisely determine the likelihood of the future outcomes. To make matters worse, there are often many different dimensions that the agent must consider when making real-world decisions, as the alternatives affect many different results that are important to selecting the "best" decision, for example, risk, convenience, and financial return. In this simple banking example, the interest rate is not constant but fluctuates, such that waiting a day might result in a higher or lower interest rate. Thus, the agent must decide not only on which bank to open an account with but also when the agent should open the account. A real-world agent, particularly a human, cannot practically assess the interest rates or likelihood of failure of all banks. Further, the agent does not know when it might need access to the money, so the agent must further select the appropriate terms for the CD. As a result, humans often use heuristics to arrive at a satisfactory solution, rather than an optimal solution, which cannot be determined, given the level of uncertainty. As a result, we, as humans, look at the interest rates for the banks we have done business with, in the past, potentially expanding this list with offers at other banks based upon recommendations from friends or media and gather some other information to understand trends in rates and trustworthiness of the considered banks. We then weigh this information and decide upon a certificate based on the reduced set of alternatives, often using our preferred and trusted bank, rather than risking a new relationship unless the potential reward is large.

Timing of a decision is often critical. Usually, waiting permits one to gather more information and reduces uncertainty through more thorough analysis. However, waiting often reduces the number of options available. Further, if we think of decisions embedded in action processes that iterate rapidly, for example, selecting the steering direction and speed when driving a car, decisions are sometimes made so rapidly that they must occur nearly continuously, limiting our ability to perceive and analyze information thoroughly as might be desired in a robust decision-making process.

9.3 HUMAN DECISION-MAKING

As we conceive of decision-making, we often reflect on the conscious thoughts that we engage in to make decisions. This is typical of decisions like the banking example in which we spend at least some time consciously thinking about the possible return on our investment. However, human decision-making must also support decisions in time-critical environments, like driving. If we think back to the time when we began driving or perhaps teaching someone else to drive, we might realize that practically every decision we made around driving at that time required conscious thought, including: "How much should we turn the steering wheel to stay in our lane" and "How much do we press the gas pedal to move at the right speed". We might also remember constantly looking at the edge lines and the speedometer to understand how our decisions on steering wheel and gas pedal position are impacting placement and speed. Over time, we learned to drive in a way that we were no longer consciously aware of some of these activities. This learning involved the formation of specialized pathways in the brain so that we no longer needed to use the limited amount of short-term memory and other areas of the brain associated with conscious thought. Instead, humans learn many of these activities to the extent that they are performed unconsciously. This statement assumes that we initially engage in conscious thought when performing novel activities and learn, and then this learning enables automaticity or intuitive decision-making. However, there is considerable evidence that humans also engage in implicit learning where they are not consciously engaged in learning or even aware that learning has occurred (Patterson & Eggleston, 2017).

This division of conscious and unconscious processing to support human decision-making is supported throughout the human decision-making literature (Patterson & Eggleston, 2017; Patterson et al., 2010). A common model representing this dichotomy is the intuitive decision-making process provided by Gary Klein in his Recognition-Primed Decision-Making model (Klein, 1993, 2008). Figure 9.1 represents this model. Klein discusses three paths through this model, depending upon the familiarity of the situation. As depicted in the white elements of this model, the human is situated in an environment and experiences or perceives the situation within the environment. If the situation is familiar, the human can recognize aspects of the situation by recognizing relevant cues and their relationships. This model links recognized cues to plausible goals, expectations of other cues, and potential actions, assuming these links exist in human memory. Based upon experience where patterns among these cues are like expectations and these cues supported previous actions, the human can implement associated actions without further analysis. This series of activities relies heavily upon individuals perceiving cues in the environment, which are like a set of cues they experienced previously and executing the associated successful actions.

A second pathway through the model further includes the light gray boxes. Under this version of the model, the individual again experiences the situation and perceives

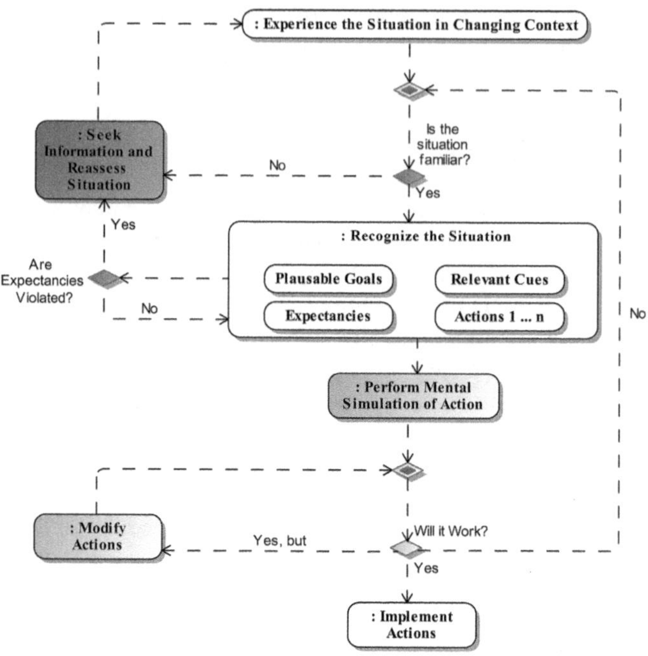

FIGURE 9.1 Representation of Klein's Recognition-Primed Decision-Making model.

familiar cues. The model assumes that the human brain links these cues to goals, expectations, and actions in human memory. However, the pattern of cues might differ slightly from those experienced previously. The individual then performs mental simulation of the actions to determine if the associated actions are likely to be successful. If the individual believes that the associated actions will be successful, the individual will implement these actions. If the individual does not believe the associated actions will be successful, the individual will modify these actions to arrive at a set of actions that mental simulation indicates will achieve the desired goals. The individual will then implement the modified actions. Depending upon the circumstances, the individual may perform these mental simulations consciously. However, the psychology literature suggests that mental simulation can occur unconsciously. Therefore, the individual might complete this process without a conscious thought. If it is determined that the individual cannot modify base actions to achieve the associated goal, then the process reverts to the recognition stage.

The final pathway through this model occurs when the situation is not familiar. Under these circumstances, humans are likely to seek additional information and to reassess. The human brain can trigger these processes using previous experience to guide actions, such as eye movements or focusing of auditory attention, to find expected cues based upon similar experienced situations. In this situation, humans are actively gathering current information to assess the situation. The human is likely taking intermediate actions, perhaps not only to gather additional information but also perhaps to test the environment in ways that might modify it, for example, a fire fighter might test the rigidity of a portion of a floor, knowing that a portion of the floor may give way, resulting in a hole that they could potentially fall into.

There are a few important points to discuss regarding this model. First, Klein refers to the Recognition-Primed Decision-Making model as a form of naturalistic decision-making. This model attempts to explain the way that individuals perform in a natural environment. Klein derived this model from examining the behavior of fire fighters. During his research, Klein noted that these individuals appeared to rely upon an array of cues, or information elements, which existed in the environment. These cues might include items such as the perception of heat, sounds, visibility or color of smoke, and proprioception of the give-in floors or movement of the building. Based upon perception of these cues, experienced firefighters would rapidly arrive at solutions that novice firefighters would not. Further, when asked why they made what turned out to be a lifesaving, instantaneous decision, the experienced firefighters were unable to recall why they made the decision or any conscious deliberation which led to the decision. In hindsight, they could develop the rationale to explain why they should have made the decision, but these explanations were not consciously available without deliberation. That is, they are performing intuitive pattern-based rather than analytic, rule-based decision-making.

A second observation regarding this model is that this mode of operation requires that the brain perform pattern matching to an array of perceptual cues to find solutions that were employed "successfully" in the past. We placed the word "successfully" in quotes as the decision was successful by the fact that the human survived to have this same experience again. This does not mean that the action is an optimal or even an appropriate action, but it was satisfactory enough that the human did not suffer catastrophic circumstances in the previous implementation.

A third observation is that humans typically evaluate very few actions according to this model. This increases the speed of decision-making, which is essential when rapid decisions are necessary for survival. Finally, this model explains the behavior of experienced individuals but does not explain the decision-making process or behavior for individuals who have not experienced the specific scenario in which they are deciding. These decisions are subject to conscious, analytic deliberation.

As stated earlier, human decision-making literature supports the concept that the human employs two separate methods for performing decision-making, including intuitive and analytic processing or decision-making. Psychologists typically refer to the mental processes captured by the Recognition-Primed Decision-Making model as ***intuitive*** processing. This processing is believed to be fast, has a large capacity, and does not require working memory or conscious executive control (Evans & Stanovich, 2013; Patterson & Eggleston, 2017). Further, this processing is believed to be capable of integrating a large number of perceptual cues to arrive at robust decisions (Betsch et al., 2001; Glöckner & Betsch, 2008; Maier, 1931). By contrast, analytic processing refers to conscious deliberation and rule-based reasoning which we usually associate with symbols and algorithms (Helie & Sun, 2010). Therefore, where perceptual patterns are likely to trigger intuitive processing, when the DSS displays information as symbols including text and numbers, the information is likely to trigger analytic processing. Analytic processing is slower than intuitive processing. Further, the human's limited working memory limits their ability to perform analytic processing, such that the human can simultaneously process only a limited number of factors. Further, analytic processing often involves cognitive decoupling. During cognitive decoupling, humans dedicate conscious thought to running thought experiments. Therefore, humans are unable to dedicate cognitive resources to process real-world stimuli. When trained to perform analytic decision-making, we will follow a decision-making process that includes assessing the situation; generating possible, alternative actions; evaluating each action; and selecting a desired action. As a human has limited working memory and ability to perform mathematical operations, the human will have limited ability to assess multiple alternatives and to consider multiple factors during the assessment. However, DSSs can evaluate multiple factors and render graphics to permit the human to compare the information elements necessary to make the decision. With this support, humans can rapidly judge the desired action that has the best combination of the outcomes.

The interaction between the intuitive and analytic systems is unclear from the literature. It is clear though that if we consider every muscle movement we make during the day as requiring control, and therefore, a decision at some level, the number of decisions we make each day is extraordinarily large, and a minor number of these decisions involve conscious thought. Patterson and Eggelston have cited evidence that humans rely on the intuitive decision-making system for more than 80% of decisions (Patterson & Eggleston, 2017). The intuitive system can influence the analytic decision-making process as it drives perception and the specific alternatives that the human will consider during analytic decision-making.

The intuitive decision-making literature makes a few additional salient points. First, as mentioned earlier, implicit learning occurs as we perform activities, and this implicit learning underpins intuitive decision-making system. When discussing teaming, simulations of future events that permit humans to interact with agents in realistic environments are likely to support implicit learning while simple discussion or textbook discussions of agent behavior are less likely to support intuitive decision-making. Additionally, the literature suggests that insight, that is, the sudden realization of a problem solution following an impasse, is largely driven by the intuitive system. This literature further suggests that the intuitive system performs this activity by restructuring the mental models of an environment, providing novel implications and solutions. Further, this process can remain engaged for extended periods of time (Patterson & Eggleston, 2017). Thus, mental models and team mental models support intuitive decision-making processes and may not be available to our conscious thought.

One last point about human decision-making is that the literature discusses the fact that different individuals exhibit different decision-making styles. Psychological literature characterizes an individual's decision-making style along five scales. These include (1) a person's desire to rely on information and methodical evaluation when making consequential decisions, perhaps indicating a desire to rely on analytic processing; (2) the person's tendency to rely on premonition, and hunches, which might indicate a desire to rely on intuitive processing; (3) the person's tendency to avoid or delay decision-making; (4) the person's tendency to make spontaneous or impulsive decisions; and, finally, (5) their desire to seek guidance or input from others during decision-making. Recent research has shown that these characteristics also impact people's interactions with team members during group decision-making (Strang et al., 2023) and, thus, might influence their interaction with DSSs. For example, a DSS might influence individuals who tend to avoid decision-making or who rely on others when making decisions more than other individuals. Further, interfaces that make the steps of this process clear might influence individuals who rely on analytic processing. Alternatively, individuals who rely more on intuition may desire interfaces that provide perceptual cues they can rapidly integrate to understand the situation.

9.4 DECISION SUPPORT SYSTEM DESIGN CONSIDERATIONS

Given the previous description of human decision-making, as system designers, we might anticipate design guidelines for DSSs. For example, to trigger rapid decision-making, the DSS should display perceptual cues that aid the user in recognizing the differences between different situations. Furthermore, the DSS should display these cues simultaneously, rather than spreading the cues across multiple interface screens. If the DSS displays all the information necessary to support a decision on a single screen appropriately, this display will permit the user to rapidly integrate and interpret the information. The design of DSS interfaces may rely on design methodologies such as ecological interface design (Bennett & Flach, 2011, 2019), incorporating design guidelines to support cognitive integration and decision-making (Patterson, 2012). Further, the DSS system should enable users to gain expertise through implicit learning to recognize situations. Therefore, training systems should simulate various situations that operators are expected to encounter and display this information as cues that are the same as the human operators will experience when using operational DSSs to permit the operators to learn the cue combinations they should associate with desired responses (Klein & Borders, 2016).

Assuming time exists for use of the analytic system, the DSS should aid operators as they generate and evaluate alternatives. Without this support from the DSS, the operators will often adopt known options with little thought or evaluation of alternatives. To support cognitive engagement, the DSS should not suggest a decision early in the process as individuals who avoid decisions, among others, will likely adopt these recommendations with little thought or consideration. Instead, to create collaboration between the AI and the humans, the DSS should permit and encourage the operators to rapidly suggest alternatives to the DSS, permit the DSS to analytically analyze the alternatives, and display these results to the operator to enable the operator to judge the utility of the various alternatives. By providing analysis of these alternatives, the DSS enhances the operator's ability to perform analysis during analytic decision-making. To support analytic decision-making, the DSS should present humans with a manageable number of decision criteria rather than individual data elements that they must integrate in their head to synthesize decision criteria. That is, it is important for the DSS to provide information relevant to the decision rather than providing data that the operator must synthesize to form the information they need to support a decision (e.g., provide the projected fuel and arrival times rather than distance, air speed, altitude, or other data which influence parameters the operator will use to make the decision). This design objective reduces the cognitive effort necessary to remember and integrate these individual data elements into criteria or information elements that are important for decision-making.

This discussion of representing information elements rather than individual data elements in this section might appear to be reminiscent of the feature space representation for AI agents in Chapter 6. Obviously, determining the information elements necessary for decision-making and providing this representation to an AI agent or the human simplifies decision-making, regardless of the agent type.

Action or production teams often operate in environments where the environment or other agents trigger the need for decisions. Thus, humans must decide and take necessary actions whenever the environment requires these actions. Therefore, humans in these types of teams often make decisions under time pressure. Additionally, regardless of the type of team, humans rarely make a single natural decision without the opportunity for subsequent decisions. Thus, there is often an opportunity for the human or other agent to decide, take an action, observe the effect of the action, and adjust. As a result, humans are often more concerned about avoiding decisions that will have negative outcomes rather than making the best or "optimal" decision. Interestingly, software development teams tend to design decision support systems to provide the "optimal" decision based upon the information available and the decision space considered by the DSS. Perhaps this is a complimentary activity in some instances, but displaying only this recommendation removes human judgment from the decision-making process.

When designing support systems for analytic decision-making, the human interface literature discusses three types of displays. These include status displays, option displays, and command displays. *Status displays* provide the user with knowledge to determine when a decision is necessary by providing knowledge of the state of the world. *Option displays* provide the user with options and often provide the trades associated with each option. *Command displays* instruct the human on the decision they should make. A DSS might provide each of these sets of recommendations to assist the human. Providing command displays increases the human's response rate as their decision-making process is primed by the input of the command display, and they often will perform the action commanded by the DSS, permitting timely responses under time pressure. Unfortunately, when the command display provides an incorrect action, humans require more time to recover from and determine a different action as they may not have actively considered the available information to arrive at a decision and will often also need to determine why the DSS failed (Sarter & Schroeder, 2001). Therefore, as indicated in Table 9.1, the designer should only use command displays if the human is operating under considerable time pressure and the AI is highly reliable both in determining status of the system or world and in recommending the proper decision. Otherwise, the DSS should generate options and provide selection support or only status, depending upon automation reliability when

TABLE 9.1
Display Recommendation Based on Time Pressure and Automation Reliability.

Automation Reliability	Time Pressure		
	Little	Moderate	High
High status and high decision	Option evaluation support with post suggestion		Command display
High status and low decision	Option evaluation support with post assessment option suggestion		Option evaluation
Low status			Status display

the time pressure is high. Designers should augment these displays with displays that provide perceptual cues, typically graphical representations, to help the humans recognize the situation they are in, permitting these cues to trigger intuitive processes.

As shown in Table 9.1, when time pressure is not high, it is often productive for the DSS to conduct a dialog with the human to jointly arrive at a decision. Layton and colleagues illustrated that the team can often arrive at a better decision when the human is able to generate options and have the DSS help to assess parameters that might affect the decision (Layton et al., 1994). In their research, both the humans and the AI were able to generate potential aircraft routes. In response, the automation generated fuel remaining, arrival time, distance, and expected maximum turbulence for each route, supporting the human's ability to assess the impact of the individual routes. Humans improved decision performance when they could suggest routes to augment routes developed by the DSS, observe the impact of these routes on performance attributes associated with each alternative, and then decide. Research has also shown that DSSs that critique human decisions after the human has selected an action can improve decision performance and do not bias the human response toward the AI recommendation in the same way that having the AI provide a recommendation prior to having the human indicate a decision (Endsley, 2015). Therefore, if the automation is highly reliable in determining status and the proper decision, but time pressure is moderate or low, we recommend that DSSs collaborate with the humans to generate options and calculate parameters associated with each option, permitting the human to make the decision. If the DSS would have selected an action different than the human, it can then share its recommendation with the human with a rationale. If the AI is less reliable, we recommend a similar display, but any post decision feedback must include the rationale regarding why the automation might have selected a different action, permitting the human to evaluate these criteria against their own criteria to consider the alternate point of view.

The following text provides one method for modeling the data and information, which is necessary to support a decision. Decisions support goals as indicated in the action loop. As noted, human and AI decisions are associated with each goal. Thus, the user interface between the DSS and the human should present the information elements that are necessary to support the human as they make required decisions. However, systems typically gather data, which must be interpreted to provide the information elements that are necessary to support decision-making (Endsley & Jones, 2012; Humphrey & Adams, 2011). In the AI literature, this transformation from data to information is also captured when describing representations (Bindewald et al., 2015).

Information elements are higher-level concepts used by agents to make decisions. *Data elements* are raw data from the environment that the agent must integrate to derive information useful for decision-making. For example, if we are controlling the flow of water into a glass to fill it, there are multiple data elements, including the amount of water in the glass, the flow rate into the glass, and the level of water we wish to achieve. We transform this data into time-to-turn-off the water. The human determines when to turn off the flow by estimating the time remaining to fill the glass. Therefore, this quite simple decision requires at least three data elements (i.e., water in glass, flow rate, and desired water level) to form one information element (e.g., time until we turn the control). In complex systems, many data elements may be required to synthesize the information we need to support a decision. Synthesis of these data elements requires cognitive effort. Therefore, we can simplify the user interface by automatically transforming the data elements into an information representation that supports decision-making, reducing the cognitive effort required by the human. The same happens in AI systems. Once again, the system must collect and transform multiple pieces of raw data into higher-level representations to support decision-making.

9.5 MODELING: DECISION-INFORMATION-DATA DIAGRAMS

9.5.1 Modeling Objective

The objective of the Decision-Information-Data Diagram is to define the information that the users of your system rely upon to make decisions. The designer applies the diagram to create the information elements by capturing the raw data available in the environment or from systems in the environment. The designer defines the items the sensors should capture and store in databases to inform decisions using the data elements captured in this diagram. The designer applies the information elements to gain insight into how to manipulate the data elements to create information useful for the decision-maker. The designer uses these information elements to determine information that the AI agent should apply or that the designer should represent in the human's user interface to support human decision-making.

To model the relationship between decisions, information elements, and data elements, AOSM uses a hierarchical block arrangement. The meta-model represents decisions as a special class of activity. Further, the meta-model represents Information and Data elements as special classes of blocks as they are items, not processes. The meta-model links these elements through the aggregation relationship, as the agent must aggregate data elements to form information elements. Further, the agent must aggregate information elements to support decision-making. The meta-model further links decisions to individual goals.

9.5.2 Diagram Components and Composition

Figure 9.2 shows a generic Decision-Information-Data Diagram. When developing this diagram, you will create a Block Definition Diagram. You will then place a hierarchical arrangement of decisions, information elements, and data elements in this diagram with the decision at the top of the hierarchy. You should position information elements in the second layer of the hierarchy. Then attach these information elements to the decision using the aggregation relationship. Although not shown in Figure 9.2, you can form additional layers of information elements between the second layer and the data elements, as you can depict the fact that lower-level information elements are aggregated into higher-level information elements. Finally, you should place data elements at the bottom of the hierarchy. Each information element must have at least one data element but will typically have multiple data elements. Information elements, such as *Information Element 1* and *Information Element 2*, can each require the same data element as input to aggregate the information element. Shown in this figure, these information elements both rely upon input from *Data Element 3*.

9.5.3 Considerations

The decision will typically be associated with controlling the system to achieve the goal. Therefore, this decision may vary depending upon the circumstances. More importantly, the precise information elements that the human relies upon to make this decision can vary depending upon the circumstances. The goal is to capture all information elements that are necessary to support the decision and the data elements that provide the base elements that form the information elements.

9.5.4 Example Application

In our mammography example, we could choose each goal and examine the decision, information requirements, and associated data elements. In this example, we examine the goal "Provide Correct Diagnosis". This goal is associated with reviewing mammograms to determine if there are anomalous regions within the mammogram. Once any anomalous regions are located, the radiologist will examine the mammogram to determine whether to dismiss the area, at least for the present time, or to recommend further, often more invasive, procedures.

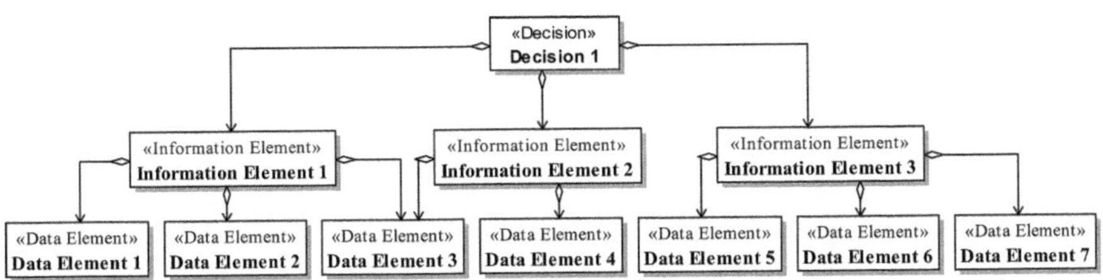

FIGURE 9.2 Generic Decision-Information-Data Diagram.

The radiologist understands that there are risks associated with misses. Misses occur when the radiologist dismisses a cancerous lesion and does not investigate it further, delaying treatment, and reducing the likelihood of patient survival. The radiologist also understands that there are risks associated with false alarms. False alarms occur when the radiologist recommends further evaluation even though the patient is healthy. These risks include mental anguish the patient undergoes as they will experience weeks of stress as they are concerned that they have cancer until the radiologist informs them that do not have cancer. Further, false alarms place strains on limited medical resources and imposes financial penalties on the patients or their insurers.

As shown in Figure 9.3, the radiologist needs to know the important areas of the mammogram to focus examination, areas with anomalous appearance, whether the characteristics of these areas are typical or atypical for this appearance, whether the areas with anomalous appearance are consistent with lesions or tumors, and whether the characteristics of the anomalous area have changed over time. These are all essential information elements that inform their decisions. Unfortunately, when picking up a patient record or mammogram, this information cannot be directly determined from the mammogram. Instead, the mammogram only contains an image of the tissue. Therefore, the radiologist must combine this image with other data elements to assess these information elements.

As shown in this figure, the radiologist derives each of the information elements from data elements. For example, the information element might be formed on the basis of data in the patient's record on the location of lumps that manual examinations have identified, areas of the mammary gland where cancerous legions are first visible, which come from the radiologist's training and experience, and the location of any previously observed anomalies that are described in the patient's record. Once again, this diagram is an example as the authors have not rigorously developed this diagram through discussions and observations with radiologists to understand if other data are also relevant. For many of the information elements, the radiologist will similarly rely on information available in the mammograms, the patient's file, and the knowledge they obtained through training and on-the-job experience.

The information element "characteristics change with time" contains some novel aspects. Specifically, this information element contains the previous mammograms in the patient's file as before. However, there is a complicating factor identified for this information element, specifically differences in imaging conditions, such as the exposure of the current and previous mammograms. If regions are inappropriately exposed, then less structure will be visible within these regions of the mammogram, complicating the comparison of the mammograms. It is also entirely possible that the radiologists obtained the two mammograms using different technologies (e.g., film vs. digital sensors) or technologies with different capabilities (e.g., digital radiographs with different resolutions or different enhancement technologies which can alter the tone scale of the image differently). Without knowing these differences, the radiologist will find it difficult to compare current and historical radiographs. Knowing these differences, the radiologist might be able to provide a comparison, using mental projections observed from other patient records, to help interpret the differences in the mammograms. Therefore, the knowledge of these technical aspects

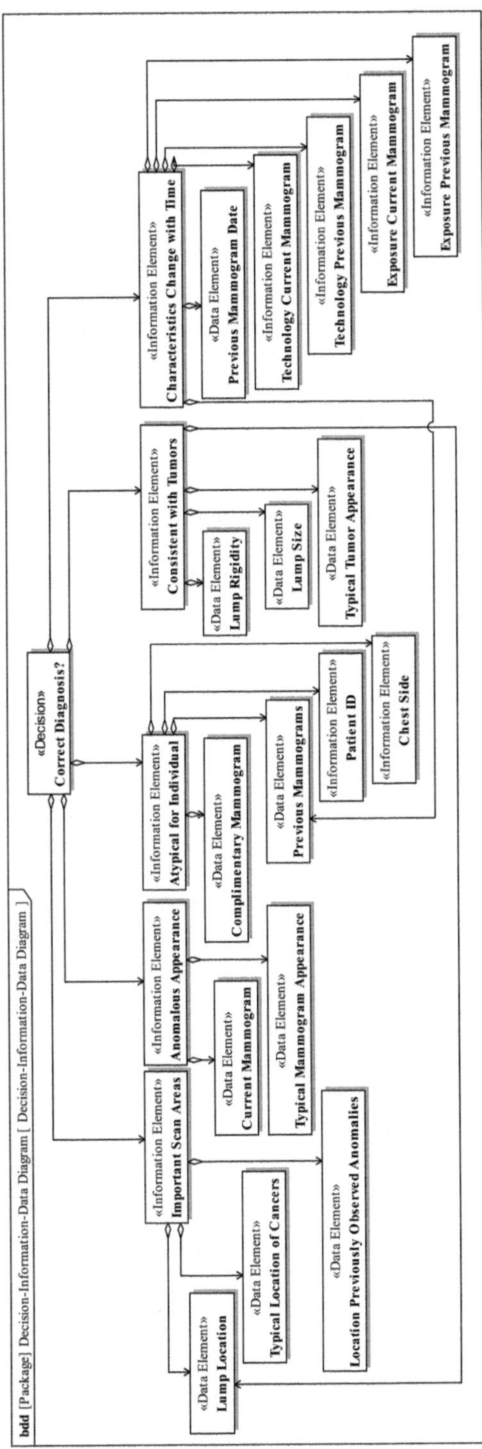

FIGURE 9.3 Example Decision-Information-Data diagram for the goal Provide Correct Diagnosis.

of mammography equipment and image processing might provide critical data that the radiologist can rely upon when attempting to assess the information they require. The design team might also consider automation to account for some visual differences between these images due to the technology differences. Such an automation tool could aid the radiologist's assessment of this information element. Similarly, the design team could develop automation aids to provide or aid the integration of the data elements into information elements. In interface design, it is useful to display either all information or data elements. However, if the system displays data elements, the radiologist will need to integrate this information to determine the relevant information. Accurate conversion of data to information can aid the radiologist's work, or the agent can use the information to provide feedback to the radiologist if an automation detects an error in the radiologist's diagnosis.

During conceptual design, it is useful to develop these diagrams for each decision that the operator must make to support relevant goals. These diagrams provide insight into data structures necessary to support the HAT. Further, they provide insight into areas where the design team can create AI to support human decision-making in later stages of development. We recommend that designers pay particular attention to those decisions that occur during atypical scenarios, as well as "pain points" and key decisions identified during the Cognitive Task Analysis, discussed in Chapter 2.

9.6 CHAPTER SUMMARY

This chapter has discussed analytic decision-making processes, human decision-making, including intuitive and analytic processes, and finally some design guidelines for Decision Support Systems (DSSs). Under time pressure, designers should design DSSs to provide perceptual cues that permit pattern matching to support the human intuitive decision-making system. However, the designer might augment these displays by command, option, or status displays, depending upon the reliability of the DSS. When making decisions over longer time horizons, the DSS can support robust decision-making by designing the system to help the human generate and evaluate options. When the AI is highly reliable, the DSS might engage in a dialogue where it provides its recommendation and rationale after the human has decided. The timing of this feedback is important to avoid biasing the human's decision-making process as some individuals are excessively influenced if the DSS provides a recommendation before the human has formed their own opinion.

9.7 QUESTIONS AND DISCUSSION TOPICS

1. Describe the difference in human activity that Klein's RPD model would predict when the cues match the cues from a previous human experience versus the human activity that Klein's RPD model would predict when the cues are similar but do not provide a match to those from a previous human experience.

2. Explain the primary differences between the intuitive and analytic human decision-making processes.
3. In AI, developers often seek to provide the "optimal" decision. Based on the description of human decision-making in this chapter, do you expect this decision to be consistent with human decision-making? Describe how you believe this might affect human–AI agent teaming?

9.8 PROJECT QUESTIONS AND ASSIGNMENTS

1. For your project, select a goal and describe a decision associated with assessing the goal.
2. For your decision, describe a situation where this decision might apply and list three or more alternatives that could be available to select from when making the decision.
3. For your decision, develop a list of information elements or parameters you should assess to select among the options you listed in the previous section and show them in a Decision-Information-Data Diagram.
4. For each information element or parameter, list the data elements that you would apply to determine the information elements, associating them with each element in your Decision-Information-Data Diagram.

REFERENCES

Bennett, K. B., & Flach, J. (2019). Ecological interface design: Thirty-plus years of refinement, progress, and potential. *Human Factors*, *61*(4), 513–525. https://doi.org/10.1177/0018720819835990

Bennett, K. B., & Flach, J. M. (2011). Display and interface design. In *Display and interface design*. CRC, Taylor & Francis Group. https://doi.org/10.1201/b10774

Betsch, T., Plessner, H., Schwieren, C., & Gütig, R. (2001). I like it but I don't know why: A value-account approach to implicit attitude formation. *Personality and Social Psychology Bulletin*, *27*(2), 242–253. https://doi.org/10.1177/0146167201272009

Bindewald, J. M., Peterson, G. L., & Miller, M. E. (2015). Trajectory generation with player modeling. In D. Barbosa & E. Milios (Eds.), *Canadian AI 2015, LNAI* (Vol. 9091, pp. 42–49). Springer International Publishing. https://doi.org/10.1007/978-3-319-18356-5

Clemen, R. T. (1996). *Making hard decisions: An introduction to decision analysis* (2nd ed.). Wadsworth Publishing Company.

Endsley, M. R. (2015). *Autonomous horizons: System autonomy in the air force—A path to the future. 1: Human-A* (p. 27). Office of the Chief Scientist, United States Air Force.

Endsley, M. R., & Jones, D. G. (2012). *Designing for situation awareness: An approach to user-centered design*. CRC Press, Inc.

Evans, J. S. B. T., & Stanovich, K. E. (2013). Dual-process theories of higher cognition. *Perspectives on Psychological Science*, *8*(3), 223–241.

Glöckner, A., & Betsch, T. (2008). Multiple-reason decision making based on automatic processing. *Journal of Experimental Psychology: Learning Memory and Cognition*, *34*(5), 1055–1075. https://doi.org/10.1037/0278-7393.34.5.1055

Helie, S., & Sun, R. (2010). Incubation, insight, and creative problem solving: A unified theory and connectionist model. *Psychological Review*, *117*(3), 994–1024.

Humphrey, C. M., & Adams, J. A. (2011). Analysis of complex team-based systems: Augmentations to goal-directed task analysis and cognitive work analysis. *Theoretical Issues in Ergonomics Science*, *12*(2), 149–175. https://doi.org/10.1080/14639221003602473

Klein, G. (1993). A recognition-primed decision (RPD) model of rapid decision making. In G. Klein, J. Orasanu, R. Calderwood, & C. Zsambok (Eds.), *Decision making in action: Models and methods* (pp. 138–147). Ablex Publishing Corporation. https://doi.org/10.1002/bdm.3960080307

Klein, G. (2008). Naturalistic decision making. *Human Factors*, *50*(3), 456–460. https://doi.org/10.1518/001872008X288385

Klein, G., & Borders, J. (2016). The ShadowBox approach to cognitive skills training: An empirical evaluation. *Journal of Cognitive Engineering and Decision Making*, *10*(3), 268–280. https://doi.org/10.1177/1555343416636515

Layton, C., Smith, P. J., & McCoy, E. (1994). Design of a cooperative problem-solving system for en-route flight planning: An empirical evaluation. *Human Factors*, *36*(1), 94–119.

Maier, N. R. F. (1931). Reasoning in humans. II. The solution of a problem and its appearance in consciousness. *Journal of Comparative Psychology*, *12*(2), 181–194. https://doi.org/10.1037/h0071361

Patterson, R., & Eggleston, R. G. (2017). Intuitive cognition. *Journal of Cognitive Engineering and Decision Making*, *11*(1), 5–22. https://doi.org/10.1177/1555343416686476

Patterson, R. E. (2012). Cognitive engineering, cognitive augmentation, and information display. *Journal of the Society for Information Display*, *20*(4), 208–213. https://doi.org/10.1889/JSID20.4.208

Patterson, R. E., Pierce, B. J., Bell, H. H., & Klein, G. (2010). Implicit learning, tacit knowledge, expertise development, and naturalistic decision making. *Journal of Cognitive Engineering and Decision Making*, *4*(4), 289–303. https://doi.org/10.1177/155534341000400403

Sarter, N. B., & Schroeder, B. (2001). Supporting decision making and action selection under time pressure and uncertainty: The case of in-flight icing. *Human Factors: The Journal of the Human Factors and Ergonomics Society*, *43*(4), 573–583. https://doi.org/10.1518/001872001775870403

Strang, M., Miller, M., Hill, R., & Elshaw, J. (2023, May). Group decision performance: The predictive role of decision making styles and cognition. *Personality and Individual Differences*, *206*, 112114. https://doi.org/10.1016/j.paid.2023.112114

10 Designing for Exceptions

10.1 VALIDATING THE "TO-BE" SYSTEM

Having developed a high-level design for our "To-Be" system, it is important to evaluate this design to determine flaws in our design, as well as to further understand our design. Figure 10.1 shows the process steps associated with validating the system. The left panel of this figure shows the overall AOSM process with the fifth step highlighted. The right panel of Figure 10.1 shows the process steps discussed in this chapter, as well as those in Chapters 11 and 12. As shown, this step begins with describing both typical and atypical scenarios for our system. As discussed in this chapter, it is important to assess not only whether the system supports common, everyday activities, but that it also supports anomalies caused by events in the environment or from within the system. To perform this evaluation, the AOSM process suggests the designer develop scenarios and apply them in an analysis and allocation process referred to as Interdependence Analysis, which we discuss in Chapter 11. This analysis method is used to evaluate system performance in each of the defined scenarios. We use information from this process to update our design. Finally, to flesh out the design, this process will lead us to consider the interfaces among the agents in our systems by detailing the items and behaviors each agent must be able to observe, predict, or direct in other agents. We will capture much of this information in Interdependence Tables and will use this information to update the diagrams in our "To-Be" system design.

10.2 EVALUATING AND REFINING YOUR HAT

In Chapter 8, you defined the basic members of your human–AI agent team (HAT) and their responsibilities within the "To-Be" system. While the HAT design must deal with everyday events, the design must also be resilient to less common perturbations, some of which designers cannot foresee during design. This is particularly important in HAT design as one of the primary HAT advantages, as compared to the typical application of AI, is the improved resilience that arises due to the human's ability to adapt to unforeseen perturbations. The design must, however, provide appropriate support to the human in this endeavor.

While we cannot understand whether a HAT will be able to respond appropriately to all unforeseen circumstances with a given design, it is important that we consider designing our HAT to respond to at least the foreseeable perturbations. This requires that we design flexibility into the system that will aid the HAT to respond to and be resilient to atypical circumstances. We introduced this discussion in Chapter 2 as we discussed the fact that we are designing soft and open systems. Because the systems are soft systems, we must remember that individual humans and situated AI differ in

DOI: 10.1201/9781003428183-10

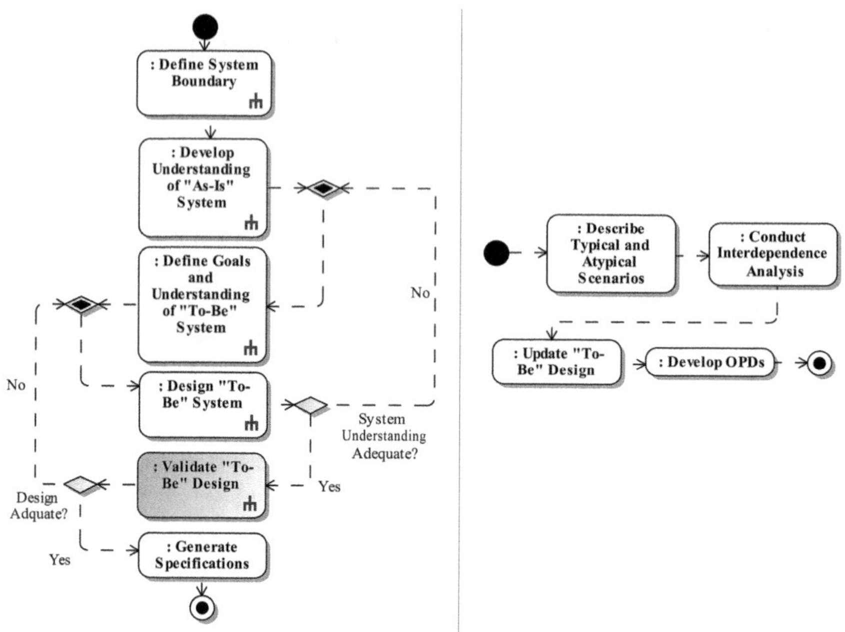

FIGURE 10.1 A depiction of the AOSM process on the left with the current sub-step highlighted. The right panel shows the sub-steps associated with Validate "To-Be" Design step.

performance and have the potential to introduce variability in system performance. The fact that these systems are open implies that variability outside the system boundary will affect the performance of the system. As discussed in Chapter 2, these facts require us to consider not only typical but also atypical scenarios during the design phase.

Anomalies, which require HAT adaptivity, originate from many sources. Anomalies can arise from components inside our system boundary due to poor or inadvertent human activities (e.g., human error), poor response by an AI agent in a particular circumstance (e.g., AI error), degradation or failure of machine components, or miscommunication among entities in our HAT. Outside our system, anomalies can also arise from many sources, including unexpected environmental variability, as well as benign or malevolent activity by external actors. Malevolent activities might include cyberattacks as well as physical effects. For example, in a military context of designing a transport vehicle, the potential to encounter anomalies in the form of traps, such as roadside bombs used by insurgents in Afghanistan to disrupt the United States' activities, might be considered anomalies that need to be detected and dealt with (Booth, 2009).

10.3 THE THREE STAGES IN EXCEPTION HANDLING

We can decompose the process of handling exceptions or anomalies into three separate stages. Each stage presents specific challenges to a HAT. Generally, these stages include detection, during which the HAT recognizes that an anomaly has or is occurring; diagnosis, during which the HAT determines error–solution pairs; and response. While detection is separate from the later stages, the later stages of anomaly response can be intertwined as actions used to diagnose the anomaly can and often will affect the time course of the anomaly and the HAT's response to the anomaly.

10.3.1 Anomaly Detection

According to Webster's dictionary, *anomaly* refers to "a deviation from the common rule or irregularity". This dictionary also states that an anomaly is "something different, abnormal, peculiar, or not easily classified" (Merriam-Webster, n.d.). Thus, anomalies are not well defined, rather they are simply the absence of the expected.

The detection of anomalies can come from multiple sources in a HAT. We can understand these sources through experiential means. Examining an instance, a particular anomaly that was difficult to detect, was described by Marstall et al. (2016). In this instance, a two-seat fighter aircraft was returning from a nighttime mission over Vietnam. The pilot in the front seat was in control of the lead aircraft in a two-aircraft formation when an encrypted message came to the crew. The copilot in the backseat stated "I've got it" but accidentally bumped the control stick with his knee. The pilot interpreted the changing pressure on the stick to indicate that the copilot had control of the aircraft, surrendered control of the aircraft, and started decoding the message. However, the copilot had intended to convey that he was going to decode the message. Thus, the anomaly occurred as neither the pilot nor the copilot was controlling the aircraft. Slowly, the aircraft banked left, which pushed the nose of the aircraft down, and the aircraft slowly began to lose altitude. However, because of the gradual nature of this change and the fact that they were both engaged with decoding the message, neither pilot noticed the change in the aircraft heading or their loss of altitude. Fortunately for the crew, the pilot in a second aircraft, who was following the lead aircraft, realized that this was not the planned route, that is, detected the anomaly and radioed to the lead aircraft asking where they were going. Both crew members of the lead aircraft quickly realized their error and began the recovery process, luckily successfully completing this process in a mountain valley, narrowly avoiding catastrophe.

This real-world incident illustrates a few important attributes of anomaly detection. First, although anomalies can and often do occur due to discrete events, for example, a faulty handover of control, the system behavior that indicates the presence of the anomaly often does not change instantaneously. Instead, the indicators of the anomaly, that is, the fact that the aircraft begins to roll and then to dive, occur gradually. Since anomaly processing requires computations that contrast current system states with common or intended behavior, they depend on relative differences in these values (Woods et al., 2002). Thus, gradual changes are often difficult to detect immediately. Instead, the anomaly is likely to affect the system over a period until the difference value becomes large enough to detect.

Second, one could employ AI agents to support detection. Had an AI on the aircraft known the flight path and constantly checked the location of the aircraft against this flight path, it could have detected the variation and alerted the pilots. However, the AI would have had to be aware of the desired flight path, the current location of the aircraft, and have the anomaly defined as a deviation between these values. Thus, the AI must be aware of the ontology of potential deviations to monitor over time (Bradshaw, 2013). Further, while these anomalies can be detected as being different from past behavior, this past behavior must be stored and understood (Ragan et al., 2015). While this alone can be difficult, without clear communication, it will be more difficult for an AI agent to understand deviations from planned behavior, which the pilot in the second aircraft used to detect the anomaly.

Third, the detection can originate from multiple locations in the team. As illustrated by this example, the pilot or copilot could have detected this anomaly if only they could see each other to know they were both decoding the message. However, the design precluded this observation as the pilots were in separate compartments within the cockpit, and it was dark. Designers anticipated errors of this type, and therefore the designers provided the pilots communication protocols where the pilot assuming control of the aircraft was instructed to clearly state they had control of the aircraft and to move the controls. The protocol also required the pilot who was relinquishing the controls to repeat that the other pilot had control of the aircraft. However, in this example, the pilots did not follow these procedures as they were experiencing fatigue as they flew the last leg of a long and stressful nighttime mission. Alternately, pilots could have detected the anomaly if either pilot viewed secondary indicators, specifically the altitude indicator, as they would have detected that they were losing altitude. Had the designers placed an AI in the aircraft to detect this anomaly, it could have provided a warning or alert to the crew, permitting detection. Finally, team members outside the controlling team can provide warnings, which in this case, was the pilot of the second aircraft. Thus, as Wiener states, detection can be performed by the primary operator, another crew member, secondary indications or displays, warning and alerting devices, or persons not in the cockpit (Wiener, 1993). Alternatively, the machine could have acted on its own, if equipment such as ground collision avoidance systems were available in the aircraft.

Overall, from the human's perspective, humans have been discussed as being relatively sensitive to anomalies (Bradshaw, 2013). However, once a human notices a deviation, it must raise suspicion enough that they transition to the next stage of diagnosis (Bobko et al., 2013). Therefore, a human operator might detect small deviations from normal system behavior, but this recognition might not raise suspicion to the point that the human recognizes the presence of an anomaly.

Further, the designer must present the human operators with information to detect an anomaly. In physical systems, we monitor many cues, vibration, sound, visual indicators, etc., which provide a background to learn and to compare during anomaly detection. Further, these signals impact multiple human senses so that an operator might detect an anomaly without requiring us to fixate our visual attention on a particular indicator. However, in remotely operated systems, many of these indicators might not be available to the human, degrading their ability to detect anomalies. Further, we might visually display much of the relevant information, requiring the operator to fixate their visual attention on the correct indicators. More concerning, the designer might design the interface such that this information displays across multiple screens, requiring the operator to page through multiple screens to gain access to the information and then mentally synthesize the information, to detect the anomaly. Lastly, rigorously monitoring visual information for rare anomalies can be classified as a vigilance task as it involves effortfully monitoring displays where targets are rare, and monitoring is mentally taxing (Warm et al., 2008). Humans are notoriously poor at this class of task with detection performance declining over a matter of minutes when performing vigilance tasks. Thus, there are many potential barriers to timely anomaly detection by humans acting alone, particularly when operating remote machines.

Detection by AI presents other challenges. First, a significant area of research is the adjustment of sensitivity to balance misses and false alarms (Artificial Intelligence Research & Development Interagency Working Group, 2019). When developers design AI specifically for anomaly detection, the priority is to ensure that the AI does not miss notable events; thus, the designer adjusts the agent's sensitivity to bias the agent to avoid misses, accepting a larger number of false alarms. In these systems, the human must discriminate between true events and false alarms. To exacerbate this problem, a single event can cascade through a system providing multiple indicators. In our example, an AI might have detected the loss of altitude, the deviation from the intended flight path, the fact that neither pilot had their hands on the controls, or other indicators of the anomaly. Different AI components could potentially examine the data from each of these possible indicators for anomalous readings. If the designer developed components to examine each of these data streams and each AI sounded a different alert to the operator, the system would sound three separate warnings to indicate the presence of the instigating anomaly. Certainly, the pilot would have become aware of an anomaly, but these different indicators now complicate the diagnosis step unless the system includes some type of integrated alerting system (Randle et al., 1980). However, as AI is narrow in scope and often not context-aware, developing AI that understands the cascading effect of both pilots relinquishing control of the aircraft to reconcile the source of the outcomes that follow this one simple anomaly is nontrivial. Once again, without general intelligence, it is difficult to differentiate correlation from causation.

Incident reports often discuss the impact of cascading alarms, as systems often employ many sensors to sense the state of a system or various processes within the system. As an anomaly can affect multiple processes, each of which triggers an alarm, individuals can become distracted by these alarms. However, the alarms themselves can fail, which leads to further confusion. An example of such a distraction resulted in the loss of Eastern Air Lines Flight 401 in 1972 as a very experienced crew became focused on diagnosing whether a burnt-out landing gear indicator light was faulty or if the landing gear had failed to deploy. During their deliberations, they accidentally and unknowingly disabled the aircraft autopilot. The aircraft gradually lost altitude and crashed, killing 101 crew and passengers. The accident investigation cited the crew's failure to monitor their flight instruments while diagnosing the faulty landing gear position indicating system (National Transportation Safety Board, 1973). While it can be difficult to determine failures in electromechanical systems, failures of more complex AI systems are likely to cause further distraction.

10.3.2 Anomaly Diagnosis

Although the example in the previous section illustrated the fact that anomaly detection by humans can be difficult, particularly when they fail to focus their attention on indicators of an anomaly, once the anomaly was detected, it was easy for the pilots to determine the source of the anomaly. However, in complex systems, diagnosing the source of the anomaly is often quite difficult. The difficulty the operator experiences in detecting an anomaly is exacerbated by multiple influences.

As indicated in the previous section, an anomaly can change system behavior in a complex fashion, with a single problem causing a cascade of indicators where each indicator could potentially result from multiple possible sources. The task of diagnosis requires the operator to observe the multiple indicators of the anomaly and to ascertain the common element or failure that might lead to the cascading events. Unfortunately, the longer the anomaly goes undetected, the more attributes of the system often deviate from normal parameters, and this variation can trigger additional failures.

For example, overpressure in a containment line, while being an indicator of something like a blockage, will eventually produce a rupture. This secondary failure then results in additional indicators of an anomaly, and it is difficult for one to understand that multiple failures have occurred or to associate each indicator of the anomaly with each unique failure. For example, a rupture might reduce the pressure in the line, eliminating the primary indicator of the blockage, which was the original anomaly. Therefore, during diagnosis, it can be important for the operator to understand not only current indicators of the anomaly but also the history or progression of the indicators. Diagnosing the source of multiple anomalies is especially challenging in unexpected situations or when under time pressure (Strauch, 2017).

It is notable that a cascade of anomalies is also possible in systems employing AI. As the AI is often designed to act under expected circumstances, during an anomaly, the inability of the AI to effectively diagnose the situation may result in cascading anomalies either because (1) the AI determines that it is unable to respond, thus unexpectedly handing control to the operator, or (2) the AI misdiagnoses and responds inappropriately, adding additional anomies (Strauch, 2017). Therefore, it is important that the operator understands the behavior of the AI such that they can reason about the actions that the automation is taking. Unfortunately, a case study based on a pair of Boeing Max 737 aircraft crashes illustrates this fact with deadly effect. Analysis of these accidents has illustrated that among the sources of failure is the fact that the pilots were not trained on the effects of AI in this system and its potential vulnerability to sensor feeds. Therefore, they were unable to diagnose the source of the anomaly or formulate a successful response, leading to the loss of two separate aircraft with their passengers and crew (Gates & Baker, 2019).

Woods and Hollnagel point out that during diagnosis, operators form hypotheses about potential anomalies (Woods & Hollnagel, 2006). The operators then often formulate actions that will have expected effects in the presence of the anomaly. The operators perform these diagnostic actions and then observe system behavior to determine if they witness the expected effect. If so, the operators witness evidence supporting the hypothesis. If the operators feel the evidence is strong enough, they may then formulate actions to overcome the expected anomaly. If not, they may formulate and perform additional diagnostic actions. During this stage, multiple failures can occur. The operators may not have the knowledge or associate their observations with the anomaly source. Second, the diagnostic actions might produce results that are consistent with the operator's hypothesis even though their hypothesis is incorrect. Third, the evidence provided might not differ from the operator's diagnosis to the degree that they abandon this hypothesis to search for alternative hypotheses. Each of these last two issues have been referred to as "garden path" problems (Woods & Hollnagel, 2006) where there is evidence that "supports" the hypothesis to the degree that the operators fail to disregard this hypothesis and continue to follow the evidence while the original anomaly is not addressed. Note that AI aids might assist, particularly less-experienced operators, to formulate alternative hypotheses such that they do not become fixated on a single, incorrect explanation. However, the AI must be designed to provide alternate hypotheses and courses of action in a way that does not overly influence the operator, potentially leading them to inadvisably rely on other incorrect hypotheses.

Woods and Hollnagel also discuss the use of "safing" actions during the diagnosis process. These are activities that reduce the likelihood or extend the time horizon of catastrophic system failures. These "safing" actions have the added benefit that they might also provide diagnostic evidence. During the diagnostic phase, effective communication is extremely important. Researchers reported that an error that led to the nuclear reactor incident at three-mile island partially resulted from the mislabeling of an indicator. Designers told the operators that a particular indicator showed the water level within the reactor. However, the indicator indicated pressure within the reactor. During normal operation, pressure increases with increases in water level. Thus, the indicator's operation was consistent with the operator's expectations. However, during the incident, water level became critically low, generating steam, which raised the pressure in the containment vessel. Based on the operator's training, they assumed the water level was too high, when it was actually too low. This then led the operators to take counterproductive actions (World Nuclear Association, 2022).

10.3.3 Anomaly Response

Once diagnosed, the operators must then take actions to overcome and correct the anomaly. This stage often involves the coordination of activities. Returning to our original example, once the pilots of the aircraft became aware of the fact that they were ever rapidly descending toward the earth, they attempted to correct the anomaly by grabbing the flight controls to regain control of the aircraft. However, given the time pressure, they responded similarly, and soon discovered they were fighting against one another. They then alternated a few times between simultaneously relinquishing control to the other pilot and reassuming control before finally only one of them assumed control and was able to recover the aircraft. The failure to coordinate their activities further delayed recovery, placing the crew and aircraft in further danger.

As we consider the design of HATs for anomaly detection, diagnosis, and response, we see the need for clear communication. The team members must be able to observe each other's behavior during these often time-critical phases, they must then be able to predict the effect of each of their actions to perform diagnosis, and then they must be able to communicate or direct each other's actions to execute the response (Johnson et al., 2014, 2018). This support for observability, predictability, and directability of agents' actions in the HAT is critically important during anomaly detection, diagnosis, and response. It is therefore important for us to consider these factors as we design team interactions in the following chapter.

Although AI algorithms are typically trained using real-world data, they can also be trained using simulated data. If robust physical models, that is, digital twins of systems, are available, environmental inputs and resulting errors can be simulated. These simulations might be used to anticipate sequences of cascading alarms. AI might be used to learn the root cause of these alarms. Such a model might be useful to support humans during overload conditions, especially if such a training method enables the AI to provide accurate predictions and suggestions of potential root causes during real world anomalies.

10.4 DEFINING ANOMALIES IN YOUR SYSTEM

During design, we envision the way we want our system to operate and design for these ***prescribed***, ***typical*** situations. In this case, the word ***prescribed*** refers to the fashion we expect the humans in our HAT to perform their work. These work processes might be based on our knowledge of the work that we expect the users to adopt as we introduce enhanced technology to the operator, helping them to conduct and simplify the work. To encourage the human adoption of these work processes, designers develop procedures and checklists, then train humans to follow these procedures as they perform work using the technology. The situations are ***typical*** as these are the situations that we assume the system will operate within once we release the system to operate in the field. Assumptions can be made about the goals that the humans will try to achieve with the system, how the humans will employ the prescribed work processes, the variability in these processes that might occur to compensate for differences in humans, the accuracy and reliability of the system, and the types of inputs or environmental changes the system will experience during normal performance. Unfortunately, we are not always aware of all the assumptions we make during design. Therefore, the system, environment, or operators can violate these assumptions during the system's life cycle, producing *atypical* situations which have numerous implications for system design.

First, the human can discover a new need they must satisfy with the existing system, and, therefore they begin to pursue new goals. To achieve these goals, humans will invent new processes and procedures that permit them to employ the system to satisfy needs that the designer never intended, a process often referred to as "learning by using" (Rosenberg, 1982). This adaptation often requires the user to access and use system functionality that the designer did not design the system to provide. We also acknowledge that the more tightly we integrate systems to optimize the intended system function, the less likely we are to make lower levels of functionality of the system available to the user. These lower-level system functions often must be accessed to permit the user to adapt the system to meet these new needs (Cox & Szanjnfarber, 2017). An example from the literature is that designers often do not design propeller-driven aircraft with multiple engines such that the pilot controls the speed of individual engines. However, in an example, the operators discovered that by controlling the engines near the fuselage of the aircraft to higher speeds than the engines farther from the fuselage, the aircraft experienced increased lift, which permitted the pilots to fly the aircraft at lower speeds than the designers intended the aircraft to fly.

Similarly, the human can find that the environment or other humans who seek to disrupt the system are able to inject variability that the system designer assumed would never occur. Once again, the humans must adapt the work processes to overcome this variability, which again may require that they access lower-level functionality. In an example, during Operation Desert Storm, the Iraqi Army hid trenches under sand, making it nearly impossible for A-10 pilots to see them. However, the A-10 crews were able to modify an infrared-guided missile, which would typically provide infrared imagery to the pilots only after the pilot fired the missile, to provide the pilots with infrared imagery while the missile was still attached to the aircraft. This modification made the "hidden" trenches visible to the pilots, providing a competitive advantage. In this example, the humans originally did not have the ability to turn on the infrared imagery from the missile. The missile automatically activated this function after the pilot fired it at a target. However, innovative maintainers were able to add access to this low-level control, permitting the system to adapt to its environment.

Variability in system inputs can also arise for many reasons, including changes in adjacent systems. For example, in our mammography example, upgrades to radiographic equipment or changes in data formats create examples that will provide changes to the inputs of the HAT. Differences in humans may occur as newly trained individuals or others join the HAT. These individuals may not use established procedures in the same fashion or communicate well with members of the team. AI agents can also perform variably as their sensors may perform better under some conditions than others, their training can change, or their embedded procedures and training may fit certain situations better than others.

Additionally, anomalies can also arise due to the intentional activities of malevolent actors. In these cases, the anomaly might arise from the behavior of the AI agent, which an external actor is able to control or manipulate to pursue behaviors the human operator does not desire. It is quite likely that this manipulation takes place under *deception*, in a manner that the malevolent actor deliberately designs to avoid detection and to preserve the ability for continued control and manipulation. The ability to effectively detect this type of anomaly will depend on the level of *suspicion* aroused in the human teammates. We can understand suspicion based on its relationship to trust. *Trust* is a belief that another agent will behave in a particular manner in a situation characterized by uncertainty and vulnerability (Lee et al., 2004). *Suspicion*, on the other hand, is a doubt in the trustworthiness of the intentions of another and thus the tendency to characterize undesirable behaviors as malicious. While trust (and distrust) is characterized by *certainty* in the positive (or negative) expectation of another's motives, suspicion is characterized by an *uncertainty* or ambiguity about the motives of another (Sinaceur, 2010). While distrust stems from knowledge and experience of previous undesirable behaviors, humans develop suspicion through sensitivity and detection of abnormal behaviors.

Each of the sources of variability can change the behavior, resulting in *atypical* behavior. These sources of variability or atypical behavior degrades performance or produces anomalies, which the HAT must detect, diagnose, and respond to in a timely manner to maintain an acceptable level of performance.

10.5 MODELING: TYPICAL/ATYPICAL SCENARIOS

10.5.1 Modeling Objective

The goal of scenarios is to provide a detailed, step-by-step description of the workflows that the HAT will need to perform. In a later chapter, we will develop the step-by-step description of each workflow. In this step, we do not provide a decomposition but only provide a description of scenarios to capture. As we describe these scenarios, we wish to capture both the most common scenarios that the team will employ, referred to as typical scenarios, as well as atypical scenarios which occur as external variability or errors occur due to variability of agents in our system occur. We will later apply a step-by-step description of each of these scenarios to provide a logical assessment of how our HAT will respond in each situation.

10.5.2 Diagram Components and Composition

These diagrams are simple tables which provide a description of each of the scenarios with a number to provide abbreviated indexing to the scenarios. For the atypical scenarios, the diagram will additionally document the source of the variability, which creates the deviation responsible for the creation of the atypical scenario.

10.5.3 Considerations

For this chapter, we begin by providing a high-level description of a series of typical scenarios. Scenarios are workflow traces through your system, which result in specific outcomes to which the system must respond. Typical scenarios are traces of activities that one could experience during normal system operation. When developing typical scenarios, it can be useful to focus on scenarios that will utilize portions of your HAT that you are redesigning as you transition from your "As-Is" to your "To-Be" system.

As we think about atypical scenarios, we consider processes that are undesirable or that include undesirable sources of variance in our system. Table 10.2 shows a total of 14 different atypical scenarios for the HAT evaluation. As we form these scenarios, it is important to consider the source of the deviation and whether we included variation from enough sources. It may be useful to return to your domain level BDD and consider how each actor or system inside your system boundary, as well as the actors outside your system boundary, might introduce unintended variability when you are creating your atypical scenarios. Note that atypical scenarios stem from deviations and variations in the expected workflows, resulting in unintended negative system performance. This differs from undesirable patient outcomes, for example, a positive cancer diagnosis, which results from typical scenario exhibiting proper system performance, even though the outcome is undesirable to the patient.

10.5.4 Example Application

Table 10.1 shows a list of nine typical scenarios that are likely to occur for the HAT in our redesign.

As shown in this table, the first three scenarios include a patient who enters the flow for general screening. Each scenario exercises various flows, which include or do not include a secondary level of screening, diagnosis, and intervention. Scenarios 4 and 5 vary the input to include patients who have physical indications which might be consistent with mammary gland cancer. Scenarios 6 and 7 exercise the system for patients where the new components of our HAT flag patients as having an elevated risk. Finally, scenarios 8 and 9 examine upgrades to the system to support the new risk assessment components of our HAT. All of these are common workflows that our HAT should be able to respond to on a regular basis.

Table 10.2 shows a series of atypical scenarios. Note the table includes the source of deviation for each of these scenarios. As shown, most of these scenarios result from a failure of one of the agents or changes in the environment or input that are outside

TABLE 10.1
Typical Scenarios for HAT Evaluation.

Scenario Number	Scenario Description
1	Patient comes in for routine screening, no areas of concern found in mammogram
2	Patient comes in for routine screening, area of concern found, follow-up screening results in positive diagnosis, intervention selected and executed
3	Patient comes in for routine screening, area of concern found, follow-up screening results in negative diagnosis, no intervention required
4	Patient with known issue comes in for screening, area of concern not indicative of cancer
5	Patient with known issue comes in for screening, area of concern found indicative of cancer, intervention selected and executed
6	Patient flagged as elevated risk comes in for screening, follow-up screening results in positive diagnosis, intervention selected and executed
7	Patient flagged as elevated risk comes in for screening, follow-up screening results in negative diagnosis, no intervention
8	Literature agent identifies a new risk indicator, to be incorporated into the risk assessment, agent incorporates this indicator in its assessment
9	Literature agent identifies a modified intervention which the radiologist and intervention agent incorporate when determining intervention options

TABLE 10.2
Atypical Scenarios for HAT Evaluation.

Scenario Number	Scenario Description	Source of Deviation
1	Radiologist disagrees with the diagnosis recommender, selects alternate diagnosis	Failure of radiologist or diagnosis recommender
2	Radiologist disagrees with the intervention recommender, selects alternate intervention	Failure of radiologist or intervention recommender
3	Risk assessment agent supports PCM to create higher patient throughput than can be supported by system	Risk assessment agent exceedingly impactful
4	Risk assessment agent increases false alarm rate in the system	Input changes affect outcomes
5	Literature agent identifies a false risk indicator, which the risk assessment agent incorporates into its assessment	Failure of literature agent for risk
6	Literature agent identifies a detrimental intervention which the radiologist and intervention agent incorporate into the intervention options	Failure of literature agent for intervention
7	Images the radiological technician captures are not of sufficient quality, resulting in incorrect diagnosis or images and a potential delay of treatment	Failure of radiology technical
8	Dissemination agent provides updates to procedures, which potentially reduce team effectiveness	Dissemination agent failure
9	Dissemination agent provides excessively frequent updates to procedures	Dissemination agent failure
10	System designers update EPR format or interface protocol	Input changes affect outcomes
11	Radiologist does not consider all the potential intervention options and prefers to recommend a suboptimal option	Radiologist failure
12	Radiologist has missed a particular type of concern across a number of patients and becomes oversensitive in future cases	Radiologist impacted by feedback
13	Radiologist is not available in a timely fashion after the radiology technician captures images	Radiologist failure
14	Radiologist is technology-averse	Radiologist failure

those expected during normal operation. While most of the examples list a failure, you should also note the presence of scenario number 3 where the risk assessment agent might indicate a large number of patients who are classified as high-risk. This agent behavior causes over subscription of the HAT services, where the system is unable to successfully provide the necessary services which leads to anomalous conditions. It is possible that anomalies do not only occur as a function of failures but might also occur if certain components or the value of the system is exceedingly positive, which will increase demands upon the system, which might also produce anomalous conditions.

10.6 CHAPTER SUMMARY

This chapter discussed the detection, diagnosis, and responses associated with overcoming anomalies. Anomalies are, by definition, abnormal and difficult-to-define events which have the potential to degrade the performance of a system or result in a system failure. AI agents can aid the detection of expected anomalies. However, the detection of the remaining anomalies and the diagnosis of the source and recovery mechanism for these anomalies often require a degree of innovation, often requiring human involvement. Designing systems with the appropriate support to aid the human in detection, diagnosis, and recovery mechanisms is essential for the creation of systems that are robust to unexpected sources of variability, which often lead to anomalies. While it is impossible to identify and purposefully design for all atypical scenarios that might involve anomalies, it is important to create and verify your design against both typical and atypical scenarios to ensure that the system is able to adapt to expected anomalies and to create a level of system flexibility that will ideally aid the human to respond to other anomalies as they are encountered.

10.7 QUESTIONS AND DISCUSSION TOPICS

1. Consider a scenario in which you are a "passenger-supervisor" in a driverless car system. Describe a potential anomaly that could occur.
    1. What external and system cues could you use to assist in anomaly detection?
    2. What are potential diagnoses for these detected cues?
    3. Provide an example of either "safing" or "garden path" behaviors that could occur.
    4. What mechanisms would exist for response?
2. How is suspicion different from distrust? Provide an example for each to clarify this difference.
3. Under what conditions is suspicion likely to be high? Under what conditions is suspicion likely to be low?

10.8 PROJECT QUESTIONS AND ASSIGNMENTS

1. Considering your project, describe at least five typical scenarios for your system.
2. Considering your project, describe at least ten atypical scenarios for your system. Identify the source of the deviation that is associated with each atypical scenario. Include at least one atypical scenario involving a malevolent external actor.
3. For one of your atypical scenarios, describe the information that an operator must observe to detect the presence of the anomaly and the information that might be required to diagnose the source of the anomaly.

REFERENCES

Artificial Intelligence Research & Development Interagency Working Group. (2019). *Advancing artificial intelligence R & D*. National Science & Technology Council.

Bobko, P., Barelka, A. J., & Hirshfield, L. M. (2013). The construct of state-level suspicion: A model and research agenda for automated and information technology (IT) contexts. *Human Factors: The Journal of the Human Factors and Ergonomics Society*, *56*(3), 489–508. https://doi.org/10.1177/0018720813497052

Booth, C. E. (2009). *Surveillance using multiple unmanned aerial vehicles*. Air Force Institute of Technology. https://scholar.afit.edu/etd/2428/

Bradshaw, J. M. (2013). From knowledge science to symbiosis science. *International Journal of Human Computer Studies*, *71*(2), 171–176. https://doi.org/10.1016/j.ijhcs.2012.10.003

Cox, A., & Szanjnfarber, Z. (2017). System user pathways to change. In A. Madni, B. Boehm, R. Ghanem, D. Ewin, & M. Wheaton (Eds.), *Disciplinary convergence in systems engineering research* (pp. 1–10). Springer.

Gates, D., & Baker, M. (2019, June 22). The inside story of MCAS: How Boeing's 737 MAX system gained power and lost safeguards. *The Seattle Times*. https://www.seattletimes.com/seattle-news/times-watchdog/the-inside-story-of-mcas-how-boeings-737-max-system-gained-power-and-lost-safeguards/

Johnson, M., Bradshaw, J. M., & Feltovich, P. J. (2018). Tomorrow's human–machine design tools: From levels of automation to interdependencies. *Journal of Cognitive Engineering and Decision Making*, *12*(1), 77–82. https://doi.org/10.1177/1555343417736462

Johnson, M., Bradshaw, J. M., Feltovich, P. J., Jonker, C. M., Van Riemsdijk, M. B., & Sierhuis, M. (2014). Coactive design: Designing support for interdependence in joint activity. *Journal of Human-Robot Interaction*, *3*(1), 43–69. https://doi.org/10.5898/JHRI.3.1.Johnson

Lee, J. D., See, K. A., & City, I. (2004). Trust in automation: Designing for appropriate reliance. *Human Factors: The Journal of the Human Factors and Ergonomics Society*, *46*(1), 50–80. https://doi.org/10.1518/hfes.46.1.50.30392

Marstall, J., Miller, M. E., & Poisson Iii, R. J. (2016). Collaboration in the cockpit: Human–system interaction beyond the autopilot. *Ergonomics in Design*, *24*(1), 4–8. https://doi.org/10.1177/1064804615572628

Merriam-Webster. (n.d.). *Confidence*. Merriam-Webster. Retrieved May 22, 2017, from https://www.merriam-webster.com/dictionary/confidence

National Transportation Safety Board. (1973). Aircraft accident report: Eastern air lines, Inc. L-1011, N310EA, National Transportation Safety Board NTSB-AAR-73-17, 1–45. Table and figure.

Ragan, E., Endert, A., Sanyal, J., & Chen, J. (2015). Characterizing provenance in visualization and data analysis: An organizational framework of provenance types and purposes. *IEEE Transactions on Visualization and Computer Graphics*, *22*(1), 31–40. https://doi.org/10.1109/TVCG.2015.2467551

Randle, R. J., Larsen, W. E., & Williams, D. H. (1980). Some human factors issues in the development and evaluation of cockpit alerting and warning systems. In *NASA reference publication* (No. 1055; Issue 1055). https://ntrs.nasa.gov/citations/19800007561

Rosenberg, N. (1982). *Inside the Black box-Chapter 6 learning by using* (1st ed.). Cambridge University Press.

Sinaceur, M. (2010). Suspending judgment to create value: Suspicion and trust in negotiation. *Journal of Experimental Social Psychology*, *46*(3), 543–550. https://doi.org/10.1016/j.jesp.2009.11.002

Strauch, B. (2017). Ironies of automation: Still unresolved after all these years. *IEEE Transactions on Human-Machine Systems*, *48*(5), 419–433. https://doi.org/10.1109/THMS.2017.2732506

Warm, J. S., Parasuraman, R., & Matthews, G. (2008). Vigilance requires hard mental work and is stressful. *Human Factors*, *50*(3), 433–441. https://doi.org/10.1518/001872008X312152.

Wiener, E. L. (1993). Intervention strategies for the management of human error. *NASA Ames research center* (Issue A-93120). http://human-factors.arc.nasa.gov/publications/Intervention_Strategies.pdf

Woods, D. D., & Hollnagel, E. (2006). *Joint cognitive systems: Patterns in cognitive systems engineering*. CRC, Taylor & Francis Group.

Woods, D. D., Patterson, E. S., Roth, E. M., & Christoffersen, K. (2002). Can we ever escape from data overload? A cognitive systems diagnosis. *Cognition, Technology & Work*, *4*, 22–36. https://doi.org/10.1007/s101110200002

World Nuclear Association. (2022). *Three mile island accident*. https://world-nuclear.org/information-library/safety-and-security/safety-of-plants/three-mile-island-accident.aspx

11 System Validation through Coactive Design

11.1 DEFINING COACTIVE DESIGN

In recent chapters, we have discussed agent capabilities where a *capability* is the extent to which someone or something possesses the knowledge, skills, and ability to perform a specified duty or task in fulfillment of a responsibility. Although we defined it earlier, we have not provided a similar discussion of capacity. Recall that we defined *capacity* as the extent to which someone or something possesses the knowledge, skills, and ability to perform a specified duty or task at a specific moment in time and within a specific context. In Chapter 7, when discussing allocation, certain responsibilities and associated capabilities were allocated to multiple agents, indicating that, depending upon the context of a situation, more than one agent could be capable of performing the necessary work. In human teams, where the humans understand how to cooperate and can innovate work methods, the humans can determine whether they have the capacity to perform the necessary work and dynamically allocate the work among themselves. Unfortunately, modern-day AI agents do not typically have the general intelligence to participate in dynamic task allocation. Therefore, it is important to design the interaction between human and AI agents to facilitate dynamic task allocation during the action phase, particularly when multiple agents can perform a task, and their ability to perform the task varies with context.

Multiple team members can take part in activities necessary to fulfill the team's common goal(s). Typically, this implies that one team member may complete a portion of the tasks necessary to achieve a goal while other team members complete other portions. Typically, these activities are not independent of one another. Instead, one agent might perform an activity which serves as an input for an activity that another team member performs. For example, on an assembly line, one person might assemble a subcomponent that then another person integrates into the final product by attaching it to the final product and verifying that the combined assemblies function as desired. It is also possible that two or more team members may need to actively work together to complete an activity—for example, two workers may need to lift a heavy assembly together to move it onto a workstation. In each of these instances, the ability of a worker to complete their portion of the work is dependent upon the actions of another worker. Johnson and colleagues clarify that a *dependence* exists when an agent "lacks the required capacity to completely perform an activity in a given context" (Johnson et al., 2014). The assembly worker is dependent upon the worker completing the subassembly as they lack the time or knowledge to complete this subassembly while the lifting worker lacks the strength to complete the lifting task.

DOI: 10.1201/9781003428183-11

An assembly line is a very specialized example of a teaming arrangement in which the process engineers design the work such that workers toward the end of the assembly line are dependent on the workers nearer the beginning of the assembly line. The fact that this dependence is unidirectional simplifies the communication among the workers, purposely streamlining the work they perform. However, in most environments, dependencies are not unidirectional among workers. Instead, a first worker may be dependent on a second worker under one circumstance, but the second worker may be dependent on the first worker in another circumstance. Additionally, the two workers may also rely upon each other simultaneously. Johnson and colleagues define this complex form of dependence as *interdependence*. Specifically, they define interdependence as "the set of complementary relationships that two or more parties rely on to manage required (hard) or opportunistic (soft) dependencies in joint activity" (Johnson et al., 2012). The examples in this chapter are examples of *hard dependencies* where a single worker cannot complete the work as they do not have the capability or capacity to perform the required work. As mentioned in an earlier chapter, *soft dependencies* are also important to consider where a single worker may have the capability and capacity to perform the necessary work activities, but the workers can perform these activities more effectively or efficiently with the help of another worker. Johnson and colleagues' definition of interdependence also includes the term *joint activity* where multiple agents perform typically interdependent activities to accomplish a common goal. To perform interdependently, each agent must understand not only the work to be performed to achieve a goal, but also the work that each agent in the team can perform under any set of circumstances, the ability to predict or determine the present circumstances, and the ability to adapt their behavior appropriately in the presence of these circumstances.

With this background, we can now define the term *coactive design* which is a term coined by Johnson and colleagues to describe an approach which they originally applied to human–robot interaction design. In their work, coactive design is the design of the interface between humans and robots to permit a human and a robot to effectively accomplish joint activities necessary to achieve common goals (Johnson et al., 2012). In our application, we are not necessarily concerned with robots alone but also the interaction between agents, both human and AI. Although the current text borrows heavily from the work of Johnson and colleagues, it is necessary to perform adaptations of the approach to coactive design as originally practiced. In the original application, the robot is a machine with an embedded AI agent that is capable of sensing and acting in an environment to perform joint work with a human. As we adapt this work to HAT design, we do not attach all AI agents to a machine. This is not a particular limitation of coactive design, however, as the original work considered both physical and cognitive activities performed by robots. In our application, AI agents may not always be capable of physical work, which places a greater emphasis on cognitive work. The larger difference between Johnson's original work and the application of coactive design to HATs is the focus on teams that are larger than simple dyads and the need to apply this approach in less-constrained environments.

It is notable that Johnson and colleagues originally intended coactive design as a method to design the interface between a human and a robot. To do so, they developed a tool called an Interdependence Table to capture and perform a subjective analysis of this design. Johnson refers to this analysis as Interdependence Analysis. While the Interdependence Table is useful to capture the designs of human–AI agent interface elements, AOSM also applies a version of the Interdependence Table and Interdependence Analysis to determine whether capabilities are missing from our Agent-Oriented Systems Model. Therefore, the Interdependence Analysis method supports the validation of the models we have created. The method serves multiple additional purposes including: (1) to ensure that specific tasks are performed by agents that have commensurate capabilities, (2) to ensure that the agent either has backup in the case of failure or that the designer purposefully designed without backup, (3) to help predict the likelihood of successful task execution, and (4) to identify important elements of communication among the team members.

Some individuals incorrectly apply the terms verification and validation as if they are synonyms. However, these terms refer to vastly different processes. System *verification* seeks to determine if the developer has constructed the system to meet its specifications or requirements. *Validation*, which is often assessed during acceptance testing that occurs just before fielding, seeks to determine if the system delivered by the developer meets the needs of the user (Buede & Miller, 2016). Similar definitions are applied to models. *Model verification* refers to the process of determining if a model represents the conceptual description of the represented entity in the requirements documentation. *Model validation* determines the degree to which a model is an accurate representation of the real world (Sargent, 2013). In our case, our modification to coactive design seeks to determine whether the model we constructed represents the work and required capabilities necessary for the system, including the "To-Be" HAT with the responsibilities and capabilities we allocated to agents, to function correctly in the environment. Therefore, this process is a validation process, although it is applied early in the development process.

11.2 DIVISION OF WORK AND LEVELS OF ABSTRACTION

Throughout this text, we have discussed levels of abstraction for commands or activity descriptions. However, we have not clearly discussed this concept and its implications. You have probably also struggled with the level of abstraction when forming your responsibilities and capabilities, wondering whether these were too broad to be useful or so narrow that you overcomplicated the ability to understand each of these elements.

Besides touching on levels of abstraction when describing work, we also touched on this concept when discussing the difference between implicit and explicit communication. Specifically, we discussed it through the example of constructing a peanut butter and jelly sandwich. When delegating work to others on a team, we can explicitly describe this work at a high level of abstraction (i.e., make yourself a peanut butter and jelly sandwich), which implies that the individual must perform the large number of activities necessary to construct the sandwich. Alternately, we can describe this work at a lower level of abstraction listing many of the sequential steps the individual must perform to form the sandwich. Hierarchies of human tasks or activities at various levels of abstraction have traditionally been used to describe human work within both the human factors engineering and training system development communities (Annett & Duncan, 1967; Stanton et al., 2010).

Johnson and colleagues provide an example of commanding robots at various levels of abstraction in their 2012 paper (Johnson et al., 2012). In the research described in this paper, the researchers asked humans to interact with virtual robots in a simulated world that included nine rooms. The simulation distributed blocks of distinct colors throughout the rooms. The experimenter required the human and robots to stack the blocks in a particular color order in a final location. The team includes an AI guided robot and a human guided robot (i.e., simulated telerobotic system). The experimenters asked the humans to interact with the AI guided robot at one of four distinct levels of abstraction for different experimental conditions. The simulation only showed each agent the blocks in the room they are currently in. The experimenters required the agents to retrieve the blocks in a particular order and carry them to the destination in the correct order. Figure 11.1 shows the activity decomposition. As shown, the experimenters asked the humans to achieve the goal using one of four sets of commands to interact with the AI-enabled robot in the simulation. These sets included commands at four levels of abstraction. At the top level, the human can only command the robot to complete the goal. At level 2, the human can command the robot to "Find a Color", "Deliver Color to Drop Zone", and "Drop off block". At level 4, these three simple commands expand into the eight separate commands of "Recognize Color", "Go to Room", "Enter Room", "Go to Block", "Pick up Block", "Go to Drop Zone", "Enter Drop Zone Room", and "Put Down Block".

In Johnson's research, he illustrates that commanding the robot at the lowest level of abstraction requires a sizable portion of the human's cognitive and physical resources such that they are distracted from their control of the telerobotic system, resulting in the lowest performance level for the team. However, at the highest level of abstraction, the human is unable to coordinate their activities with the robot, resulting in the second lowest level of performance. Team performance was highest

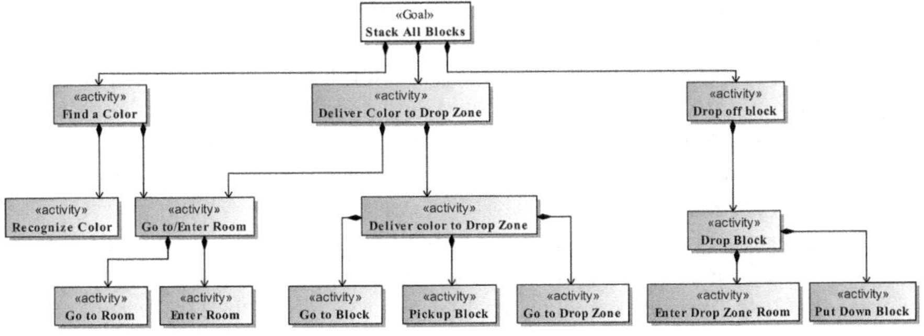

FIGURE 11.1 Activity decomposition illustrating the available commands at distinct levels of abstraction.

when commanding at the third level of abstraction and only slightly lower for the second level of abstraction. This experiment illustrates a need to provide commands at a level to support coordination of activities within Human–Agent teams but that providing commands at too low a level can also reduce team performance. As we move to our next design stage, it is thus important to select a level of decomposition that permits the sharing of work between humans and AI agents in the human–agent team without overburdening the human during typical work scenarios. This does not mean that the interface should not provide information or permit commands at lower levels of abstraction, since, as we discussed in the previous chapter, these lower-level commands may be necessary to support anomaly diagnosis and recovery.

11.3 CAPACITY AND THE IMPORTANCE OF BACKUP BEHAVIORS

It is useful to further consider the fact that although an agent might be capable of performing a task in the abstract sense, it may not be possible for the agent to perform a task reliably in a specific scenario. As noted earlier, AI team members may not be able to adapt to a specific scenario, and, therefore, they may not be able to accomplish the task in that instance. Ideally, a human teammate will be able to innovate or bring different knowledge, skills, and abilities to this task that permits them to complete tasks that the AI agent cannot. On the other hand, humans often experience high levels of workload, as their capacity is limited. Thus, they are unable to perform specific tasks in a timely manner. These capacity limitations are particularly salient for novices, individuals experiencing fatigue, or individuals who are otherwise cognitively impaired. Further, humans do not always distribute their attention as required by the task and therefore may not consider all valuable information during decision-making. AI agents consistently have access to specific pieces of information and perform in a consistent manner. Thus, these agents can perform tasks to assist the human when the human experiences capacity limitations. However, the designers must design the system to support effective collaboration, enabling agents to understand and compensate when their teammates experience capacity limitations.

The AOSM adaptation of coactive design supports the design of appropriate backup behaviors through a process that begins with defining scenarios, such as the typical and atypical scenarios we discussed in the previous chapter. The method decomposes these scenarios into tasks at a level of abstraction that permits us to examine the distribution of work among agents in the HAT. Often, we will arrange these tasks temporally to permit us to understand how information needs of one task are related to the information in subsequent or previous tasks. However, to simplify this process, we represent any repeated or cyclic tasks only once within the task decomposition for each scenario.

Once we form the task list, we assign Agent(s) to each task. We assign the agent that is "Responsible" for performing the task under typical conditions as the "Primary" agent for this task. We then consider whether other agents in the HAT could provide useful backup behaviors to support the primary agent as they complete this task. We assign these agents as "Secondary" Agent(s). The secondary agents might include agents that are either "Accountable" for the task or able to "Support" the task. We then perform an evaluation, wherein we estimate the primary agent's capacity for performing the task, as is the capacity for each of the "Secondary" agent(s). Through this process, we assess the team's ability to perform this task. As we imagine this work, we need to define the information that the agents must transfer to other agents to support successful teaming behaviors. In coactive design, we capture this information in terms of information that is observable, predictable, and directable.

11.4 DESIGNING FOR OBSERVABILITY, PREDICTABILITY, AND DIRECTABILITY

To be able to team, each agent must be aware of the status of relevant elements of the team's activities to plan and execute their portion of the work. Therefore, it is important to define the communication necessary to establish that situation awareness. The coactive design method specifically enumerates three diverse types of communication that must exist among agents that engage in joint work. These include observability, predictability, and directability or OPDs (Johnson et al., 2012, 2018).

The term *observability* refers to the ability of an agent to make their status, the status of the work they are performing, and their understanding of the state of the team and environment known to others. This is not a unidirectional flow of information from one agent to another. Instead, for information to be observable, each agent in the team must also be able to observe and interpret these same signals from other team members. In traditional AI agent systems, it is common to discuss the AI agent's ability to observe information from the environment to pursue a goal. However, the concept of observability also requires that the AI agent be able to observe and interpret the actions of other agents on the team, as well as the humans being able to observe and interpret the actions of AI agents. Since AI agents do not have general intelligence, this ability is not possible to create in the general sense. Therefore, to support appropriate design, it is important to define the information that each AI agent will need to make available to others as well as the information that each AI agent will need to receive from other members of the team, which will permit the team to achieve their function.

Observability permits one to understand the status of other members of the team, which is important as it permits other agents to react to this status. However, to function efficiently, team members often must anticipate certain activities of others so that they can either plan their own work or intervene to avoid errors. This implies that the agent's actions should be predictable. *Predictability* refers to the need to be able to predict or anticipate other agent's actions to a degree that one can consider and plan their current or subsequent actions. This predictability might come from prior performance or training with agents to understand their likely behaviors. However, in human–AI interaction, we might use anticipatory displays that help the human understand the information that the AI is considering in its decision-making and the likely behavior. As an example, an automotive display might show a graphical rendition of the road ahead with rendering of obstacles that are known to the autonomous driving agent and a line showing the anticipated path the car will follow. Once again, predictability does not refer to a unidirectional flow of information. To support predictability, the AI agents will also need the ability to consider and anticipate other agent's actions when planning their own actions.

Directability refers to an agent's ability to direct the behavior of others and similarly to be directed by others. Therefore, during coactive design, it is necessary to understand the directions that a human will need to provide to AI agents they interact with, as well as consider advice or direction that the AI agent should provide to the humans or other AI agents on the team. This direction includes explicit commands, task allocation, guidance, suggestions, and warnings. Johnson and colleagues also include less explicit direction by including requests for assistance or asking for input to support decision-making under the umbrella of directability. Once again, this includes the ability to provide and receive direction from others.

It is important to realize that during design, the goal is not to maximize or minimize OPD. If we design too much communication into the interaction, this communication will increase the workload of the human to unnecessary levels, and we will burden the AI agent designer with implementing functions that are not necessary to support the team behavior. On the other side of this balance, if we design too little communication by not enabling adequate OPDs, the team will not be able to effectively coordinate their knowledge and actions. Therefore, it is important to understand the baseline of information that team members will need to transfer within the HAT on a regular basis, as well as additional information team members will wish to transfer upon request and during anomalous conditions.

11.5 USING OPD ELEMENTS TO BUILD TRUST AND ALERT SUSPICION

Effective teaming relationships depend on *trust*—the belief that your teammates will perform as expected within a specific context. For human teams, this sense of trust stems from the combination of the trustee's ability, integrity, and benevolence, whereas in human–AI teaming, this depends on (1) the AI agent's performance, (2) transparency of behavior, and (3) the relevance of the capabilities designed into the AI agent for the task at hand (Hoff & Bashir, 2015).

Communication using the Observability, Predictability, Directability elements provides an opportunity to increase trust through greater transparency (Johnson & Bradshaw, 2021). As cautioned earlier, designers should recognize that there is a trade-off between providing complete transparency and information overload. Judiciously presenting OPD elements that provide critical, actionable information while withholding irrelevant information is essential for optimal performance of the human team members. However, it is recognized that to build trust in the AI, humans may need insight into an AI agent's rationale or behavior and that this trust-building information may go beyond what the human needs for typical system operation.

Communication using the Observability, Predictability, and Directability elements also provides an opportunity to design to alert suspicion. As discussed in Chapter 9, internal factors as well as malicious external actors can alter system and AI agent behavior. The ability to accurately arouse ***suspicion***, a doubt in the trustworthiness of the intentions of another, is critical to early detection, diagnosis, and response.

Intentional communication of specific system behaviors can increase the visibility of anomalous behaviors, whether due to natural degradation or deceptive activities, increasing the human team members' awareness to potential failures.

11.5.1 OPDs to Communicate AI Agent Performance

Ideally, the AI agent would be designed to have perfect performance, but the reality is that all contexts cannot be accounted for in a design, and when an AI system encounters complex real-world scenarios, it will inevitably be presented with an edge case that was not part of its design parameters. When this happens, the system will either "fail" or will respond in a way that is suboptimal to team performance. The human team members will experience this as the AI being "unreliable", "unpredictable", or "erratic". One potential mechanism to minimize this experience involves enabling the AI agent to communicate a level of confidence in its selected action or response—in a sense confessing its own estimated level of reliability. This communication can help calibrate the human team members to the AI agent's inputs and allows for the humans to appropriately account for the confidence level in deciding whether to accept or override their AI teammates' actions (McGuirl & Sarter, 2006). To minimize information overload, the interface does not always need to display the confidence level, but the confidence level should be made salient early enough in the sequence of activity so that the human operator can reengage with the system and build situation awareness before they must make a critical decision when the AI agent's confidence is low.

Additionally, undesirable AI agent performance should be an indicator that the human should become suspicious of the AI agent or system performance. The ability of the human to differentiate between reliability degradation due to out-of-bounds parameters versus malicious behavior is critical to producing an appropriate response. This is especially the case if a malicious actor is involved because rapid detection and response are essential to minimizing damage (Gay et al., 2017).

11.5.2 OPDs to Communicate Transparency of Behavior

Transparency of AI agent behavior can serve multiple functions. First, transparency through observability can build trust by increasing the comprehension of system behavior. Over time, human team members will recognize patterns in the AI's behavior, increasing predictability as well.

Second, in cases of handoffs, especially emergency handoffs or handoffs that occur because the automation is incapable of responding, the human must be capable of processing the current state of the environment and any mechanical systems that the AI agent is controlling. Transparency of the current state and the AI activities that lead up to the state requiring the handoff will enable the human to have the requisite situation awareness to effectively receive the handoff.

Third, transparency can make it more difficult for a bad actor to successfully gain control of the system without detection. Additionally, as human team members recognize patterns in the AI agent's behavior, it will be easier to detect when the AI is not performing as intended.

11.5.3 OPDs to Increase the Understanding of Designed Capabilities

Humans are more likely to trust automation when it is functioning within the scope of its intended purpose and capabilities. With traditional automations (toasters and dishwashers), the intended purpose and capabilities are often intuitable without the need for detailed education of the human operators. In the case of AI agents, humans are often unaware of the boundary of the AI's capabilities and are also often unaware of the scope of the dataset used to train the AI. Furthermore, use of and experience with the AI aid do not necessarily produce a transparent understanding of the AI's capabilities and limitations.

Increasing the observability and predictability of the AI agent's activities will enable human teammates to more rapidly and accurately understand the designed boundaries of the AI agent's capabilities. Effective interface design with respect to the completeness and limits of the AI agent's directability will also increase this understanding for the human teammates.

As we discuss coactive design in this chapter, you should recognize that we are discussing the teaming behaviors that we are instantiating into our system, similar to the human teaming behaviors discussed in Chapter 4. It is useful to recall some of these important behaviors, many of which were summarized in a paper describing ten behaviors for making automation a team player (Klein et al., 2004). These include the following behaviors.

1) The AI agent should engage in common grounding or the construction of team mental models to include information regarding the agent's ability to successfully fulfill their responsibilities.
2) The AI agent should model other team member's intentions and actions to determine alterations to their behavior in response to changes in intentions and actions. This likely requires the construction of teammate's mental models.
3) Consistent with item 2, the agents must be able to observe and interpret important signals necessary for determining the coactive agent's status and intentions, which requires the inclusion of sensors to enable this observation.
4) The system should be constructed to make the behavior of coactive agents predictable to one another.
5) The system should be constructed to permit agents to direct one another's behavior in response to anomalies.
6) The agents should be able to make pertinent aspects of their status and intentions available to teammates. However, it is important to recognize that providing irrelevant data can cause data overload. Thus, the system must be able to distinguish relevant from irrelevant data.
7) The agents should be able to negotiate goals and their relative importance. This activity should include planning responsibilities during

the preparation stage and the replanning of tasks and responsibilities during the action phase.

8) The agents should be able to participate in planning to enable collaboration. This objective is particularly important in time-critical environments involving action and production teams.
9) The agents should be able to participate in managing the attention of their teammates to aid the construction of consistent team mental models.
10) All agents should work to control the cost of coordinated activity, assessing the consistency of team mental models held by teammates and engaging in explicit communication when necessary to realign team mental models.

While these are all desirable teaming behaviors, today's AI agents typically lack the ability to participate in these activities. Thus, today, it is often necessary to design the interface between humans and AI agents in significant detail to support the desired interaction. The design of Interdependence Tables, and the capabilities and interface behaviors associated with these capabilities, should attempt to create the illusion of these AI agent behaviors to the degree possible. As we progress into the future, one would expect that AI agents will be constructed that will enable these behaviors. Work in this direction includes the development of Joint Activity Graphs to describe and plan joint work within AI agents (Vignati et al., 2022), the development of structures for AI agent mental model development (Lapso et al., 2023), and efforts for determining or specifying human intentions (Miller, 2017; Miller & Parasuraman, 2007; Schneider et al., 2023).

11.6 MODELING: COACTIVE DESIGN PROCESS AND TABLE

11.6.1 Modeling Objective

In AOSM, a slightly modified version of Johnson's coactive design method is employed. To conduct a thorough analysis, we develop Interdependence Tables for the typical scenarios as well as the foreseeable, atypical scenarios for our system. We assume that by designing the system to support interdependence between agents in our system for the foreseeable, atypical scenarios, we will design sufficient OPDs into the system to support appropriate teaming behavior during unforeseen scenarios. As stated earlier, the process objectives are to define the teaming opportunities between agents as they perform tasks, enumerating and exploring the backup behaviors to be instantiated in your system, providing a prediction of system success for each task to support potential redesign, providing validation of the capabilities assigned to agents, and finally to provide an understanding of the OPDs that must be supported by the system and individual agents to enable proper interdependence.

11.6.2 Diagram Components and Composition

Figure 11.2 provides a generic Interdependence Table as implemented in AOSM. As shown, the table in this figure comprises four major sections. These sections include the Scenario Description, Primary Agent, Secondary Agent(s), and OPD sections.

The first section is the *Scenario Description* section. This section includes the scenario and the tasks that we expect the HAT to conduct during the scenario. We list the tasks in the order we expect the HAT to conduct them during the scenario. It is not necessary to include any task multiple times within this list. Therefore, we will not list repeated tasks for expediency, which may cause breaks in the temporal order.

The second section is the *Primary Agent* section. This section includes the name of the agent that is *Responsible* for conducting the task, or if there is no *Responsible* agent then the agent that is *Accountable*. In addition to the agent's name, this section includes the capabilities that the agent must have to support the task. While listing these capabilities, it is important to ask whether the capabilities available for this agent in the model are sufficient or whether we need to add capabilities to support the task. If you define new capabilities, then you should add these capabilities to your agent's capabilities in the model, including the Responsibilities-Capabilities Diagram and Agent Capability Assessment Table. Additionally, you should list a performer rating for your primary agent. The performer rating uses a scale from 1 to 4 as shown in Table 11.1. As shown, the primary goal of this rating is to capture the degree to which the primary agent requires backup behaviors from other agents. A rating of 4 indicates the need for reallocation, training requirements, or innovative design for an AI agent's capabilities.

The Secondary Agent(s) section includes the name and capabilities of the backup agents. When the primary agent is Responsible, then we include the Accountable agent in this section. Additionally, we list any agents that are Supporting in this column. You do not need to capture agents that we have rated as simply informed or consulted in this section, but you do need to consider these agents when describing OPDs. Although not shown in Figure 11.2, you can also include a scale value for your secondary agent as shown in Table 11.1.

The OPD section comprises an overall Team Rating as well as the OPD descriptions. The team rating scale is also described in Table 11.1. Also, we can provide the final element of the OPD section as a single column or decompose this section into multiple columns. In either case, this section includes a description of the information transfer mechanisms that the design team must create to support appropriate teaming behaviors and achieve the team rating.

11.6.3 Considerations

The construction of the Interdependence Table for each scenario can be time consuming. For a class exercise, it is sufficient to complete the Interdependence Table for select scenarios. From an educational viewpoint, completing these tables for at least a couple of typical as well as atypical scenarios is sufficient. However, in a real-world project, it is important to fully specify and understand the communication mechanisms that the system must support to enable appropriate teaming behaviors.

Scenario Description		Primary Agent			Secondary Agent(s)				OPD Section	
Scenario	Tasks	Capability	Agent Name	Performer Rating	Capability	Agent Name	Capability	Agent Name	Team Rating	Observability, Predictability, Directability
S1	Task 1	Capability 1	Agent 1	1	Capability 1	Agent 3			1	
	Task 2	Capability 2	Agent 1	2	Capability 2	Agent 3			1	
	Task3	Capability 3	Agent 1	2	Capability 3	Agent 3	Capability 3	Agent 4	1	Agents 2, 3, and 4 need to observe and predict Y
	Task 4	Capability 4	Agent 1	2	Capability 4	Agent 3			1	
	Task 5	Capability 3, 7	Agent 1	2	Capability 3, 7	Agent 3			1	
	Task 6	Capability 6	Agent 1	2	Capability 6	Agent 3			2	
	Task 7	Capability 7	Agent 1	3	Capability 7, 9	Agent 3			2	
	Task 8	Capability 8	Agent 2	3	Capability 8, 14, 17	Agent 3	Capability 8, 14, 17	Agent 4	1	
	Task 9	Capability 3, 11, New	Agent 2	2	Capability 3, 11, New	Agent 1			1	
	Task 10	Capability 10	Agent 2	2	Capability 10	Agent 1			1	
	Task 11	Capability 11	Agent 2	2	Capability 11	Agent 1			1	
	Task 12	Capability 12	Agent 2	2	Capability 12	Agent 1			1	

FIGURE 11.2 Generic interdependence table containing generic elements.

TABLE 11.1
Rating Scales to Apply in the Interdependence Table.

Rating	Performer Rating	Secondary Agent Rating	Team Rating
1	Capable of performing task alone	Capable of improving task efficiency	Team can perform task efficiently and effectively
2	Capable but task reliability less than 100%	Capable of improving task reliability	Team can perform the task with acceptable reliability
3	Capable of performing part of task but need assistance	Capable of supporting remaining task activities	Team can perform task, but backup is not present, providing potential reliability issues
4	Incapable of performing a significant portion of the task	Incapable of providing assistance on task	Team cannot reliably perform task

To minimize the effort, it is important to construct your scenarios following a modular approach where scenarios are a relatively constrained set of tasks that the HAT can sequence with or embed into larger scenarios. This organization reduces the need to replicate this process for a series of tasks.

One expects that if the primary agent receives no support from any other agent, the team rating would be equal to the performer rating. However, as the primary agent receives support, it is reasonable to expect the team rating to improve.

We expect that there will be significant overlap between the capabilities that each agent has for any task. If the capabilities for the secondary agents are the same as for the first, one can simply apply shorthand, such as "same". However, it is important to verify that the secondary agents have these capabilities in the model. In areas where the performer rating is 3 or below, they may lack capabilities to perform the task on their own. Under these circumstances, it is reasonable to assume that the secondary agents will require capabilities beyond those available to the primary agent. Figure 11.2 exemplifies the need for these additional capabilities for Task 7.

11.6.4 Example Application

Figure 11.3 shows an example Interdependence Table for Scenario A2. Scenario A2 is the atypical scenario in which the radiologist disagrees with the intervention agent's recommendation. Although this scenario could begin with the patient arriving for a screening mammogram, which is the first encounter with this patient, the scenario focuses on the intervention selection processes and thus begins with the radiologist providing a positive diagnosis. This allows us to narrow this scenario to the unique elements for intervention selection. This contained module of this scenario ends after the radiologist forms an agreement regarding the intervention with the patient and documents this agreed intervention.

As shown in Figure 11.3, we provide a designation for the scenario and a sequential list of the tasks that we expect the HAT to perform in the left two columns of the table within the Scenario Description section of the Interdependence Table. We then list the primary agent in the Primary Agent section of the table. As shown, the Intervention

Scenario Description		Primary Agent			Secondary Agent(s)		OPD Section	
Scenario	**Tasks**	**Capability**	**Agent Name**	**Performer Rating**	**Capability**	**Agent Name**	**Team Rating**	**Observability, Predictability, Directability**
A2	Receive Positive Diagnosis	Ability to Access Patient Records	Intervention Recommender	1	Ability to Access Patient Records	Radiologist	1	Intervention agent should observe radiologist accessing patient record and begin activities
	Retrieve Patient Record	Ability to Access Patient Records	Intervention Recommender	1	Ability to Access Patient Records	Radiologist	1	Radiologist should observe the intervention agent retrieve correct patient record and begin process
	Determine Intervention Options	Knowledge of Intervention Options, Knowledge of Outcomes of Interventions	Intervention Recommender	2	Knowledge of Intervention Options, Knowledge of Outcomes of Interventions	Radiologist	1	Radiologist should be able to observe all intervention options considered
	Evaluate (Score) Intervention Options	Knowledge of Intervention Options, Knowledge of Outcomes of Interventions, Knowledge of Financial Cost, Knowledge of Invasiveness, Knowledge of Unintended Consequences of Interventions, Ability to Synthesize Indicators	Intervention Recommender	2	Knowledge of Intervention Options, Knowledge of Outcomes of Interventions, Knowledge of Financial Cost, Knowledge of Invasiveness, Knowledge of Unintended Consequences of Interventions	Radiologist	1	Radiologist must be able to observe scores for all options with rationale (e.g., weights on factors for patient record)

FIGURE 11.3 Interdependence table for example scenario.

Scenario Description		Primary Agent			Secondary Agent(s)		OPD Section	
Scenario	Tasks	Capability	Agent Name	Performer Rating	Capability	Agent Name	Team Rating	Observability, Predictability, Directability
	Select a Recommended Intervention	Knowledge of Intervention Options, Knowledge of Outcomes of Interventions, Knowledge of Financial Cost, Knowledge of Invasiveness, Knowledge of Unintended Consequences of Interventions	Intervention Recommender	2	Knowledge of Intervention Options, Knowledge of Outcomes of Interventions, Knowledge of Financial Cost, Knowledge of Invasiveness, Knowledge of Unintended Consequences of Interventions	Radiologist	1	Intervention agent shall be able to present the recommended intervention
	Select Alternative Intervention Options	Knowledge of Intervention Options, Knowledge of Outcomes of Interventions, Knowledge of Financial Cost, Knowledge of Invasiveness, Knowledge of Unintended Consequences of Interventions	Intervention Recommender	2	Knowledge of Intervention Options, Knowledge of Outcomes of Interventions, Knowledge of Financial Cost, Knowledge of Invasiveness, Knowledge of Unintended Consequences of Interventions	Radiologist	1	Radiologist should be able to observe rationale and differentiating factors for all alternate options
	Populate Patient Record with Intervention Information	Ability to Populate Patient Records	Intervention Recommender	1	Ability to Populate Patient Records	Radiologist	1	Radiologist should be able to observe that patient record has been populated

Radiologist Disagrees with Intervention Recommender, Substitutes Different Options (True Cause: Human Error or AI agent Error)	Knowledge of Intervention Options, Knowledge of Outcomes of Interventions, Knowledge of Financial Cost, Knowledge of Invasiveness, Knowledge of Unintended Consequences of Interventions	Radiologist	1		1	Radiologist directs adoption of alternative option
Radiologist Annotates Rationale	Ability to Populate Patient Records	Radiologist	2		2	
Discuss Intervention Options and Expected Outcomes with Patient	**Ability to Communicate with Patient**	Radiologist	1		1	
Decide on an Intervention	**Ability to Communicate with Patient**	Radiologist	1		1	
Document Selected Intervention	Ability to Populate Patient Records	Radiologist	1		1	Intervention agent should be able to observe the final selection and any rationale for change

FIGURE 11.3 (Continued)

Agent is responsible for the first seven tasks while we show the Radiologist, who is accountable, as the secondary agent for these tasks. In the revised system, the Radiologist is accountable for the remaining tasks but receives no support from other agents in the HAT. We list the capabilities for the primary agent for each task. This analysis indicates that our original design did not capture the Radiologist's *Ability to Communicate with Patient*. Therefore, we show this capability in bold and add this capability to the required capabilities for the Radiologist. We then capture performer ratings. As shown, we expect that the agents will be able to perform the tasks with ratings of 1 or 2, indicating that reliability will not be near 1 for some conditions.

We then capture the secondary agent. In this case, our HAT does not include other supporting agents, so only the Radiologist appears as the Secondary Agent. The capabilities required of the Radiologist in this role are the same capabilities required from the Intervention Agent, even though we expect that the method to complete these capabilities will be materially different. For example, we might expect that the Intervention Agent might form an objective function in which it assesses the risk and reward of each intervention using multiple criteria and computes an expected value score based on the relative importance of these criteria. We do not expect that the Radiologist will perform a similar computation but weigh risks and rewards qualitatively before arriving at a recommendation.

Finally, in the OPD Section, we form the team rating. In this case, we assess most ratings to be 1, indicating that we expect the team will be capable of successfully completing these tasks if the right communication protocols are established with the desired capabilities. The one outlier in this assessment is the Radiologist's annotation of the rationale as we expect that the Radiologist may not enter meaningful information when experiencing time pressure, and there is no agent capable of providing backup in this example. As shown, the OPDs capture information that the Radiologist needs to know from the system or the Intervention Agent, as well as information the Intervention Agent needs to receive from the Radiologist for the system to function appropriately. When forming these OPDs, the team may find themselves discussing the details of the interaction. For example, when discussing the task "Evaluate (Score) Intervention Options", a significant discussion occurred as to whether it was adequate for the Radiologist to simply observe the rationale of the Intervention Agent or whether it would be useful to direct the agent to explore additional attributes or weights. In the end, we decided that given the time pressure, we expect the Radiologist to experience, permitting them to adjust these values to better understand that the trade-space was a feature that we would not expect them to use. Therefore, we did not include this directive information. In our example, the Intervention Agent is primarily a decision support agent, and therefore it does not observe, predict, or direct the human to a measurable amount. However, it does need the ability to observe the Radiologist's final selection which the designers might use in future training of the underlying AI within the agent.

11.7 CHAPTER SUMMARY

This chapter has provided a discussion of coactive design and the importance of designing teaming behaviors within human–AI teams, including the need to design interaction by enabling observability, predictability, and directability. Within AOSM, we use a modified version of coactive design to determine these teaming behaviors using a modified Interdependence Table. We discussed the importance of applying this process during both typical and atypical scenarios, as well as the importance of appropriate observability, predictability, and directability to support the humans in the HAT as they build appropriate levels of trust and suspicion. Finally, we discussed and illustrated the application of Interdependence Tables as a component of AOSM.

11.8 QUESTIONS AND DISCUSSION TOPICS

1. Imagine that an adversary used a cyberattack to modify the weights assigned to the value rating for the Intervention Agent in the example. How could we design the system to permit the Radiologist to gain this information? Do existing OPD elements support this assessment? If not, what additional ones do we need to add to the system?
2. What must agents be able to do to fully support predictability?
3. Describe a concrete recommendation from this chapter for improving human trust of AI agents.
4. In your own words describe why it is important to include both typical and atypical scenarios during coactive design as applied in AOSM.
5. Compare and differentiate coactive design from interdependence analysis or interdependence tables.

11.9 PROJECT QUESTIONS AND ASSIGNMENTS

1. For your project, select a pair of typical scenarios and develop a list of tasks for each of these scenarios, which are appropriate for use in an Interdependence Table.
2. Select a pair of atypical scenarios which are variances on the selected typical scenarios from the previous step. Develop a list of tasks for each of these scenarios, which are appropriate for use in an Interdependence Table.
3. For one of your typical scenarios, develop a full Interdependence Table. Discuss any learnings from the construction of this table.

REFERENCES

Annett, J., & Duncan, K. (1967). Task analysis and training design. *Journal of Occupational Psychology*, *41*(211–221).

Buede, D. M., & Miller, W. D. (2016). *The engineering design of systems: Models and methods*. John Wiley & Sons, Inc.

Gay, C., Horowitz, B., Elshaw, J., Bobko, P., & Kim, I. (2017). Operator suspicion and decision responses to cyber-attacks on unmanned ground vehicle systems. *Proceedings of the Human Factors and Ergonomics Society*, *2017-October*, 226–230. https://doi.org/10.1177/1541931213601540

Hoff, K. A., & Bashir, M. (2015). Trust in automation: Integrating empirical evidence on factors that influence trust. *Human Factors: The Journal of the Human Factors and Ergonomics Society*, *57*(3), 407–434. https://doi.org/10.1177/0018720814547570

Johnson, M., & Bradshaw, J. M. (2021). The role of interdependence in trust. In *Trust in human-robot interaction*. In C. S. Nam & J. B. Lyons (Eds.), *Trust in human-robot interaction* (pp. 379–403). Academic Press. https://doi.org/10.1016/b978-0-12-819472-0.00016-2

Johnson, M., Bradshaw, J. M., & Feltovich, P. J. (2018). Tomorrow's human–machine design tools: From levels of automation to interdependencies. *Journal of Cognitive Engineering and Decision Making*, *12*(1), 77–82. https://doi.org/10.1177/1555343417736462

Johnson, M., Bradshaw, J. M., Feltovich, P., Jonker, C. M., Van Riemsdijk, M. B., & Sierhuis, M. (2012). Autonomy and interdependence in human-agent-robot teams. *IEEE Intelligent Systems*, *27*(2), 43–51. https://doi.org/10.1109/MIS.2012.1

Johnson, M., Bradshaw, J. M., Feltovich, P. J., Jonker, C. M., Van Riemsdijk, M. B., & Sierhuis, M. (2014). Coactive design: Designing support for interdependence in joint activity. *Journal of Human-Robot Interaction*, *3*(1), 43. https://doi.org/10.5898/JHRI.3.1.Johnson

Klein, G., Woods, D. D., Bradshaw, J. M., Hoffman, R. R., & Feltovich, P. J. (2004). Ten challenges for making automation a "team player" in joint human-agent activity. *IEEE Intelligent Systems*, *19*(6), 91–95. https://doi.org/10.1109/MIS.2004.74

Lapso, J. A., Peterson, G. L., & Miller, M. E. (2023, February). A hybrid cognitive model for machine agents. *Cognitive Systems Research*, *81*, 1–10. https://doi.org/10.1016/j.cogsys.2023.02.007

McGuirl, J. M., & Sarter, N. B. (2006). Supporting trust calibration and the effective use of decision aids by presenting dynamic system confidence information. *Human Factors*, *48*(4), 656–665. https://doi.org/10.1518/001872006779166334

Miller, C. A. (2017). The firefox fallacy: Why intent should be an explicit part of the external world in human automation interaction. In P. J. Smith & R. R. Hoffman (Eds.), *Cognitive systems engineering: The future for a changing world* (pp. 269–294). CRC Press. https://doi.org/10.1201/9781315572529

Miller, C. A., & Parasuraman, R. (2007). Designing for flexible interaction between humans and automation: Delegation interfaces for supervisory control. *Human Factors*, *49*(1), 57–75. https://doi.org/10.1518/001872007779598037

Sargent, R. G. (2013). Verification and validation of simulation models. *Journal of Simulation*, *7*(1), 12–24. https://doi.org/10.1057/jos.2012.20

Schneider, M. F., Miller, M. E., & McGuirl, J. M. (2023). Assessing quality goal rankings as a method for communicating operator intent. *Journal of Cognitive Engineering and Decision Making*, *17*(1), 26–48. https://doi.org/10.1177/15553434221131665

Stanton, N. A., Salmon, P. M., Walker, G. H., Baber, C., & Jenkins, D. P. (2010). *Human factors methods*. Ashgate Publishing.

Vignati, M., Johnson, M., Bunch, L., Carff, J., & Duran, D. (2022, December). Interdependence design principles in practice. *Frontiers in Physics*, *10*, 1–18. https://doi.org/10.3389/fphy.2022.969544

12 Human Capabilities and Capacity

12.1 DIFFERENTIATING HUMAN AND ARTIFICIAL INTELLIGENCE

Presently, deep learning algorithms are state of the art in AI and in some ways mimic the human brain as these models involve complex arrangements of artificial neurons. AI designers weave layers of these artificial neurons together, with connections that involve statistical parameters that are tuned through training. Advanced deep learning networks, specifically large language models, might consist of arrays of up to 540 billion artificial neurons that are organized in around 100 interconnected layers (Raiaan et al., 2024). This is a truly impressive number, as compared to early neural network models.

However, the similarity in form between these AI algorithms and the human brain is irrelevant, given the difference in scale as the human brain is a network of approximately 80 billion neurons and an almost equal number of glial cells (Azevedo et al., 2009). These neurons are connected to subsequent layers in a feedforward network as modern-day deep learning networks but contain significant feedback loops as well. These cells are highly interconnected with large cortical pyramidal cells having as many as or more than 10,000 excitatory inputs. Some cells, such as Purkinje cells, have as many as 100,000 inputs (Klopf, 1972). Further differentiating is the complexity of this organ which is awash in a liquid bath that significantly affects its behavior. Nutrients and oxygen may be plentiful at one time and sparse at another, allowing the neurons to function well when sufficient nutrients and oxygen are present but perform with degraded performance as high levels of metabolism within these cells deplete these materials. Further, the brain generates and modifies hormones within our body. The circulatory system then circulates these hormones into the liquid bath surrounding the neurons, having a significant impact on the performance of these neurons. These hormones provide control of the neurons in the brain and body in near real time at a more global level. While we usually associate activities of certain centers in the brain with motor movements, activity in the brain directly drives visceral, vascular, and glandular activity as well. Each of these can have significant effects on our performance and motivation. An example is the release of adrenaline to increase the flow of oxygen and nutrients to the extremities, which permits temporary boosts in strength, endurance, and speed. A second example is the production of melatonin, which reduces metabolic activity to prepare the body to sleep. As learning affects the strength of connections within the brain, the variation in nutrients and hormones further regulates the function of the human brain. Thus, despite the amazing advances in AI, the complexity of this human organ far exceeds the complexity of modern-day AI, providing general intelligence that is only present in the artificial intelligence imagined in science fiction.

DOI: 10.1201/9781003428183-12

The complexity of the brain provides the ability to perform tasks that are uniquely human, including understanding what others know, to imagine ourselves acting as another (permitting us to gain insight to their physical and emotional states), to produce large numbers of phonemes permitting complex speech, and to quickly assimilate and mimic the actions of others (Calcagno & Fuentes, 2012). When we couple these abilities with the fact that the human brain requires longer to fully develop than the brain of any other species, providing a longer time for children to learn from adults, these abilities not only permit enhanced learning but also allow us to readily pass knowledge across generations. Researchers have argued that it is our cognitive flexibility that provides humans with a unique advantage. This flexibility allows us to store vast amounts of knowledge; adapt to and induce change; and to craft patterns of behavior, concepts, and cultures (Lieberman, 2013). As a result, we have remade our existence, allowing us to live in a world that is unlike the world that existed hundreds or even thousands of years ago. On the other hand, the lives of all other animals in the wild on our planet are much the same as they have been for the last thousand years.

It is also important that unlike computers, each human is unique from every other individual. This difference is not preordained by an individual's genetic composition, while differences exist at this layer. Our brains grow and store information differently based upon our experiences throughout life. These differences between individuals not only exist as simple behavioral differences, but also these differences exist physically in the form of the different neural connections that our brain forms and prunes. Therefore, humans are able to innovate, making us unpredictable not only as a species but unpredictable at the individual level as well. Again, this ability to innovate and adapt is uniquely human, and the uniqueness among individuals enables different innovations as unique individuals harness their knowledge and abilities to inspire others.

In his book, Brigadier General Y.S. relates a fable in which a pair of scholars agree to learn from one another (Y.S, 2021). After the death of one of them, their smartest student assumes the role of the deceased scholar. After a time, the remaining living scholar expresses the desire to die, explaining to the student that "for every new idea that I had and then discussed with [your mentor], he gave me 24 reasons that I was wrong, and then we had to think together to improve our thoughts and create new knowledge together. In your case, I give you an idea and you give me twenty-four reasons that I am correct, there is no new knowledge". This fable illustrates that diversity of thought and conflict which can arise from this diversity is important to learning and the growth of knowledge.

Information from the world not only modifies our knowledge, but it affects humans from an emotional perspective as well. It is well known that increasing the intensity of stimulation, resulting in an emotionally charged stimulation, increases the speed with which our brains encode memories making these memories more vivid. This mechanism, which permits humans to store vivid memories of important, life-changing, or novel situations, is an important mechanism for filtering and reinforcing impactful events from typical, everyday events. This mechanism is absent in all artificial intelligence systems, forcing them to filter large amounts of information for storage. Researchers have argued that a similar mechanism, wherein individual areas of the brain are excited more than others, is responsible for driving the "spotlight" of human attention. The brain balances a desire for new stimulation with a desire to maintain homeostasis, resulting in curiosity modulated by a degree of caution to maintain survival (Klopf, 1972). The drive to experience novel situations, which we term curiosity, facilitates learning and adaptation of the brain to additional stimuli. The human brain is constructed to learn stimuli associated with novel events, encode these events within the complex structure of neurons, and to do so in a way which is supportive of social connections in the presence of emotion.

While we might use terms such as emotion, curiosity, and motivation to describe human properties, as engineers, software developers, and designers, too often we ignore these properties during system design. In fact, the design of our systems will affect a person's level of curiosity, motivation, and emotional response. Leveraging design to influence these items has propelled the use of social media, gaming, and other consumer-facing software but is less commonly discussed during the construction of commercial software to be deployed in work environments, instead focusing on aiding task work (Eyal & Hoover, 2014). This is unfortunate as it has been suggested that the passion of highly motivated individuals coupled with curiosity is more important to innovation than intelligence (Friedman, 2006). Thus, if we can use AI to free individuals to perform the work activities that they are passionate and curious about, while removing the boring and tedious work activities, we might better harness unique human contributions (Bates, 2018). Therefore, as we examine the division of work and the Interdependence Tables, it is useful to consider the question of whether we have designed the system to encourage human curiosity and innovation where useful or is the system designed to consume the human's attention without encouraging novel thought and unique contributions. For example, in the Coactive Design Table for Scenario A2, the sequencing of the intervention decision can affect the thoughts of the Radiologist. If the Innovation Recommender provides the Radiologist recommendations without providing stimuli for them to consider their response, then they are likely to review and accept the recommendation much of the time. If, instead, the Radiologist could explore the solution space and the medical evidence for similar cases in an efficient fashion, they might find a novel solution that the Intervention Recommender did not identify. The decision to construct the system using one approach or the other will trade the time required for the Radiologist to perform this analysis without access to the Intervention Recommender's recommendation with the potential benefit of deriving potentially better innovations. This can be an important design decision. In this instance, given the current shortage of Radiologists, we may decide that speed is more important in most everyday applications. This may mean, however, that individuals suffering from unique circumstances may not receive the best possible care.

12.2 HUMANS AND ANALOGY

It is important that the human brain does not encode raw sensory information but readily converts this sensory information to concepts and encodes them. For example, we do not walk into a room and remember the exact pattern of cones and rods which fire within our eye in response to seeing a chair. We may not even perceive or be aware of subtle differences between one particular chair and another chair. Often, we walk into a room and see "a chair". It is the concept which we remember rather than the pattern of neurons which fire. Instead, when a similar pattern of neurons activates in the brain, we associate this with chair and then collapse this information into even higher levels of information. We might walk into a room with artfully arranged rows of chairs, and rather than perceiving chairs, we simply say that we have walked into a classroom. We may not be aware of the number of chairs in the room or their general design, let alone the dab of paint on the fourth chair from the right in the third row in the room. We simply integrate these stimuli to provide a useful description of the room, a process often referred to as "chunking".

As we access this memory of "classroom", we instantly have access to substantial amounts of information. We know this is a location where a professor or teacher is likely to be present, where multiple students congregate, and we are aware of the social norms associated with spaces such as this classroom. These concepts are not present in the room. They are concepts we have associated with the category we call "classroom", what Don Norman would refer to as "knowledge in the head" (Norman, 2013). How do we make the association between the classroom we are perceiving and past classrooms we have associated with? Are the chairs all the same color? Are they aligned in precisely the same fashion? Is the chalkboard always aligned precisely with each of the chairs? None of these are true. However, we can generalize from one like set of symbols to another set of symbols. This classification is referred to as analogy-making (Hofstadter, 2001).

The categories of information are themselves learned over time, however. We form some categories simply because we are shown examples and given labels (i.e., named as words). These broad categories we share with others, although we often learn distinct categories based on our experiences.

As humans, we do not always share these categories; in fact, the same categories are not always distinct among individuals; for example, when does a classroom become large enough to become an auditorium? For individuals, this difference may occur at a number not much beyond 20 chairs. For others, this number may be as large as 100 chairs. Further, it is not this single variable which differentiates these spaces. Perhaps if the chairs are more comfortable, the seating organized in a raised or stair-step pattern, an expensive electronic projection device is present, carpet is on the floor, or the podium is located nearer the edge of the projection screen, we might be more likely to classify a room as "classroom" or "auditorium". Thus, each category, whether named or not, is complex and imprecise. Yet, each category is also associated with a dynamic array of associated multi-attribute information.

One can then argue that it is this ability to associate disparate information which permits us to link and generalize concepts to new situations. Thus, we use past experiences to guide behavior in novel situations, not relying on only one such experience but accessing many different prior experiences to anticipate successful behaviors. This ability to create these associations, not just at low levels of abstraction but across distinct levels of abstraction, gives us access to an array of models of behavior (e.g., plans) we might employ in a particular situation. For instance, if we could only employ low levels of abstraction, we might be able to assume that because a chair is present, we should sit. However, at a higher level of abstraction, I might see that chairs are present within a classroom and associate this with the knowledge that sitting in a chair in the front row might encourage the instructor to be more likely to ask me a question. Thus, I arrive at the plan that I should not just sit in a chair, but I should sit in a chair near the back to avoid the effort necessary to answer anticipated questions.

As today's AI technology is incapable of automatically clustering experiences across multiple abstract levels of description, it is unable to recall information in any similar fashion, although this is an area of active research (Forbus et al., 2017). Thus, this ability to generalize information from one situation to another is uniquely a human capability within today's human–AI agent teams. This is not only a critical skill in developing and crafting language but also a critical skill when trying to develop novel solutions to novel problems.

In summary, the human brain allows us to conceptualize and understand information through analogy, permitting us to fill in beliefs about the function of systems by understanding how they are like other well-understood systems. On the other hand, we can extrapolate beyond these models by merging these analogies with other analogies. As an example, we might have a dimension for animals which includes a category for birds and a category for humans. We might also have a dimension which represents distinct types of motion. Thus, we might know that the bird can walk, and the bird can fly. We might then associate these two motions with all other categories of animals. Therefore, we might imagine that a human can walk, and a human can fly. It is this type of analogy-making that we might try to construct in AI algorithms. However, as humans, we can determine that one of these categories does not inherently make sense. That is, humans (without machine aids) cannot fly. This is the knowledge we have at an early age. As humans, we not only perform these analogies, but can also construct models in our heads which permit us to evaluate whether these analogies are likely to be possible or how we might need to modify an element within the analogy (i.e., providing humans with wings) to make the analogy possible. Therefore, analogy is not the only human trait that today's AIs do not possess. Our ability to use mental models to assess and associate attributes provides humans the ability to innovate and rapidly assess whether combinations of attributes "make sense". It is also important that humans are not always consciously considering these mental models. Instead, there is evidence that we are constantly applying these mental models, often unconsciously to aid decision-making.

On the other hand, this ability to use knowledge to constrain the solution space can lead humans to discount solutions that are possible. This is especially true in situations where there are multiple interactions that we might not anticipate or are so complex that we are unable to estimate their combined effect. In these situations, the designer can develop AI agents and interfaces to them to help us investigate alternate solutions. Designers can apply techniques such as genetic algorithms or generative adversarial networks to generate and potentially evaluate alternatives that human operators might discount. These algorithms use stochastic processes to randomly modify design parameters or process steps and then evaluate these solutions using a specified criterion. If it is possible to formulate a value function to evaluate each permutation that these algorithms generate, the algorithms can evaluate large numbers of permutations of system design or process parameters, returning viable solutions that the human can evaluate further. The human can interact with these algorithms, for instance, by modifying the value function or adding restrictions to the parameter space, to obtain novel solutions which expand upon the human's use of analogy, searching for creative solutions. This concept provides an alternate method for the Radiologist to interact with the Intervention Recommender in our example.

12.3 HUMAN ATTENTION AND SITUATION AWARENESS

As we consider factors which limit human capacity, it is important to understand that humans are far from being omniscient. Instead, our perceptual systems are bandwidth limited requiring us to focus our attention on items to fully perceive their attributes. We can easily demonstrate these attributes through the discussion of the human visual system. Our eyes contain a region called the fovea, which allows us to perceive information in a circular patch with a diameter of about 2 degrees of visual angle (about the size of the width of our thumb when held at arm's length). Most of the color sensors in our eye are in the fovea, and the resolution of our eye decreases rapidly as the distance from the center of our fovea increases. To compensate for the lack of visual resolution across our visual field, we move our eyes and our head, permitting us to attend to and sample visual information from our environment. We gather visual information from across our visual field, but to see the detail at any point requires that we move our eyes to place a desired point to view on our fovea. Other sensors in our auditory or somatosensory systems permit us to hear sound or feel sensations from our entire surroundings, permitting us to guide our attention and our visual system to gather information from the direction of sound or touch. Our sensory systems, therefore, function together to permit us to typically gather enough information about our world to permit us to prosper. However, we can only attend to and access limited amounts of visual, auditory, somatosensory, proprioceptive, or other signals at any moment in time. As such, we are only aware of a small portion of our environment at any moment in time.

To overcome limited attention, we construct mental models of our environment and search our environment for changes or unexpected items. Human cognition guides attention through both "top-down" and "bottom-up" processes. In top-down processes, humans gather information to support goal-directed behavior. That is, as we pursue goals, we realize the need for information; and as we understand our environment, our brain directs attention to the areas of the environment that contain this information. In a well-designed operator workstation, the designer assigns visual information to specific locations so the operator can quickly guide their eyes to the location containing the desired information, perceive this information, and then move their eyes to the next piece of information. This not only enhances user performance under most conditions but also permits the operator to ignore some information within the environment for periods of time, which might, unbeknownst to the operator, suddenly change in an important and unexpected way. Thus, while experienced operators gather information more efficiently under standard operating conditions, they can fail to perceive anomalous information.

In bottom-up processes, human cognition processes unexpected cues to drive attention. For example, suddenly flashing a colored light in the human's visual field or sounding an auditory alarm that apparently comes from a specific location will likely draw the human's attention to that location (Wolfe, 2007). Evolutionarily speaking, this is an important feature of human perception which, as Marr indicates, is important to permit us to detect a predator within a natural environment (Marr, 1982).

The limited bandwidth for attention has significant implications for human understanding of the environment and the state of systems we interact with. In human interface design literature, this awareness is typically referred to as *Situation Awareness (SA)* (Endsley, 1995). In her landmark paper, Endsley discusses SA as perception of information in the current situation, comprehension current situation, and projection of the future status. Robust decision-making requires the human operator to maintain SA. It is important to realize that there are many potential barriers to obtaining and maintaining good SA, as illustrated in Figure 12.1. As shown in this figure, humans may not be able to perceive the real world in many of these systems. Elements of the system may not enable the user to access important information from the systems or the real-world environment, the system may not provide all available information available to the user, and, finally, the user may not perceive all displayed due to limited perceptual bandwidth (Endsley & Jones, 2012). Additionally, converting the information we perceive to understand our situation and project the future requires cognitive resources. However, limitations in cognitive resources can impede these processes in the presence of time pressure. Furthermore, the individual must have knowledge of the system and environment to understand the state and implications for the future. The designer can support or impede SA based on the decisions that are made during the design and construction of agents and system interfaces.

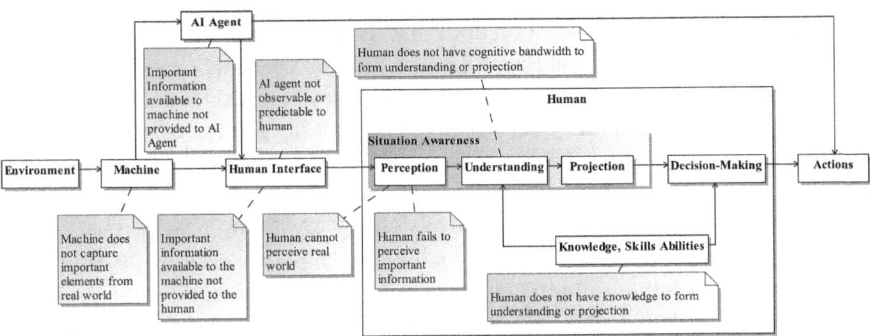

FIGURE 12.1 An illustration of the SA process with labels indicating potential barriers to human SA when interacting in a human–AI agent team.

You may recognize the association between SA and the OPDs we discussed in the last chapter. The ability of the agents to provide observability is aiding the perception of the state of the team while providing predictability is supporting the human's projection of the state of the system. The human–AI agent literature also refers to *team SA* when discussing similar knowledge (Gorman et al., 2006). This term is also related to a similar concept, *common ground,* in the communications and cognitive systems engineering literature (Clark & Brenan, 1991).

Thus, as we review the Interdependence Tables, it is important to consider areas where we might filter or alter information, asking whether we are designing these alterations to remove information that is necessary for the human(s) to maintain SA. The formation of SA in time-constrained environments can also be hindered by displaying too much data to the operator that is not useful to support decision-making or that has to be altered in some way to support decision-making, resulting in data overload (Woods et al., 2002). Therefore, we must also ask whether we present the data in a way that the human has the information they need to support decision processes.

12.4 VIGILANCE

We have mentioned vigilance many times throughout the text. *Vigilance* refers to a person's ability to maintain sustained attention when asked to detect temporary and infrequent signals, often in the presence of distractors, over prolonged periods of time (Warm et al., 2008). A problem, referred to as the *vigilance decrement*, often occurs when work requires humans to monitor the performance of an automated system without performing active tasks. In these tasks, the ability for a human to detect a signal drops from a value in excess of 90% to values on the order of 60–70% within a matter of 20 to 30 minutes (Jeroski et al., 2014; Kamrud et al., 2021). The reason for this decrement in performance remains a topic of discussion. One hypothesis is that the individuals experience boredom due to lack of stimulation, which reduces blood flow to parts of the brain, resulting in reduced oxygen and nutrient delivery and loss of mental acuity (Pattyn et al., 2008). Others have argued that at least in certain tasks, blood flow to the brain does not decrease, but oxygen is depleted due to high cognitive workload (Funke, 2009).

While the reason for the vigilance decrement is unclear, the vigilance decrement is less likely to occur if the human is actively engaged with the system. That is, developers should design the system so that the human interacts with it by providing input requiring physical motion, routinely performing important cognitive work, and participating in tasks which vary over time. It is also unreasonable to expect humans to rapidly detect a failure of the automation without interaction. The designer can help to avoid failed handoffs by providing the operator insight into the operation of the automation so that they are aware of the state of the system and the AI with respect to its expected operating envelope (Billings, 1996). This information permits the operator to anticipate or predict potential failures and the potential need to override or assume control of the automation before the operator must assume control. Another approach to enabling handoffs is to embed reliable indicators of the AI's reliability or certainty in the system (McGuirl & Sarter, 2006). From a design perspective, review your Interdependence Tables for areas where humans are only monitoring the performance of the AI. Consider whether there is a better design solution that enables the operator to be more involved or permits the operator to better understand the state of the AI.

12.5 WORKLOAD

Workload is the amount of effort that results when a human performs physical or cognitive tasks. Perceived workload is related to the perceived effort required by a user to respond to a specific task load. Although we can view physical tasks performed by a human to ascertain their physical workload, cognitive or mental workload is typically not observable but is generally believed to be inversely related to available cognitive capacity (Szalma, 2009). We are generally interested in workload as it is thought to follow a similar relationship as the relationship between arousal and human performance, as explained by the Hebb–Yerkes–Dodson Law (Hebb, 1955; Teigen, 1994). According to this law, human performance is related to workload according to an inverted-U function, which qualitatively might appear as shown in Figure 12.2. As illustrated by this function, human performance is low when the human is not actively engaged and experiencing a minimum level of workload. Performance then increases as workload increases until a point where human capacity limits human *performance*, that is, how well the human accomplishes a task, then decreases with further increases in workload. Therefore, we would ideally design systems to impose a task load on the human to provide a level of workload that is near the peak in this function. This goal was a primary motivator for adaptive automation systems, which attempted to dynamically allocate tasks in time-sensitive environments between humans and AI agents, depending upon an assessment of human workload or activity (Goodman et al., 2016; Rusnock & Geiger, 2014).

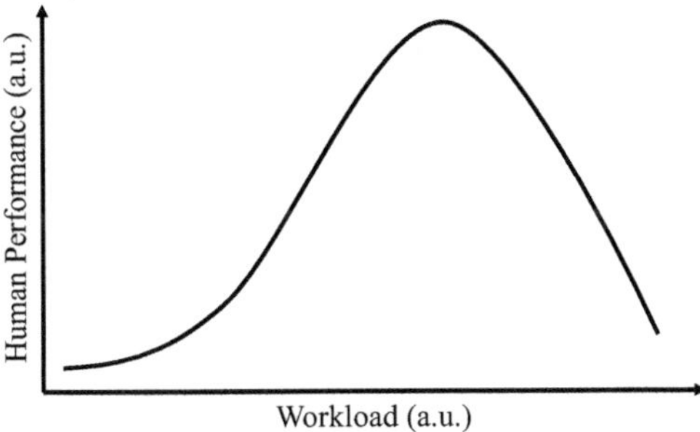

FIGURE 12.2 Illustration of representative relationship between workload and performance based on the Hebb–Yerkes–Dodson Law.

Unfortunately, real-time assessment or prediction of human workload can be difficult. From a physical standpoint, humans cannot control more than two controls with their hands at a time or look in more than one location at a time. Thus, any system which would require more physical resources than we have available at any moment of time would certainly exceed our capacity and, by definition, result in excessive workload. Similar constraints exist for most of our perceptual systems. For example, although we can switch our attention between different conversations in a crowded room, we cannot simultaneously process two different conversations at the same time. We can simultaneously receive stimuli from different perceptual systems (i.e., auditory, and visual) at the same time and process these signals; however, our perceptual systems may be less sensitive as compared to times where we receive only a single signal. Thus, there is less conflict between signals' input to two perceptual systems than between signals' input from a single perceptual system. Finally, we can encode and respond to signals of diverse types, processing linguistic, symbolic, and subconscious information simultaneously, but we might not be able to process two linguistic or two symbolic signals simultaneously. Figure 12.3, which shows Wickens' Multiple Resource Theory model graphically, depicts the idea that we can process some types of information simultaneously but not others. This model assumes that multiple signals can be processed in different boxes, but only one signal can be processed in any box at one time without exceeding human capacity.

While Wickens' Multiple Resource Theory helps to explain the conflict that can occur within the human cognitive system, potentially resulting in excessive workload, this graphic does not explicitly show the effect of experience. When individuals are new to a task, they employ deliberate reasoning to perform the task using conscious processes, shown as the symbolic and linguistic reasoning blocks in Figure 12.3. However, as we gain experience and skill, our brain forms new connections, permitting these activities to be performed more automatically such that over time we perform these activities subconsciously, usually reducing the time required to complete processes than is required for conscious processes. As it is difficult to understand which tasks the human will perform consciously versus subconsciously, it is difficult to estimate the impact of these tasks on human workload. Thus, when

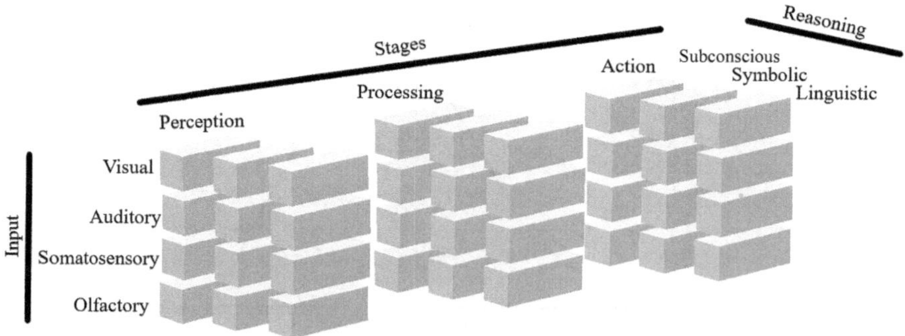

FIGURE 12.3 Illustration of independent human information processing channels based on Wicken's Multiple Resource Theory.

designing human–AI teams, it is important not only to understand the task load we are imposing on individuals to assess whether this will result in excessive workload but also to understand the expertise of our users. If they are novices, designers should develop AI agents to provide additional support to permit the humans to successfully complete certain activities. Experts may require overall less support, with that support being specifically amenable to unconscious processing, which often means displaying information in a way that facilitates pattern recognition.

During this discussion, we have discussed workload as though it is a quantity that changes moment to moment at the individual level. This viewpoint is valid and often applied in the human factors literature. However, we can also think about and discuss the balance of workload across team members. If workload is not well-balanced across human team members, then some individuals will be overloaded, experience long-term stress, and may become burnt out while other members of the team may suffer from boredom. This aspect of workload is also important to consider as we design the distribution of work in our team.

With each of the human advantages and capacity limitations in mind, it is useful to revisit the RASCI table from earlier. Regarding workload, this table shows responsibility allocation across agents. It is reasonable to ask if individual human team members are experiencing similar levels of workload such that the system design balances workload across the team members. As perhaps you noticed before, the Radiologist in this table is accountable for a larger number of the responsibilities. Therefore, we might expect the Radiologist will be overworked and experience fatigue. This table does not indicate the number of Radiologists on this team, and one could theoretically increase the number of Radiologists as compared to other humans on this team, but as this is a specialized agent, this is unlikely. Another approach would be to examine the items that the Radiologist is accountable for and ask whether workload balance could be achieved by adding another agent to the team. Figure 12.4 shows an updated RACSI table, which now includes a Data Scientist. Accountability for the responsibilities associated with the Risk Assessment Agent which implements the new functions associated with our "To Be" HAT has been allocated to the Data Scientist so they are responsible for these items. However, we understand that the Data Scientist may require expertise from the Radiologist to support their judgments; thus, we now assess the Radiologist as Consulting for these responsibilities. We can also examine our Interdependence Tables to ask whether humans are overly tasked and are likely to experience high workload to attempt to redesign the work. We acknowledge, however, that the exact interface implementation can have a significant effect on workload, and, therefore, it is premature to provide a thorough workload analysis.

12.6 FATIGUE

Fatigue refers to a physiological state of reduced mental or physical capacity, often resulting from sleep deprivation, circadian rhythm disruption, or long periods of high task demand (Wilson et al., 2021). A designer should expect that individuals experiencing fatigue will make poorer decisions, be less able to process socio-emotional

Responsibilities	Diagnosis Recommender	Dissemination Agent	Intervention Recommender	Learning Database Maintainer	Literature Agent	Radiologist	Radiology Technician	Data Scientist	Risk Assessment Agent
Alert to Anomaly	S			I		A	S	A	S
Calibrate Sensitivity of Detection Criteria	R		I			A	S		
Coordinate with PCM to Implement High-Risk Patient Procedures		R				A			S
Develop Shared Understanding of Mammogram Quality						A	R		
Differentiate Normal from Abnormal	R					A			S
Establish High-Risk Patient Procedures		S			S	A			
Evaluate for Likely Errors	S					A	R		
Identify Abnormal Patterns	S					A	R		
Identify High-Risk Patients	I	I	I	I		C		A	R
Identify Useful Interventions			R		S	A			
Implement Identification Criteria into Screening Systems						A	S		
Interpret Health History for Risk Level				S	S	C		A	R
Interpret Physical Indicators	R					A	R		
Maintain State-of-Art Knowledge	I		I	I	R	A	S		I
Obtain Complete Mammograph History	R		R	S		C	R	A	R
Obtain Patient Health History	R		R	S		C	I	A	R
Obtain Quality Biopsy	I		I	I		A			
Obtain Quality Images	I		I	I		C	A		
Perceive Anomaly Alert	R		R	I		A			
Perform Intervention						A	R		
Perform Risk Assessment	I		I	S	S	C		A	R
Provide Diagnosis	R		I	I		A			
Recommend Further Diagnostics	R		R			A			
Train for Likely Errors						A	R		
Update Anomaly Detection Criteria	I			R	S	C		A	I
Update Electronic Patient Record	R		R			C	R	A	R
Update Intervention Evaluation			R		S	A			
Update Intervention Implementation						A	S		
Update Risk Assessment	I		I	S	S	C		A	R

FIGURE 12.4 Revised RASCI Table illustrating reallocated responsibilities after including the Data Scientist agent.

cues from others, be more likely to commit errors, and otherwise suffer from decreased performance. While vigilance decrements usually appear within minutes of beginning a task, task-induced fatigue occurs after longer periods of performance.

We can expect fatigue to occur in environments which involve routine schedule changes, for example, environments where workers rotate between day and night shifts or work across time zones as is typical for large transport aircraft. These environments disrupt circadian rhythms, making it difficult for individuals to sleep as their "internal clock" needs to reset to changes in their sleep/wake schedule. Exposure to light, particularly blue light, or sunlight, during the sleep hours exacerbate our inability to sleep. We can expect fatigue if workers perform long shifts or if we require workers to maintain elevated levels of performance over extended periods without rest. In these environments, it is particularly important to provide backup to aid job performance, provide active cognitive aids that monitor for potential errors, engage the operator in error checking, and provide emotional support to help avoid conflict with coworkers. When designing systems to support operators in environments that are likely to induce fatigue, designers should consider the incorporation of agents which actively engage with humans to provide appropriate aids. Alternately, the designers can incorporate AI agents which can purposefully monitor the human to determine fatigued or drowsy states and enable enhanced error checking and support for operators who are determined to be in such a degraded state.

12.7 RESPONSE TIME AND ACCOUNTABILITY

While AI agents may be capable of responding to certain stimuli within milliseconds, humans require time to acquire and process information. For example, simply making an eye movement may require up to 100 ms, making the cognitive selection for where to turn a rotary dial may require on the order of 700 ms, physically turning the rotary dial may require 730 ms, and somewhat more complex tasks, such as determining your own location on a map, may require around 7 s (North Atlantic Treaty Organisation & Research and Technology Organisation, 2009). Even more complex tasks, such as diagnosing a fault, could require several minutes to tens of minutes for a human to complete.

These time limitations impose constraints on the human's ability to perform with an agent. Chapter 4 discussed the fact that the agents could simply respond at a rate that could disrupt human processing, causing them to simply react to the actions of the AI agent, rather than constructively contribute to the work of the team, and this is one implication of having AI agents that can respond at rates that exceed human processing and response times. This also has implications for human accountability. As we indicated earlier, typically, the society will legally and morally hold humans accountable for the performance of the AI agents they interact with or supervise. However, because the AI agents can take actions at a rate that exceeds the human's ability to understand and override these actions, a conflict can occur. Therefore, designers should ask whether an action an AI agent is responsible for poses significant risk to the lives and property of others. If so, it is reasonable to consider whether this risk is so high that the AI agent should not be permitted to make the decision without human intervention. If human intervention is required, the system must be designed to provide the accountable human proper time to analyze and agree or disagree with the agent's assessment. Alternately, the designer should consider methods to place the human in this control loop in a way that they have the capacity to intervene should the AI agent select a course of action that imposes more risk than the accountable person is willing to accept. However, the designer should be cautious about placing the human in the control loop without active engagement, as vigilance errors can occur, degrading the human's ability to respond, and leading to undesirable outcomes (Hawley et al., 2005). Once again, we can examine the Interdependence Tables for tasks that have an AI agent as the primary or responsible agent that also imposes the potential to harm others and ask whether the OPDs are designed such that the accountable human is truly able to understand decisions, intervene and override the AI agent should it intend to take an undesirable action.

12.8 CHAPTER SUMMARY

We intend this chapter to help you recognize the advantages of humans in human–AI teams, as well as capacity limitations that the AI can assist with. The chapter began by discussing attributes of human cognition, which further differentiate the general intelligence provided through the complex human brain from modern-day AI. Although our technology-centered culture espouses the virtues of AI, it is important to recognize that this technology is not capable of general intelligence. We followed this discussion with a brief discussion of the use of analogy to support human innovation but recognized that designers can develop AI techniques which may further support innovation.

We also discussed human limitations, including bottlenecks in attention which affect our situation awareness, our limited vigilance, workload, and response time, as well as our limited endurance which can lead to a loss of performance due to fatigue. With each of these sections, we discuss the impact of these limitations on our thinking when performing allocation or when considering work using Interdependence Analysis. We intend this knowledge to aid you to further consider aspects of your design that we capture in these modeling artifacts.

12.9 QUESTIONS AND DISCUSSION TOPICS

1. Recalling a time that you experienced excessive workload or fatigue, describe the way that you coped with this event and how an AI agent might have interacted with you to help in your coping strategy.
2. Think of a time that you had to innovate to overcome a situation. How did your knowledge of similar situations either aid or hinder your ability to innovate in this situation?
3. Compare and contrast vigilance decrement and fatigue. Discuss the preconditions for each of these phenomena as well as the effect they have on your ability to perform tasks.

12.10 PROJECT QUESTIONS AND ASSIGNMENTS

For your "To-Be" system, review your RACSI table and at least one of your Interdependence Tables to complete the following:

1. Identify any steps in your Interdependence Table where you expect a human might perform a vigilance task. If one is present, discuss how you might revise your design to increase human activity during this step.
2. In your RACSI table discuss whether you feel the work is well-balanced across the human agents. If not, describe how you might overcome this imbalance.
3. In your Interdependence Table, identify any tasks that you would consider the human responsible for where they may not be able to participate in the decision-making. Discuss whether you believe that you should redesign this task based on the material in this chapter.
4. In your Interdependence Table, identify any task where human innovation might be useful. Discuss whether you believe the implied interaction is appropriate for this task, given the discussion in this chapter, and, if not, discuss how you would change the interaction.

REFERENCES

Azevedo, F. A. C., Carvalho, L. R. B., Grinberg, L. T., Farfel, J. M., Ferretti, R. E. L., Leite, R. E. P., Filho, W. J., Lent, R., & Herculano-Houzel, S. (2009). Equal numbers of neuronal and nonneuronal cells make the human brain an isometrically scaled-up primate brain. *Journal of Comparative Neurology*, *513*(5), 532–541. https://doi.org/10.1002/cne.21974

Bates, A. (2018). *Augmented mind: AI, humans and the superhuman revolution*. Neocortex Ventures LLC.

Billings, C. E. (1996). Human-centered aviation automation: Principles and guidelines. In *NASA technical memorandum* (Issue A-961056). Ames Research Center, National Aeronautics and Space Administration.

Calcagno, J. M., & Fuentes, A. (2012). What makes us human? Answers from evolutionary anthropology. *Evolutionary Anthropology*, *21*(5), 182–194. https://doi.org/10.1002/evan.21328

Clark, H. H., & Brenan, S. E. (1991). Grounding in communication. In L. B. Resnick, J. M. Levine, & S. D. Teasley (Eds.), *Perspectives on socially shared cognition* (pp. 127–149). American Psychological Association. https://doi.org/10.1201/b17574

Endsley, M. R. (1995). Toward a theory of situation awareness in dynamic systems. *Human Factors: The Journal of the Human Factors and Ergonomics Society*, *37*(1), 32–64. https://doi.org/10.1518/001872095779049543

Endsley, M. R., & Jones, D. G. (2012). *Designing for situation awareness: An approach to user-centered design*. CRC Press, Inc.

Eyal, N., & Hoover, R. (2014). *Hooked: How to build habit-forming products*. Penguin.

Forbus, K. D., Ferguson, R. W., Lovett, A., & Gentner, D. (2017). Extending SME to handle large-scale cognitive modeling. *Cognitive Science*, *41*(5), 1152–1201. https://doi.org/10.1111/cogs.12377

Friedman, T. L. (2006). *The world is flat: A brief history of the twenty-first century*. Farrar, Straus and Giroux.

Funke, M. E. (2009). *A comparison of cerebral hemovelocity and blood oxygen saturation levels during vigilance performance*. University of Cincinnati.

Goodman, T. J., Miller, M. E., Rusnock, C. F., & Bindewald, J. M. (2016). Effects of agent timing on the human-agent team. *Cognitive Systems Research*, *46*, 40–51. https://doi.org/10.1016/j.cogsys.2017.02.007

Gorman, J. C., Cooke, N. J., & Winner, J. L. (2006). Measuring team situation awareness in decentralized command and control environments. *Ergonomics*, *49*(12–13), 1312–1325. https://doi.org/10.1080/00140130600612788

Hawley, J. K., Mares, A. L., & Giammanco, C. A. (2005). *The human side of automation: Lessons for air defense command and control* (Tech Report ARL-TR-368). U.S. Army Research Laboratory, Aberdeen Proving Ground, MD. https://apps.dtic.mil/sti/pdfs/ADA431964.pdf

Hebb, D. O. (1955). Drives and the C. N. S. (conceptual nervous system). *Psychological Review*, *62*(4), 243–254. https://doi.org/10.1037/h0041823

Hofstadter, D. (2001). Analogy as the core of cognition. In D. Gentner, K. J. Holyoak, & B. N. Kokinov (Eds.), *The analogical mind: Perspectives from computer science* (pp. 499–538). The MIT Press.

Jeroski, J., Miller, M. E., Langhals, B., & Tripp, L. (2014). Impact of vigilance decrement upon physiology measures. *Proceedings of the 2014 Industrial and Systems Engineering Research Conference*, 1184–1193.

Kamrud, A., Borghetti, B., Kabban, C. S., & Miller, M. (2021). Generalized deep learning EEG models for cross-participant and cross-task detection of the vigilance decrement in sustained attention tasks. *Sensors*, *21*(16). https://doi.org/10.3390/s21165617

Klopf, A. H. (1972). *Brain function and adaptive systems—A heterostatic theory* (AFCRL-72–0164). http://www.dtic.mil/dtic/tr/fulltext/u2/742259.pdf

Lieberman, P. (2013). *The unpredictable species: What makes humans unique*. Princeton University Press.

Marr, D. (1982). *Vision*. W.H. Freeman and Company.

McGuirl, J. M., & Sarter, N. B. (2006). Supporting trust calibration and the effective use of decision aids by presenting dynamic system confidence information. *Human Factors*, *48*(4), 656–665. https://doi.org/10.1518/001872006779166334

Norman, D. (2013). *The design of everyday things* (Revised and expanded ed.). Basic Books.

North Atlantic Treaty Organisation, & Research and Technology Organisation. (2009). Human behavior representation in constructive simulation. In *Final report of task group 128, consolidating the findings of task group 143* (Vol. 323, Issue September). ISBN: 978-92-837-0074-6

Pattyn, N., Neyt, X., Henderickx, D., & Soetens, E. (2008). Psychophysiological investigation of vigilance decrement: Boredom or cognitive fatigue? *Physiology and Behavior*, *93*(1–2), 369–378. https://doi.org/10.1016/j.physbeh.2007.09.016.

Raiaan, M. A. K., Mukta, S. H., Fatema, K., Fahad, N. M., Sakib, S., Mim, M. M. J., Ahmad, J., Ali, M. E., & Azam, S. (2024, February). A review on large language models: Architectures, applications, taxonomies, open issues and challenges. *IEEE Access*, *12*, 26839–26874. https://doi.org/10.1109/ACCESS.2024.3365742

Rusnock, C. F., & Geiger, C. D. (2014). Simulation-based assessment of performance-workload tradeoffs for system design evaluation. *Proceedings of the 2014 Industrial and Systems Engineering Research Conference*, 394–403.

Szalma, J. L. (2009). Individual differences in performance, workload, and stress in sustained attention: Optimism and pessimism. *Personality and Individual Differences*, *47*(5), 444–451. https://doi.org/10.1016/j.paid.2009.04.019

Teigen, K. H. (1994). Yerkes-Dodson: A law for all seasons. *Theory & Psychology*, *4*(4), 525–547. https://doi.org/10.1177/0959354394044004

Warm, J. S., Parasuraman, R., & Matthews, G. (2008). Vigilance requires hard mental work and is stressful. *Human Factors*, *50*(3), 433–441. https://doi.org/10.1518/001872008X312152

Wilson, M. D., Ballard, T., Strickland, L., Amy Boeing, A., Cham, B., Griffin, M. A., & Jorritsma, K. (2021, March). Understanding fatigue in a naval submarine: Applying biomathematical models and workload measurement in an intensive longitudinal design. *Applied Ergonomics*, *94*, 103412. https://doi.org/10.1016/j.apergo.2021.103412

Wolfe, J. (2007). Guided search 4.0: Current progress with a model of visual search. In *Integrated models of cognitive systems* (pp. 99–119). Oxford University Press. https://doi.org/10.1093/acprof:oso/9780195189193.003.0008

Woods, D. D., Patterson, E. S., Roth, E. M., & Christoffersen, K. (2002). Can we ever escape from data overload? A cognitive systems diagnosis. *Cognition, Technology & Work*, *4*, 22–36. https://doi.org/10.1007/s101110200002

Y.S, B. G. (2021). *The human machine team: How to create synergy between human and artificial intelligence that will revolutionize our world*. eBook Pro Publishing.

13 Completing System Specifications in the AOSM

13.1 GENERATE SPECIFICATIONS

The last step in the AOSM process is to generate specifications to aid other team members and developers to understand the work required to instantiate the "To-Be" design. This chapter provides an overview of this step. This step involves the development of requirements or user stories that aid developer's understanding of the system components they are responsible for designing and constructing. As discussed earlier, these specifications can include the capabilities of each agent, including their training requirements, actuators and sensors for each AI agent, and the requirements for the interfaces among the agents.

13.2 PHILOSOPHY OF REQUIREMENTS

As systems become more complicated and complex, larger numbers of developers are required to complete the design. With this increase in the number of developers, there is an increased opportunity for developers to pursue development strategies that are counterproductive with the activities of other team members. To maintain alignment, it is important to create a common vision of a system and align the activities of individual development groups with this vision. *Requirements*, a description of "what" the system will need to be capable of doing including operating conditions and constraints, were intended to capture the vision of the system and convey this vision to large teams of developers, permitting them to align their design activities with the common system vision. Systems engineers develop requirements for individual components, which they expect subteams to design by decomposing system level requirements into specific design requirements for individual development teams, clarifying the role of each team and enabling them to achieve the system vision. In general, requirements support planning of activities, ideally defining the role of each team in the design process.

An example comes from the development of the Geospatial Positioning System (GPS), a system that we all use daily. The technology that provided the basis of this system was developed over decades without a clear vision and development path (O'Brien & Griffin, 2007). The GPS we know today began with the vision that the space-based system shall enable an aircraft to "drop five bombs in the same hole"

DOI: 10.1201/9781003428183-13

anywhere on the globe. Although this is a simple statement, it restricts the decision space for this system in meaningful ways. Specifically, this statement implies:

1) The system shall provide signals to an aircraft anywhere on the globe. This statement required components of the system to cover the earth, implying a satellite constellation.
2) The system shall provide control of the satellites, implying the need for ground control stations.
3) The system shall provide position in three dimensions as the aircraft's altitude as well as the position over the earth affects this capability.
4) The designer shall incorporate a GPS receiver into individual vehicles. The original requirement stated that this receiver shall cost less than $10,000.
5) The system shall provide location data with enough accuracy for a pilot to drop five bombs within the blast radius of a common aircraft bomb.

Therefore, the developers could decompose this simple system level vision statement or stakeholder requirement to provide insight into development for satellite constellations, ground control stations, and the GPS receiver (like the ones in your cellular phone). Thus, we can see that this single statement implies deliverables for the three major components in the final system. Thus, three independent development teams could simultaneously develop the three components. Additionally, the systems engineers working with these development teams could define the interfaces between these major components to create an integrated system. Philosophically, the requirements should align the activities of large organizations with the need of the system stakeholders and improve cooperation across large development organizations. In the development of GPS, this vision of the systems coalesced development efforts after decades of technology demonstrators that showed promise but failed to project a single vision that excited the stakeholder community to maintain a system development effort. This vision not only permitted others to understand and value the envisioned system but also constrained the problem to define a technology solution.

Realistically, requirements-driven approaches suffer from three issues. First, requirements can change over time. Returning to the GPS example, the stakeholders and developers intended this system to support military requirements. However, an accident involving the drowning of a civilian passenger aircraft, which accidentally traveled into restricted airspace, motivated the United States Government to require the Department of Defense to make the system available to the civilian world to avoid similar accidents. Although this is an extreme example, system requirements can change well into the development of a system or after the system deployment. Second, it is difficult to understand the implications of all requirements. For example, a requirement might require the system to complete an activity that requires excessive processing, reducing the responsiveness of the system. In our Intervention Recommender example, if this system required ten minutes to provide a recommendation, it would be useless to a radiologist who must form a recommendation within a couple of minutes. Therefore, if a functional requirement imposed the need for excessively lengthy calculations, then this requirement may not be desirable. Instead, there might be more desirable alternatives that do not comply with this requirement. Often, these types of system interactions or constraints are discovered during the development process, and the stakeholders and requirement writers cannot anticipate these interactions or constraints when forming initial requirements. Third, requirements can be overly restrictive, narrowing the solution space, such that the most desirable solutions do not comply with the stated requirements. As a result, requirements must be viewed within a requirements management process where the development team can question or suggest changes in requirements and where the development team or a requirements change board can assess the effect of these changes on other parts of the system before adopting the change.

As a result of the failure of purely requirements-driven processes, processes such as the Agile software process have become popular. For example, the software development community historically employed a requirements-driven process referred to as the Waterfall Software Design and Development process. However, this process often fails to deliver useful software within schedule and cost limitations. As a result, many organizations have, in recent years, adopted Agile software processes which rely far less on initial requirements. While these software development processes are beneficial in certain contexts, they can also suffer from deficiencies (Hochmüller & Mittermeir, 2008). As a result, many organizations employ hybrid approaches, such as the so-called "Water-Scrum-Fall" or "Wagile" approaches, which often involve requirements-driven techniques that drive the big picture of the system during development, with embedded Agile activities to quickly understand development risks, leading to more robust solutions (Kuhrmann et al., 2017).

The Waterfall Process is a nearly linear process beginning with requirements development, followed by design, development, test, and deployment, with little iteration. This process relies on developing stable requirements early in the software development process and includes little-to-no end user interaction between requirements development and deployment. Additionally, a strict waterfall process assumes a fixed budget, scope, and time. The Agile processes are conceived of as continuously looping processes from design, development, test, to deployment. In Agile, the development team scopes their work during each loop to respond to customer feedback from the previous deployment to deliver value to the customer. The development team establishes the scope to deliver a minimum viable product and then provides improvements to this product in subsequential releases, each based on user feedback from previously deployed versions. The terms "Water-Scrum-Fall" and "Wagile" represent a hybrid development approach. Although different projects can design this hybrid to meet their needs, the "Water-Scrum-Fall" name indicates the adoption of early and later processes from the Waterfall method, with Agile processes being employed for the actual software development phases including multiple releases to test, rather than releases to the field. Therefore, these approaches might incorporate an initial requirements development and analysis process as applied in the Waterfall development process, followed by software development using Agile concepts, followed by robust test and documentation for maintainability. Both the initial requirements and analysis process and documentation processes have traditionally been associated with the Waterfall process. Therefore, development teams pursuing a pure Agile process typically do not perform these process steps.

The "Wagile" techniques appear to share some attributes with traditional User-Centered Design (UCD) methodologies, which often involve upfront requirements analysis, combined with the use of rapid prototypes of increasing fidelity to iterate and refine the user interface requirements. The UCD methodologies often result in a prototype which helps to serve as a portion of the software specification (Endsley, 2015; Norman & Draper, 1986). Thus, large software systems often involve the use of requirement sets with processes to maintain and update these requirements. Regardless of the development method employed, all large software systems will require documented requirements to ensure stakeholder alignment, enforce development accountability, and enable cooperation and integration across development teams.

We intend the design modeling process in this book to develop a clear understanding of the user needs and some of the important logical trade-offs that drive design of the human–AI team. We can use this model to develop specific requirements or user stories for the next stage of development, which we can expect will provide additional insight that will also change our understanding of the system trade-offs. However, this approach specifies the big picture for the system, potentially feeding a "Wagile" process.

User or Human-Centered Design is based upon three principles that are critical to the AOSM process. First, the process focuses on the people that the product impacts. Second, the process focuses on the function of the system, recognizing that optimizing some parts of the workflow by using AI to automate segments of the workload might not optimize the larger process employed by the team or organization. Third, the process focuses on the system and all its components, including humans, AIs, machines, and integrating software to deliver the final desired result.

13.3 WRITING USEFUL REQUIREMENTS

Major development efforts employ large development teams, sometimes spanning multiple organizations or corporations. Requirements should align the goals of these large development teams and often form contractual obligations among these organizations. As such, requirements often include the terms "shall", "will", and "should" which have specific legal implications. Of these terms, only statements including "shall" are legally binding obligations that the legal establishment requires the contracted firm to deliver. Statements of fact use the term "will". The term "should" is not obligatory, and statements using "should" are rightfully given lower priority or even ignored by prime and subcontractors as they seek to achieve budget and timeline requirements (Endsley, 2015).

The AOSM method decomposes requirements as discussed in the previous section. The organization or the stakeholders for the system develop the original requirements. These requirements are used to derive requirements that are specific to the system, that are then used to derive requirements for system components, which are used to derive requirements for the interfaces between these components (Buede & Miller, 2016). Within the systems described in this text, the system will include the HAT and machines applied by the HAT. We consider members of the HAT, for example, AI agents and humans, to be components of the system. The description of the human as being a component of the system or discussing the requirements for humans might seem objectionable as humans are not simply objects that we can design during a product design process. However, we can specify the requirements for humans, including possessing specialized skills or abilities, the training that these humans will undergo, and the prescribed procedures they are to follow, to function effectively in the system.

Within SysML, all requirements have an identification number, a title, and a text-based statement of the requirement. This text statement includes the subject which represents the relevant system or component followed by the word "shall" and the specifying statement. SysML includes various types of requirements, including those shown in Table 13.1, the descriptions of which are defined on the basis of the specification by Lightsey (2001). As the developers work with the system

TABLE 13.1
Extended SysML Requirement Types with Descriptions.

Requirement Type	Description
Business	Specifies specifications and baselines developed prior to design, system cost, team assignments and structure, control mechanisms, required project metrics, standards and guidelines, policies, and procedures, as well as the physical, financial, and human resource allocations to the project.
Usability	Specifies the learnability, efficiency, memorability, error rate, and satisfaction of the system for a specified human
Functional	Defines what the system or component must be able to do. In AOSM, these are often associated with the capabilities assigned to each agent.
Interface	Specifies the functional or physical interfaces between systems or components. Could include the specification of electrical, thermal, data, information, control, procedures, or other interactions. The transfer of information between agents to support decision-making is of particular concern in AOSM.
Performance	Specifies the effectiveness, efficiency, error rates, etc., which are acceptable for the system or a component in the system. In AOSM, these are often associated with quality goals.
Physical	Identifies the required physical characteristics of the system.

stakeholders to create the HAT team using AOSM, the developers should consider business requirements early in our design. However, as we have defined our system, we are specifying functional and performance requirements based on the capabilities we have defined. Further, we define interface requirements based primarily on our OPDs. Finally, the specification may include usability requirements, including specifying learnability, efficiency, memorability, error rate, and satisfaction for specified humans when interacting with our system. Because we are interested in total system performance, including the performance impact of the AI and human agents on the system's performance, it may be more desirable to specify the performance of the human–AI team to include efficiency, effectiveness, or error rate of the entire team rather than specify efficiency and error rate for the human as usability requirements.

When writing requirements, use active voice and avoid the use of negative predicates (e.g., avoid stating the AI agent *shall not*). It is also important to avoid the use of compound predicates or requirements where a single requirement specifies more than one attribute (e.g., the AI agent shall be capable of *reading and writing data* from and to the patient record). Instead, your development team should decompose these requirements further and write two independent requirements (e.g., REQ1: the AI agent shall be capable of *reading data* from the patient record and REQ2: the AI agent shall be capable of *writing data* to the patient record).

We should write succinct requirements with measurable criteria. If you cannot measure or determine precisely whether a system or agent satisfies the requirement, then it is impossible to enforce this requirement upon the design team. Therefore, when writing each requirement, always ask whether you can determine whether the system fulfills this requirement. This includes avoiding verbs such as "optimize", "maximize", and "minimize" as it is impossible for us to know when we have achieved these verbs. Further avoid imprecise or judgmental adjectives, such as "better", "best", "easy", "flexible", "robust", "sufficient", or "user-friendly" as adjectives such as these are typically not measurable. Your team should also specify how your team will assess each requirement (e.g., inspection, demonstration, simulation) within the model. This assessment method is part of the test process.

It is also important to understand the origin of each requirement. This is particularly important when modifying requirements as a change in a requirement may necessitate changes in other requirements that are associated with elements of the system impacted by the changed requirement. Unless you can trace the requirement to its source, you will not be able to quickly determine the associated requirements.

Buede and Miller provide a list of desirable attributes for a set of requirements (Buede & Miller, 2016), including:

- Each item in the list of requirements is necessary to fulfill the purpose of the system.
- The set of requirements together describe all system inputs and responses.
- Prioritized, your team can compare or rank the requirements with one another such that the available timeline and budget can be allocated to focus on the most important requirements.
- Each requirement should accurately state the true need.

- The requirements should be consistent with one another such that no requirement conflicts with another requirement.
- Each requirement should be unambiguous and thus clear to the user and developer.
- Requirements should be feasible, so that is it must be possible for these requirements to be satisfied on the basis of the available resource constraints.
- It must be possible to directly map requirements to desired functions and features of the system.
- Each requirement must be measurable (quantifiable), and we should define them so that one can objectively determine whether the developed system fulfilled the requirement.

It can be useful to review the set of requirements for an individual agent or machine in your design, as well as the larger system to ensure that the set of requirements complies with these guidelines.

Although not provided as an example, many of the earlier discussions around design goals could be captured as requirements in our model. We could capture the performance requirements, such as "The HAT shall have a false alarm rate during initial screening of less than or equal to 10%" and "The HAT shall have a miss rate less than or equal to 2%". Requirements such as these frame the overall performance of our system. Of course, we will need to develop methods to determine that the final system achieves these performance levels, perhaps using historical mammograms with known anomalies or making this determination based on a trial period using a specified number of assessments. Note that these requirements are for the HAT, not the radiologist or the diagnostic aid, but for the combination of the two entities working in the prescribed teaming arrangement. This concept of measuring the performance of the team, with additional measurements of individual team members' performance, permits us to assess the design of the HAT rather than the design of individual components within the system, while the additional measures illustrate the contribution of individual components.

13.4 AN ALTERNATE SPECIFICATION THROUGH USER STORIES

Although system developers suggested the use of requirements to aid communication of duties during development, the use of requirements in legal specification implies that, once adopted, a developer or stakeholder cannot change the requirement, without renegotiating a legally binding contract, making these changes difficult. To overcome this issue, the Agile community has adopted user stories. User stories describe functionality that provides value to a stakeholder, providing the ability to specify and allocate functionality among team members, but are intended to be negotiable between the development team and the stakeholders over time, permitting the evolution of the specification as learning and trade-offs among cost, schedule, and functionality are required (Cohn, 2004). User stories can include four parts, including a (1) a written description of the story, (2) conversations about the story that provide context and other implementation details, (3) constraints, and

(4) tests that the development team can use to evaluate the implementation of a story. Large stories are referred to as "Epics" where each epic can include multiple user stories (Cohn, 2004). An example Epic might include an item such as "The Radiologist can utilize the patient record to view ancillary information to inform their diagnosis". The user story development team can decompose this item by into multiple user stories. Like functional requirements, epics and user stories specify functionality that is valuable to the user and should not be technology specific. Typically, members of the software team and stakeholders collaborate to write and develop user stories.

The format of the written description of a user story is highly structured, similar to a requirement. However, it is not specific to a hardware or software element but specifies value to the user. The format of the written description as typically applied in Agile Software processes follows the form "As a [role], I want to [means], so I can [ends]" (Lucassen et al., 2015). In AOSM, we replace the term "role" with "Agent", and we replace "means" either by the "task" from a scenario or by "capability" from the agent specification. The "ends" represents the result or advantage the user story provides to the stakeholders. As a result, we can leverage the capabilities and interdependence analysis tables to derive the written description of the user stories. An example user story within the epic discussed previously is "The *radiologist* [Agent] can *view the source of all updates to the patient record* [means] to *determine the reliability of each information element* [ends]". Unique to AOSM is the fact that the functionality has been allocated to specific agents in our system, and thus user stories are specific to specific AI agents or to the User Interface between a human and AI agent. Therefore, your team should trace this allocation between user stories and the relevant software component(s) within the model. In the example user story, AI agents and the user interface might affect the user story. Therefore, all items need to be associated with the user story, so the teams associated with developing each component know they need to derive an integrated solution. However, other user stories, such as "The radiologist can view the patient identification for the recommended intervention to verify they are currently viewing the recommendation for the correct patient", will only impact the design of user interface and database for the patient record system and not impact any of the AI agents.

> Traditionally, the development team develops user stories from discussions with end users about the needs of the envisioned or "To-Be" system. However, the users may not be able to envision dramatic and desirable changes in their workflow. Therefore, alternate methods are often required for deriving user stories that involve significant changes in the goals of the system or the user's workflows. For instance, the literature discusses the use of tabletop exercises and real games as methods (Dorton et al., 2023). AOSM provides another method for defining these user stories as the agent capabilities and OPDs derived in the AOSM process can be rephrased to form user stories.

Returning to the four components of a user story, the description does not have a structure and provides a free-form capture of relevant context from discussions with stakeholders. Constraints do not have a specific structure in Agile development either, but they capture limits from the context with the user. For example, if only specific diagnoses are possible, one might capture the specific diagnoses phrased in the language of the Radiologist, as possible entries in the diagnosis fields. The development of test cases is an art of their own, which is beyond the scope of the current text. These test cases will often involve inspection, analytic analysis and simulation, instrumented testing, or demonstrations. Further, discussion of test methods and system qualification have been described by Buede and Miller (Buede & Miller, 2016).

Like requirements, guidelines guide the development of the written user story description. These guidelines have been summarized in a framework by Lucassen and colleagues (Lucassen et al., 2015). The framework provides three categories of guidelines, including syntactic, semantic, and pragmatic. According to syntactic guidelines, the description should be (1) atomic, specifying exactly one feature; (2) minimal, containing no more than the role, means, and ends; and (3) well-formed, such that it includes at least a role and a means. The semantic guidelines include: The user story should be (1) conflict free so that it is not inconsistent with other user stories; (2) conceptually sound so that the means express a feature, and the ends express the rationale; (3) problem-oriented, expressing the problem but not a solution; and (4) unambiguous, avoiding terms or abstractions that have multiple interpretations. Pragmatic guidelines include (1) completeness, such that the user stories create a feature-complete application so that no steps are missing; (2) explicit dependencies, such that the user stories link all nonobvious dependencies between stories; (3) providing full sentences; (4) independence such that no user story is dependent on another; (5) scalability so that they are not so broad as to be difficult to plan and prioritize; (6) uniform such that all stories follow the same template; and (7) unique so that duplicate user stories do not exist within the set.

The decision of whether to use requirements, user stories, or a combination of the two will depend upon the team and the structure of the effort. As indicated earlier, projects are likely to use hybrid approaches such that the development team specifies a portion of the elements of the project with requirements and the remaining elements of the project with user stories. Software teams often use Agile processes. Although agile processes originated in the software community, development teams can use similar methodologies to specify and design the behavior of components that are a combination of hardware and software. The AI agents will consist of software embedded in hardware systems that include sensors and actuators. The structure, expectation, and management framework used by the development team will typically dictate the types of specifications. To reduce confusion, we use requirements to illustrate the specification process throughout the remainder of this chapter. However, your development team can apply a similar process to develop user stories.

13.5 SPECIFYING FUNCTIONAL REQUIREMENTS FROM CAPABILITIES

When discussing the design of our "To-Be" system in Chapter 5, we could have, and in fact a true development team should have, captured the high-level requirements for the system in the model as we discussed them. At this stage of model development, we can establish a set of requirements for specific elements within the team based on the information we have specified in the model. For an AI agent, such as the Innovation Recommender, we can begin this process by examining the final set of capabilities allocated to the AI agent. Figure 13.1 provides an example. This figure displays a requirements diagram for the Intervention Recommender Agent by displaying the agent, its capabilities, and then applying Functional Requirements to express the functions associated with each of the capabilities. Note that these requirements specify the functions, which include the inputs and outputs associated with each of the capabilities. Each capability will require at least one functional requirement, but some, such as "ability to populate record", will require multiple functional requirements to represent the implied ability assigned to the agent.

Although not relevant in this example, we might also specify the ability to sense certain features or actuate items within the environment as we review these capabilities. Also, as some of the agents are likely to need further training over time, we may need additional requirements to provide training requirements for these AI agents or even the data sets that might be necessary to provide this training.

13.6 HUMAN PROCEDURES AND TRAINING

Your design team can perform a similar exercise for the human. As you examine the capabilities for the human, you will form training system requirements or procedures that provide the human the ability to perform the functions that are indicated by the human capabilities in the model. In our example, the radiologist likely learned many of the capabilities that are specified through formal education and residency. Therefore, in this example, internal training and procedures are unnecessary for many capabilities. However, items such as "Coordinate with PCM to Implement High-Risk Patient Procedures" indicate the need to update procedures for this activity, and, therefore, one would expect a requirement to provide this procedure. There is also the need to train the Radiologist to expect and respond to errors that are committed by any of the agents in the system or through software or hardware failures. Therefore, a training process should train the radiologist to respond to errors as well as on methods to avoid committing errors within the system. Therefore, we can capture requirements to address these needs.

13.7 TRANSLATING OPD ELEMENTS TO INTERFACE REQUIREMENTS

Another significant source of requirements is the OPD elements from within our Interdependence Tables. The Requirements Table shown in Figure 13.2 lists not

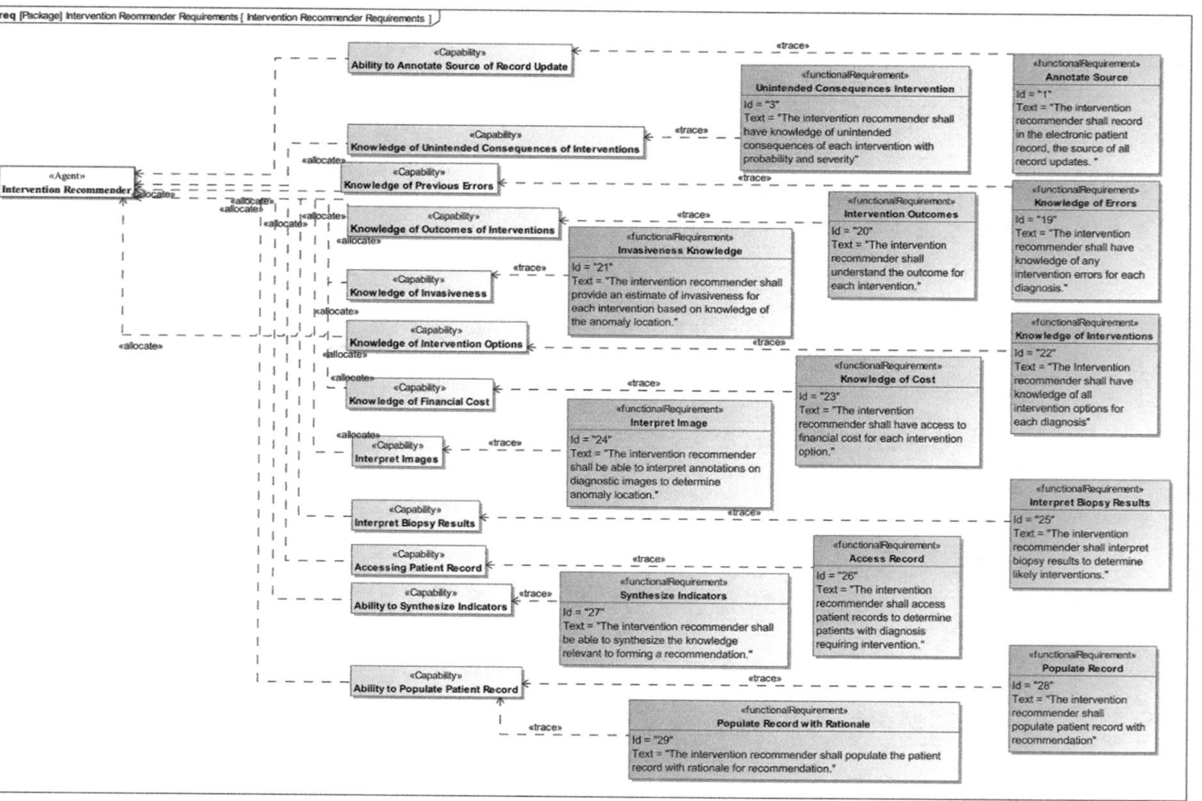

FIGURE 13.1 A requirements diagram showing the Intervention Recommender agent with its associated capabilities. Requirements are written to capture the implied functionality of the capability, with each functional requirement linked to the capability using the "trace" relationship.

#	△ Name	Text	Traced To
1	1 Annotate Source	The intervention recommender shall record in the electronic patient record, the source of all record updates.	Ability to Annotate Source of Record Update
2	3 Unintended Consequences' Intervention	The intervention recommender shall have knowledge of unintended consequences of each intervention with probability and severity	Knowledge of Unintended Consequences of Interventions
3	4 Begin Intervention Evaluation	The intervention recommender shall begin the intervention evaluation when a patient record has a diagnosis with no corresponding intervention	
4	5 Display Patient Identification	The intervention recommender shall display patient identification and diagnosis for the current recommended intervention	
5	6 Display Intervention Options	The intervention recommender shall display intervention options for evaluation	
6	7 Display Scores	The intervention recommender shall display scores and rationale for all intervention options	
7	8 Compute Scores	The intervention recommender shall compute scores for all intervention options	
8	9 Select Recommendation	The intervention recommender shall determine the preferred intervention	
9	10 Present Recommendation	The intervention recommender shall present the preferred intervention	
10	11 Populate Record with Intervention Information	The intervention recommender shall populate the patient record with intervention information	
11	12 Display Record Updates	The intervention recommender shall display confirmation of patient record updates	
12	13 Compute Score Speed	The intervention recommender shall compute the scores for all intervention options in less than 30 s (T), preferably less than 10 s (O)	
13	14 Begin Intervention Reliability	The intervention recommender shall automatically begin Intervention Evaluation for at least 95% of interventions (T), preferably at least 99% (O)	
14	16 Manual Intervention Activation	The software system shall provide a manual method to initiate the Intervention Recommender	
15	18 Observe Final Intervention Selection	The intervention recommender shall update its data model to include the final selection and rationale for the patient-diagnosis	
16	19 Knowledge of Errors	The intervention recommender shall have knowledge of any intervention errors for each diagnosis.	Knowledge of Previous Errors
17	20 Intervention Outcomes	The intervention recommender shall understand the outcome for each intervention.	Knowledge of Outcomes of Interventions
18	21 Invasiveness Knowledge	The intervention recommender shall provide an estimate of invasiveness for each intervention based on knowledge of the anomaly location.	Knowledge of Invasiveness
19	22 Knowledge of Interventions	The Intervention recommender shall have knowledge of all intervention options for each diagnosis	Knowledge of Intervention Options
20	23 Knowledge of Cost	The intervention recommender shall have access to financial cost for each intervention option.	Knowledge of Financial Cost
21	24 Interpret Image	The intervention recommender shall be able to interpret annotations on diagnostic images to determine anomaly location.	Interpret Images
22	25 Interpret Biopsy Results	The intervention recommender shall interpret biopsy results to determine likely interventions.	Interpret Biopsy Results
23	26 Access Record	The intervention recommender shall access patient records to determine patients with diagnosis requiring intervention.	Accessing Patient Record
24	27 Synthesize Indicators	The intervention recommender shall be able to synthesize the knowledge relevant to forming a recommendation.	Ability to Synthesize Indicators
25	28 Populate Record	The intervention recommender shall populate patient record with recommendation	Ability to Populate Patient Record
26	29 Populate Record with Rationale	The intervention recommender shall populate the patient record with rationale for recommendation.	Ability to Populate Patient Record

FIGURE 13.2 A requirements table showing the requirements derived in the Requirements Diagram in Figure 13.1 and additional requirements derived from the OPDs in Figure 10.3.

only the earlier requirements, which we traced to the earlier capabilities, but also an additional set of requirements, which we derived from the OPDs in Figure 10.3. As shown, these requirements include functional, interface, and performance requirements. We used Interface Requirements to capture the specific information elements that our model specified in terms of OPDs. For example, we derived the requirement "The intervention recommender shall display patient identification and diagnosis for the current recommended intervention" from an observability element. We derived the interface requirement "The software system shall provide a manual method to initiate the Intervention Recommender" from the need to direct the intervention agent to perform its evaluation if the system does not initiate it automatically. Thus, it embodies a directability element. This requirement is related to the functional requirement that "The intervention recommender shall begin the intervention evaluation when a patient record has a diagnosis with no corresponding intervention". This indicates our desire that the agent determine when it needs to recommend an intervention on its own without a human request, particularly when the Radiologist implies this action by accessing the patient record after all information necessary to determine the recommendation has been collected. Finally, this set of requirements includes performance requirements such as "The intervention recommender shall compute the scores for all intervention options in less than 30 s (Th), preferably less than 10 s (O)". This requirement captures the idea that to be useful, the intervention recommender will need to provide the radiologist with information to drive their decision-making within a reasonable time, in this example in 30 s or less. We indicate this value as the threshold (Th), meaning that the system must create this information within 30 s to be useful to the radiologist. This requirement also specifies an objective (O), in this case 10 s. The objective indicates that while 30 s would be acceptable, the function would be more valuable if the system could provide this information in as little as 10 s. Thus, it signals the development team that, if possible, they should attempt to reduce the processing time required down to 10 s.

> Although not explicitly captured in the requirements set that we have derived, the development team, with input from the stakeholders, should prioritize the requirements. To better align the process with Agile principles, this prioritization might include prioritizing the development of specific agents, such as the Diagnosis Recommender and Risk Assessment Agent, over the Intervention Recommender and Learning Database Maintainer, which we might prioritize over the Learning Agent and Dissemination Agent. Similarly, we might prioritize specific features in each agent over others. We can design such a prioritization to support the definition of a minimum viable product for initial releases, followed by releases with increased functionality assuming the timeline and budget permits.

13.8 CHAPTER SUMMARY

This chapter discussed the need and use of requirements, guidelines for requirements writing, and the application of the Agent-Oriented Systems Model to derive specifications (i.e., requirements and/or user stories) for an AI agent, the System in which they are embedded, the user interface, and human requirements for HATs. Although examples are provided using requirements, the chapter discussed the use of user stories as an alternative method to requirements when specifying systems using Agile approaches. Early parts of this process established goals and requirements for the HATs. As we have used AOSM to define the high-level interaction of the system and HAT, this high-level design can be used to derive more specific system and AI agent requirements or user stories. We illustrate a process to convert the information from within an AOSM model into requirements to guide the development of components within the HAT, although we could use an analogous process to derive user stories for some system components.

13.9 QUESTIONS AND DISCUSSION TOPICS

1. The specification of an agent requires the specification of both sensors and actuators. How can you apply information from the agent's capabilities to determine the information the agent must be able to sense and what it needs to be able to manipulate or output?
2. When defining the system, what are the critical components that your development team must specify through requirements, assuming that your development team is applying only requirements to specify your system?
3. How does one apply OPDs to derive requirements?
4. Why is it important to be able to trace the origin of your requirements or user stories to items within your model?
5. When using Agile software development processes, when do you think it is appropriate to apply user stories versus requirements? What specific components of your system do you think you should specify using requirements versus user stories?
6. Determine an epic for your system and draft user stories that are components within this epic. Indicate the elements in your model that illustrate the need for this epic.

13.10 PROJECT QUESTIONS AND ASSIGNMENTS

1. For one of the AI agents involved in one of your Interdependence Tables, apply the agent's capabilities to develop a Requirements Diagram for that agent, tracing the derivation of each requirement to the capabilities for that agent. Highlight or discuss any requirement in this diagram, which applies to a portion of the system outside of the AI agent.
2. For the Interdependence Table you developed for your atypical scenario, examine your OPDs and develop a set of corresponding requirements, indicating the type of requirement as highlighted in the text.
3. From your Agent Capabilities Assessment Table, develop five training requirements or training system requirements to support the humans within your system.

REFERENCES

Buede, D. M., & Miller, W. D. (2016). *The engineering design of systems: Models and methods*. John Wiley & Sons, Inc.

Cohn, M. (2004). *User stories applied for agile software development*. Addison-Wesley.

Dorton, S. L., Barrett, E., Fersch, T., Langone, A., & Neville, K. J. (2023). RAD-XP: Tabletop exercises for eliciting resilience requirements for sociotechnical systems. *MDPI Systems*, *11*(10). https://doi.org/10.3390/systems11100487

Endsley, M. R. (2015). Human systems integration requirements analysis. In D. A. Boehm-Davis, F. T. Durso, & J. D. Lee (Eds.), *APA handbook of human systems integration* (pp. 63–80). American Psychological Association. https://doi.org/10.1037/14528-005

Hochmüller, E., & Mittermeir, R. T. (2008). Agile process myths. *Proceedings—International Conference on Software Engineering*, 5–8. https://doi.org/10.1145/1370143.1370145

Kuhrmann, M., Diebold, P., Münch, J., Tell, P., Garousi, V., Felderer, M., Trektere, K., McCaffery, F., Linssen, O., Hanser, E., & Prause, C. R. (2017). Hybrid software and system development in practice: Waterfall, scrum, and beyond. *ACM International Conference Proceeding Series, Part F1287*, 30–39. https://doi.org/10.1145/3084100.3084104

Lightsey, B. (2001). *Systems engineering fundamentals*. Defense Acquisition University. https://apps.dtic.mil/sti/tr/pdf/ADA387507.pdf

Lucassen, G., Dalpiaz, F., Van Der Werf, J. M. E. M., & Brinkkemper, S. (2015). Forging high-quality user stories: Towards a discipline for agile requirements. *2015 IEEE 23rd International Requirements Engineering Conference, RE 2015—Proceedings*, 126–135. https://doi.org/10.1109/RE.2015.7320415

Norman, D., & Draper, S. (1986). *User-centered system design: A new perspective on human-computer interaction* (D. Norman & S. Draper, Eds.). Lawrence Erlbaum Associates.

O'Brien, P. J., & Griffin, J. M. (2007). *Global positioning system systems engineering case study*. Air Force Institute of Technology. https://scholar.afit.edu/docs/34/

14 Future of Agent-Oriented Systems Modeling

14.1 MAINTAINING A ROBUST MODEL

This text has focused on modeling human–AI agent teams to support initial team design and to derive requirements for the overall systems, AI agents, humans, and machines to support the human–AI agent team. The model that results provides a high-level conceptual design for the system. The creation of this model would typically be the responsibility of the design team, and, if applicable, of the systems engineers associated with the design team. However, the design team should share this model and the model artifacts with stakeholders across the target community, involving them in the model's development, discussing and obtaining feedback from the stakeholders before progressing to the next stage of design. The model and process discussed in this text present only the first stage of development, resulting in a conceptual design for the human–AI agent team and the supporting artifacts. The next phase of development is to design each of the components of the system. The design team should retain this model and incorporate additional design of these components using standard SysML or UML, maintaining traceability from the conceptual design and its justification to detailed design of each system element. If the design team maintains and updates the model to include the system components they develop, the model can provide a source of documentation for the overall project, as well as support the test and evaluation of the system as it progresses through the system life cycle. The reader is referred to standard texts on SysML and UML to support these model additions (Delligatti, 2013; Friedenthal et al., 2014).

As the design team develops the model, the team should focus on describing and understanding the exchange of information between entities. By formally defining the interfaces between the agents and system components, the design team will improve the likelihood that the information that an agent, human or AI, needs to support decision-making will be available during typical scenarios, as well as during many off-nominal conditions, improving the resilience of the system to anomalous conditions. By developing these interfaces, the design team will also help define necessary information flows within the human interfaces to the system. We began to describe a process for defining information to support decision-making in the team in Chapter 9. As we discussed in this chapter, besides designing the structure of the system, it is critical to define the data structures to support the appropriate sharing of information among team members. Having the appropriate information available supports both human and AI agent decision-making, as well as the interaction

DOI: 10.1201/9781003428183-14

between these agents. The development of data structures for human–AI teams has been discussed in more detail by Sterling and Taveter, among others (Sterling & Taveter, 2009).

Although the previous chapter discussed requirements development, this phase of the project should also consider testing and evaluation of the resulting HAT and system. If your project has a team dedicated to testing the software and system to determine if the system meets the requirements, this team should also develop test methods, metrics, and procedures for each of the requirements. Some of these requirements might be evaluated through the observation of the functionality, physical characteristics, or connectedness of the interfaces in the resulting hardware and software and, thus, not require formal test procedures. However, the performance and usability requirements, as well as some functional and interface requirements, are likely to require formal test procedures. Therefore, the test team can begin developing requirements and procedures for formal tests, and these tests and test results can be maintained in the model, providing a sole source of truth for the development and operations teams to access throughout the system life cycle.

We have evolved and used this modeling method within courses and research projects over the past decade. During this time, we and our students have demonstrated the ability to help design and development teams think through the design of complex human–AI agent teams. Frequently, the users of this modeling language and method generate new concepts for including AI in these human–AI teams. As is common, these innovative ideas are exciting to the developers. In the spirit of user-centered design, the design team should create early-stage prototypes around these concepts to gain feedback from users before progressing too far in development. Providing opportunities for user insight into the concepts permits the intended users to envision real-world constraints that the designers may not have considered. It is necessary to understand not only the advantages but also potential disadvantages of each concept before committing significant resources toward detailed design and development. Among the most exciting concepts, commonly observed by students, as they apply this approach are the relatively simple AI instantiations that, at least initially, help the humans on the team overcome constraints that they have simply learned to live with. Often, these instantiations give the humans the ability to meet goals that they were accountable for achieving in their "As-Is" system but were never able to reliably achieve within the available time limits. A second set of exciting concepts students often observe are tools that support teaming, enhancing the interaction among humans to achieve their work more efficiently. Traditional education too often trains technical personnel to perform tasks independently. However, as these individuals encounter real-world work environments, they must interact with others, often in large teams, as most real-world problems require teams of people to address the complexity and time demands of the work environment. As a result, these individuals often attempt to solve problems that involve teamwork by employing tools that were designed to support individual work. Understanding the constraints of these tools has led to significant innovations within some of these domains (Henry, 2020). The design team can only realize these innovations by maintaining communication with the target user communities as the development of the tools progress. As the design team develops and evaluates these concepts through early-stage prototypes, it is important to update the model to account for new learnings and design changes (Oryschak, 2020).

14.2 LOOKING BEYOND DESIGN TO SUSTAINMENT

Maintenance of the model is important as it provides a description of the system and maintains connections to earlier design decisions. While a multidisciplinary team has ideally taken part in the definition of the model as it exists, such that members of each discipline understand the design and the decisions that led to the current design, this team is likely to expand and change as the project progresses to later stages of development, deployment, and sustainment. Although the current approach has not focused on the sustainment of the AI agents, it is important to understand that effective human–AI teams will require significant expenditures of time, money, and effort to sustain the AI agents if the system is to be successful. Developers and program managers too often do not consider the cost of sustainment when envisioning and selling the vision of systems employing artificial intelligence. This sustainment includes all aspects required to maintain the relevancy and operation of the system. These updates include: (1) training data, (2) algorithms, (3) system hardware, (4) interfaces with other internal AI agents, (5) interfaces with external systems as those systems are updated, and (6) the user interface.

Although traditional software engineering practice considers software system sustainment as a function that a remote group of software developers perform, the deployment of AI can change the nature of sustainment. While traditional software could often remain relatively stable over time, systems employing human–AI agent teams are often taking part in real-time decision-making in dynamic environments under real-world constraints and pressures. This environment often drives the need for near real-time updates of the AI models and algorithms as discrepancies occur or are discovered between the real-world needs and the decisions provided by the AI. In our example system, we envisioned the embedding of a data scientist directly in the radiology team to aid the update of AI models in response to new data. This individual was required to update the models in response to newly discovered patterns in the data to improve decision-making. It may be necessary to provide similar support that is tightly integrated with the human–AI team, permitting the AI to be updated in response to rapid environmental changes.

The underlying need is for the AI to be responsive to rapid changes in the environment. Embedding part of the technology team with the operations team provides a deeper understanding of the challenges and needs of everyday decision-making. This need might be supported in ways other than embedding data scientists or other AI developers in the operational team. However, the system will need to be able to rapidly adapt to changes in the environment, where some of these changes will be initiated by intelligent entities outside the system boundary who are acting in an adversarial or opportunistic role. As a result, the cost of sustainment for these systems is likely to be higher than is typical of other software development efforts.

A commonly-claimed advantage of AI agents is their predictability compared to humans. However, all benefits have associated drawbacks. Thus, as the system becomes more predictable, it foregoes adaptability. For AI agents, this inability to adapt occurs not only as a result of software but also because of hardware, as well as the interaction between the two. For example, individuals have found they can exploit this deficiency by placing cones on the hoods of certain autonomous vehicles to block the sensors and disable the automobiles. Once discovered, this fact spread rapidly over social media, leading to numerous individuals interfering with these vehicles using this approach. Human drivers would simply look around the cone, get out of the car and remove the cone, use changes in the car's acceleration to remove the cone from the hood, or apply other novel approaches to overcome the "obstacle". However, the decision space for the AI is limited due to hardware constraints, as the sensor does not move, preventing it from looking around the cone. The agents cannot simply get out of the car to remove the cone, and the AI has not been trained to use the acceleration of the vehicle to remove objects from its shell, if this would be a permissible activity for the AI. Thus, this relatively simple intervention on the part of opportunistic individuals can significantly disrupt the function of the AI agents. If the development team is tightly coupled to operations, there is a chance they could produce innovative ideas to overcome this issue. However, if the operations team is trying to service many customer needs, as is often the case for remote software teams, these teams may require substantial periods of time to prioritize, update, and overcome the challenges imposed by these opportunistic individuals. Thus, these individuals have the potential to effectively defeat the function of at least some of the systems for substantial periods of time unless sustainment is highly coupled to system operation.

14.3 MODELING TO PREDICT OPERATIONAL EFFECTS

A limitation of the current approach is the fact that the model developed thus far is a descriptive, qualitative model of the human–AI team. The modeling language provides a framework to discuss shared work between the humans and AI agents within a team structure. When applied within the described method, the model describes the work and the work environment performed by individuals responsible for the work. The model captures this knowledge in the "As-Is" model. The development team can apply the model's representation of this work and the interactions between team members to guide the development of the "To-Be" model. The model and method permit the team to discuss and apply their intuition regarding the types of capabilities that can be embodied in AI agents, given the current technology. The development team should combine this information with the knowledge of human capabilities, capacities, and team behaviors discussed in this text to guide the development of possible, and ideally desirable, "To-Be" systems. This approach relies heavily on the designers' intuition for the capabilities they can imbue in AI agents, as well as methods to take advantage of the capabilities and capacities possessed by humans.

Importantly, the AOSM language and method do not currently provide a quantitative, analytic approach to the formation of human–AI agent teams.

Digital Engineering, which is a process that applies Model-Based Systems Engineering, and SysML, ideally applies quantitative models to understand system design trade-offs early in the development process. Similarly, it is desirable to apply quantitative methods to evaluate trade-offs during the design of human–AI agent teams. However, such models require effective models of both human task performance and AI performance. Methods for analytic modeling of human performance using SysML have been explored (Goodman et al., 2017; Kim et al., 2018; Watson et al., 2017); however, more robust models of human task performance and its relation to system performance are needed to provide tools to develop early estimates of the benefit of various AI implementations in human–AI teams.

Although the previous literature discusses modeling HAT performance or system performance that includes HATs, modeling performance alone does not illuminate design improvement opportunities. Some recent papers have begun to at least enumerate desirable teaming behaviors (Dorneich et al., 2023; Sparrow et al., 2023). Dorneich and colleagues list the desirable behaviors in the context of what they refer to as the teaming action loop, which includes preparation, execution, evaluation, adjustment, and chemistry. It is worth noting that many of these categories align well with the work of Marks and colleagues as discussed in Chapter 3 (Marks et al., 2001). Within preparation, the team should be performing mission analysis, specifying goals, and planning both the desired actions, as well as actions the HAT should execute during contingencies or off-nominal conditions. This planning step should define the roles that each agent will play and how these roles might shift to accommodate off-nominal conditions. This planning step is particularly important for action or production teams due to the temporal demands these teams experience during the execution stage. During execution, the team should be capable of employing both explicit and implicit coordination with implicit coordination requiring the use of team mental models. During evaluation, the team members should be able to critique the actions of team members and aid each other in performing teaming activities, specifically aiding each other with attention allocation, workload transfer, and shifting agency (i.e., responsibilities). Evaluation also includes evaluating the state of systems that are employed and monitoring the process to detect any anomalies that might disrupt the process. Adjustment includes performing backup behaviors, coaching, problem solving, and reviewing the actions of others. Finally, the chemistry category includes trust, collective efficacy or team performance, cohesion, and adherence to the culture of the organization. The resulting human–AI team should exhibit these behaviors, which the design or test and evaluation team can assess to determine the overall effectiveness of the HAT as well as to understand areas for AI agent, system, or team performance.

14.4 CHALLENGES IN MODELING EVOLVING SYSTEMS

There are many issues with the current state of AI for supporting human–AI teams. As discussed throughout the text, developers typically develop AI, intending to replace, rather than augment humans. As we discussed in Chapter 3, humans have evolved to be social beings, and as such, we have developed behaviors that permit us to function as team members. The ability to create mental models of each other to attempt to understand the other human's goals and intent is a skill that we, as humans, develop from an early age. Unfortunately, there is limited research on attempting to construct AI that has similar capabilities or even to develop tools that support the development of AI agents with these capabilities (Johnson et al., 2018; Schneider et al., 2022, 2023). There is certainly a need for further technology in this area. Ideally, such technology would reduce the need for the tedious effort that was required to perform the Interdependence Analysis discussed in Chapters 11 and 12. However, without this technology, it is necessary to develop detailed design of the interaction between the humans and AI agents. The process we discussed only attempted to understand the information and interface requirements for this interaction. The design team must augment this initial stage of design with human user interface design to include methods of communication, for example, voice or visual, as well as the design of the interaction in terms of speech or graphical user interfaces.

Another significant issue is that human's mental models of the world and of other agents, as well as their behavior, are not stationary over time. Instead, humans continuously learn and develop new methods for performing work (Rosenberg, 1982). As a result, if humans find an AI teammate to be useful, they will employ this teammate, and their interaction with this teammate will evolve as the humans learn the AI's capabilities and limitations. Therefore, humans are likely to use AI agents in ways that the designers might never have anticipated to fulfill needs that were not obvious or even understood by the users at the time of system design. Ideally, the AI agents would coevolve their behaviors as well. However, designing AI agents to successfully coevolve requires trust in the agents to successfully undergo this evolution without verification or validation. Additionally, these agents should coevolve not only task behaviors but also teaming behaviors, which require them to construct mental models of their teammates (Lapso et al., 2023). Given the state of AI today, as current algorithms learn, they sometimes lose the ability to perform tasks they could perform successfully in the past. Therefore, it is unlikely AI will be employed in some applications, including safety-critical applications, without test, verification, and validation processes to assess its performance.

14.5 WHAT IS THE FUTURE?

More than 60 years after the invention of AI, a debate which began between Marvin Minsky and J.C.R. Licklider over whether AI will be successfully applied to replace or augment humans is ongoing (Licklider & Clark, 1962; Minsky, 1961). Like many dichotomous decisions, the question of which view will dominate is irrelevant. A better question is: "Under what circumstances should designers pursue each viewpoint to successfully further human goals and desires?" While popular literature often emphasizes the ability of AI to replace humans and the fear of lost employment that would accompany this future, current technology-focused software development processes lead to AI that is designed to replace a portion of each human's activity. As a result, these predictions of replacement are often not practical, as the design process for AI systems leaves much of the work that a "replaced human" performs to fall on other humans who must adapt their work practices to accommodate the AI. This fact is illustrated in our example as decades of AI development to "replace" radiologists who perform mammography have only attempted to replace the radiologist's task of image interpretation, without considering other portions of their responsibilities (Zorpette, 2023).

Despite the incessant discussion of human replacement by AI, there is significant evidence and ongoing discussion of the fact that designing AI to augment humans can have significant advantages (Bates, 2018; Cummings, 2014; DeCremer & Kasparov, 2021), as can AI-based training or education (Dyulicheva & Glazieva, 2022; Rosser Jr et al., 2007). There is a need, however, to increase investment in technologies and design methods that can aid the design of these cooperative systems. However, examples of compelling, interactive human–AI teaming technology are not generally available beyond gaming environments (Nathan, 2023), where support systems augment humans to make them more effective in these environments. The AOSM method and SysML language extension provided in this text, or at least the point of view provided in the text, is intended to provide an approach to system design that will augment human cognition while reducing effort. This design method and language extension is intended to support the development of practical, real-world technology solutions that enhance and augment human performance through the application of AI agent teammates. Until we begin to focus on augmenting humans, providing AI that leverages each of our unique abilities and our passion to overcome different challenges, fear and distrust of AI are likely to grow. We should all, therefore, work toward creating a day where AI developers do not seek to replace humans but apply AI to challenge us, leading to faster and greater personal and organizational learning, growth, and satisfaction.

Until recently, AI developers worked to improve the credibility of this technology by demonstrating AI that could outperform humans. Examples such as International Business Machine's ability to construct AI that could defeat chess grand masters or champions on the game show Jeopardy provided very public demonstrations of this capability. It is important that these applications leverage the strength of AI in games that are challenging for humans and yet have defined rule sets and system boundaries. These characteristics of the games limit the need for the AI to adapt. An important, arguably negative, side effect of these demonstrations was to establish the view in the minds of individuals across the globe that AI would compete with humans. Now that technology such as deep learning has demonstrated the ability to perform equally and sometimes exceed human performance in complex tasks such as image recognition or to some degree image and text creation, there is no longer a need to demonstrate credibility through replacement demonstrations. It is important to design AI to support humans, freeing humans from dangerous, dirty, and repetitive work. It is important that future AI developers design AI to support, rather than marginalize the characteristics that make us all human.

14.6 CHAPTER SUMMARY

This chapter discusses advantages and limitations of the approach provided in this text. Included is a discussion of not only the modeling approach but also the challenges associated with sustaining AI systems, the need to understand the impact of the design of these systems, and associated challenges. The chapter also provides some thoughts around implementing this type of system as part of a team where the human's approaches are likely to evolve as they experience interaction with any tool, including an AI agent. Finally, the chapter provides a brief discussion of the limitations of current AI technologies for teaming and concluding thoughts on the replacement–augmentation dichotomy.

14.7 QUESTIONS AND DISCUSSION TOPICS

1. Non-playable characters proliferate modern video games. Discuss your impressions of how well these characters fulfill the requirements of teammates as discussed in this text. How could game developers improve these characters to permit them to rival human teammates?
2. Discuss your current impressions of human replacement versus augmentation using AI agents. When do you believe each is appropriate?
3. Given the content of this book, how would you assess your current understanding of self-driving technologies? Do you believe the current approach to develop self-driving vehicles will be successful in widespread adoption? Why or why not? What changes might lead to faster or more successful adoption of these basic technologies?

REFERENCES

Bates, A. (2018). *Augmented mind: AI, humans and the superhuman revolution*. Neocortex Ventures LLC.

Cummings, M. M. (2014). Man versus machine or man + machine? *IEEE Intelligent Systems*, *29*(5), 62–69. https://doi.org/10.1109/MIS.2014.87

DeCremer, D., & Kasparov, G. (2021). AI should augment human intelligence, not replace it. *Harvard Business Review*. https://hbr.org/2021/03/ai-should-augment-human-intelligence-not-replace-it

Delligatti, L. (2013). *SysML distilled: A brief guide to the systems modeling language*. Addison-Wesley.

Dorneich, M. C., Gilbert, S., Francis, R. F., Talyat, M., & Didic, E. (2023). Team skill metrics that span human-human and human-agent teams: An initial assessment. *Proceedings of the Human Factors and Ergonomics Society Annual Meeting*, *67*(1), 123–130. https://doi.org/10.1177/21695067231192892

Dyulicheva, Y. Y., & Glazieva, A. O. (2022). Game based learning with artificial intelligence and immersive technologies: An overview. *CEUR Workshop Proceedings*, *3077*, 146–159.

Friedenthal, S., Moore, A., & Steiner, R. (2014). *A practical guide to SysML*. Elsevier Science and Technology.

Goodman, T. J., Miller, M. E., Rusnock, C. F., & Bindewald, J. M. (2017). Effects of agent timing on the human-agent team. *Cognitive Systems Research*, *46*, 40–51. https://doi.org/10.1016/j.cogsys.2017.02.007

Henry, W. C. (2020). *Analytic provenance for software reverse engineers.* Air Force Institute of Technology. https://scholar.afit.edu/etd/3882

Johnson, M., Bradshaw, J. M., & Feltovich, P. J. (2018). Tomorrow's human–machine design tools: From levels of automation to interdependencies. *Journal of Cognitive Engineering and Decision Making, 12*(1), 77–82. https://doi.org/10.1177/1555343417736462

Kim, S., Miller, M. E., Rusnock, C. F., & Elshaw, J. J. (2018, February). Spatialized audio improves call sign recognition during multi-aircraft control. *Applied Ergonomics, 70*, 51–58. https://doi.org/10.1016/j.apergo.2018.02.007

Lapso, J. A., Peterson, G. L., & Miller, M. E. (2023, February). A hybrid cognitive model for machine agents. *Cognitive Systems Research, 81*, 1–10. https://doi.org/10.1016/j.cogsys.2023.02.007

Licklider, J. C. R., & Clark, W. E. (1962). Online man-computer communication. *AFIPS Proceedings—1962 Spring Joint Computer Conference, AFIPS 1962*, 113–128. https://doi.org/10.1145/1460833.1460847

Marks, M. A., Mathieu, J. E., & Zaccaro, S. J. (2001). Remember me a temporally based framework and taxonomy of team processes. *Academy of Management Review, 26*(3), 356–376.

Minsky, M. (1961). Steps toward artificial intelligence. *Proceedings of the IRE, 49*(1), 8–30. https://doi.org/10.1109/JRPROC.1961.287775

Nathan, B. (2023). *Artificial intelligence in gaming: Revolutionizing the player experience NPC behavior.* AI Accelerator Institute. https://www.aiacceleratorinstitute.com/artificial-intelligence-in-gaming-revolutionizing-the-player-experience/

Oryschak, J. E. (2020). Specifying space defense operator interfaces through the application of cognitive systems engineering and prototyping, Air Force Institute of Technology. https://scholar.afit.edu/etd/4543/

Rosenberg, N. (1982). *Inside the black box: Technology and economics* (1st ed.). Cambridge University Press.

Rosser Jr, J. C., Lynch, P. J., Cuddihy, L., Gentile, D. A., Klonsky, J., & Merrell, R. (2007). The impact of video games on training surgeons in the 21st century. *Archives of Surgery, 142*(2), 181–186. https://doi.org/10.1001/archsurg.142.2.186

Schneider, M. F., Miller, M. E., Ford, T. C., Peterson, G., & Jacques, D. (2022). Intent integration for human-agent teaming. *Systems Engineering, 24*(4), 291–303. http://doi.org/10.1002/sys.21616

Schneider, M. F., Miller, M. E., & McGuirl, J. M. (2023). Assessing quality goal rankings as a method for communicating operator intent. *Journal of Cognitive Engineering and Decision Making, 17*(1), 26–48. https://doi.org/10.1177/15553434221131665

Sparrow, D. A., Fealing, C. A., Wilkins, J., Ferguson, K. A., Wojton, H., & Vickers, B. D. (2023). A team-centric metric framework for testing and evaluation of human-machine teams. *Systems Engineering, 27*(3), 466–484. https://doi.org/10.1002/sys.21730

Sterling, L. S., & Taveter, K. (2009). *The art of agent-oriented modeling.* MIT Press.

Watson, M., Rusnock, C. F., Miller, M. E., & Colombi, J. M. (2017). Informing system design using human performance modeling. *Systems Engineering, 20*(2), 173–187. https://doi.org/10.1002/sys

Zorpette, G. (2023). Just calm down about GPT-4 already. *IEEE Spectrum.* https://spectrum.ieee.org/gpt-4-calm-down

Appendix A: AOSM Profile, Language Extension

Within our model, we have created some specialized entities that we have used to extend SysML. Specifically, we have created entities that include agents, sensors, actuators, information elements, data elements, goals, responsibilities, roles, capabilities, and decisions. These modeling constructs do not exist as part of the standard SysML language or the UML language which SysML is built upon.

To create these entities, we have created a profile within a profile diagram. Our profile diagram is shown in Figure A1. As shown in this figure, the AOSM profile contains each of these specialized items, referred to as stereotypes. The items in this diagram are generally formed into two separate groups. The first group includes agents, sensors, actuators, information, and data elements.

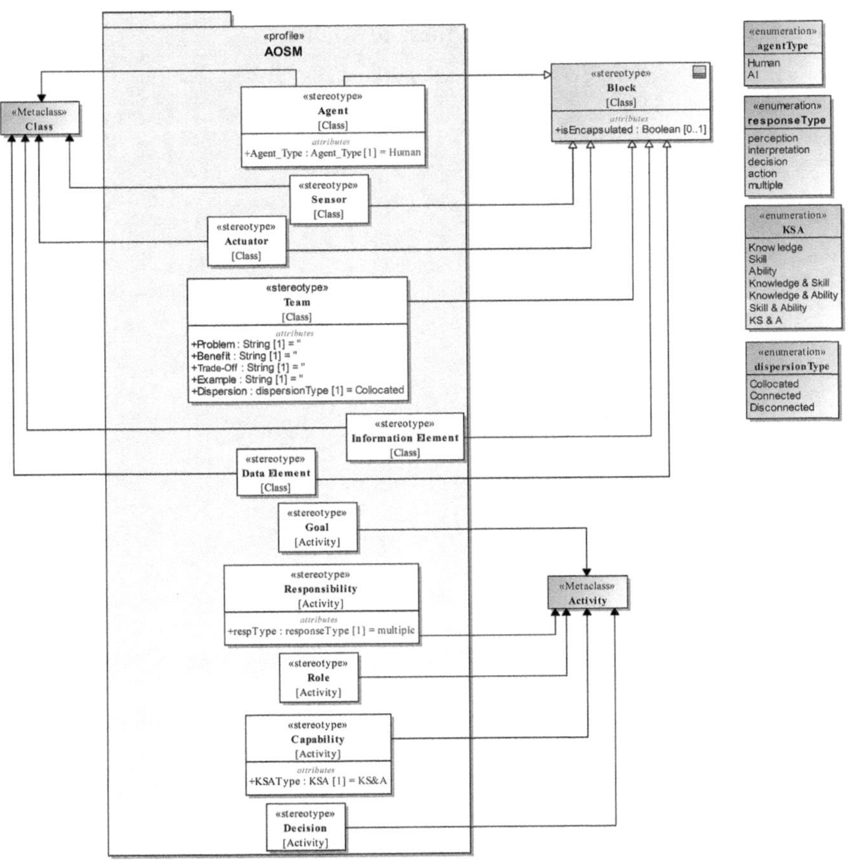

The items in this first group are all objects, and, as shown, the stereotype for each of these objects extends the base UML item of Class. These items are also shown to be specific types of blocks or that a block is a generalization of each of these objects. Therefore, these items can generally be used in SysML diagram types that would normally contain blocks, and their behavior is inherited from the Block stereotype.

The items in the second group, including goals, responsibilities, roles, capabilities, and decisions, generally describe behavior that is associated with the objects in the first group. Therefore, the items in this second group are all shown as extensions of the Activity metaclass that is defined in UML. Therefore, these items all assume the behavior of an activity and can be applied as one applies activities in UML or SysML.

Also shown are two enumerations. The first is agentType. This enumeration has been applied to Agent, illustrating that an agent can assume the attribute tag of either human or AI. The second enumeration is of responseType. This enumeration can assume the values of perception, interpretation, decision, and action, which are the four stages of the action process we used within this text. This enumeration also includes the value of multiple, implying that the responseType may also apply to more than one of the four stages of the action process. This enumeration is applied to the responsibility stereotype, indicating that the attribute of responseType within each responsibility can assume one of these five values.

Glossary

***Accountable*:** bearing the risk or consequences of an outcome; expected to have the capability to justify decisions

***Action*:** a discrete step taken by an agent to execute a higher-level activity

***Action Phase*:** period of time when teams conduct activities to accomplish the goal or objective

***Action Process*:** a series of process steps where teams conduct activities to accomplish the goal or objective

***Action Team*:** a team that is highly integrated in their environment such that the environment often establishes the pace of the team's work and that comprises agents with a broad range of specializations to overcome the fact that they perform in highly unpredictable environments. The need for different specializations can additionally drive the need for dynamic adjustments to team composition

***Activity*:** a type assigned to an action to enable sequencing

***Adaptive Automation*:** a type of automation in which tasks can be allocated to either the human or the machine. The allocation adapts to the condition of the human, with increased allocation to the machine to compensate for predicted performance declines due to perceived human overload

***Advice Team*:** a type of team that is typically not highly integrated with the environment, permitting the team to drive the pace of work, and that typically comprises group members with similar specialized knowledge in a narrow area

***After-Action Phase*:** a period of time in which a team engages in transition and interpersonal processes to examine their past behavior during the action phase to create future looking plans, goals, strategies, and methods to address interpersonal conflicts and preparation for future activities

***Agent*:** a persistent entity capable of perceiving its environment, processing the perceived environment to determine a desired action, and performing the desired action in response to the environment

***Agent-Based Software*:** a computer programming design methodology centered on autonomous agents; it consists of modeling agent behavior according to rule sets

***Agent-Oriented Systems Modeling (AOSM)*:** a graphical modeling language that extends SysML for the purpose of designing systems that support the teaming interaction between humans and AI agents

***AI agent*:** a software-based automated intelligent agent

***Allocation*:** a step of the engineering process in which particular functions are assigned to physical resources

***Analytic*:** a type of mental processing or decision-making that requires conscious deliberation using established rules or relationships

***Anomaly*:** a deviation from the expected; something different, abnormal, peculiar, or not easily classified

Artificial Agent: a persistent, man-made artifact, capable of perceiving its environment, processing the perceived information, and taking some form of action in response to changes within its environment without being activated by any other entity

Artificial Intelligence: qualities of intelligence such as perception, interpretation, prediction, and action that are exhibited by a man-made entity

Atypical: situations containing unexpected characteristics; an abnormal set of conditions from what we expect the system to operate in

Authority: the power or right to control someone or something

Automation: the process of substituting an activity originally performed by a human with an activity performed by a man-made artifact or system

Autonomous Agent: agents that create and pursue their own goals, as opposed to being controlled by another agent

Autonomy: the authority, capability, and capacity to decide upon or perform an action, free from external control or influence

Behavior: description of "how" the system will perform, including the interaction between structural elements, changes in states, and sequencing of activities

Capable/Capability: the extent to which someone or something possesses the knowledge, skills, and ability to perform a specified duty or task

Capacity: the extent to which someone or something possesses the knowledge, skills, and ability to perform a specified duty or task at a specific moment in time and within a specific context

Certainty: a set of conditions in which each action that may be taken by the agent is known and will invariably lead to a specific, known outcome

Closed System: a system that is unaffected by external influences; where nothing passes across its boundary

Coactive Design: a design method for the interface between humans and nonhuman agents to permit the agents to effectively accomplish joint activities necessary to achieve common goals

Cognitive Task Analysis: a set of tools used to describe the cognitive aspects required to perform a task

Command Display: a type of display used in a decision support system where the display instructs the human on the decision that they should make based upon the system's analysis of the situation

Commands: nonnegotiable directives that the sender expects the receiver to carry out as soon as possible

Common Ground: information, perspectives, opinions, or mental constructs that are agreed to by team members

Communication: the conveyance of information between two or more entities; involves actions to both send and receive information

Communication Overhead: the coordination cost due to the effort required to manage a handoff or to initiate a required activity by another team member

Complex Environment: a decision-making environment where the agent must consider numerous known inputs that interact often in a nonlinear fashion to affect an outcome

Complexity: consisting of many interacting components

Complicated Environments: a decision-making environment where the agent has a large number of options to choose among where these options can result in a large number or range of outcomes

Concept Map: a diagram showing the relationships between concepts, usually items, ideas, or information.

Consulted: providing opinions, guidance, feedback, or information as assistance in accomplishing task activities

Control: the ability to direct a course of action

Controlled Autonomy: a state in which an agent acts autonomously only to the extent that it is beneficial to the physical entity (e.g., human) who is responsible for controlling it

Coordination: the cyclical communication process that enables synchronized actions of teammates who are working on interdependent tasks

Cybernetics: the study of control and communication in "teleological mechanisms" in both animals and machines

Data elements: raw data from the environment that an agent must integrate to derive information useful to support decision-making

Deception: an intentional manipulation designed to create a mistaken belief about something that is not true; often accomplished covertly with the intent to gain personal advantage

Decision Support System: a system designed to provide cognitive support to a human during decision-making

Deep Learning: a class of statistical machine learning algorithm that utilizes a multilayered network of nodes and connectors, the values and weights of which are tuned as the algorithm "learns" the relationship between known inputs and outputs

Dependence: a condition in which an entity relies on another entity or resource to accomplish a goal, lacking the required capacity to completely perform an activity

Diagnosis Overhead: the coordination cost due to the time spent diagnosing a problem that occurs because of interrelated activity

Directability: the ability to require or request direction from your teammates to perform a desired activity, as well as the ability to receive requests and directions

Disembodied Agent: an agent that does not exist in a single location, as a single physical entity, but rather is distributed across a physical space and only exists as a logical entity

Domain: the part of the environment in which the system operates

Domain-Specific Vocabulary: terms, acronyms, slang, and short phrases that express commonly occurring complex concepts within a specific domain

Embodied Agent: a single physical entity, for example, a robot, capable of meeting the requirements of an agent

Emergence: a phenomenon which occurs when interactions among objects generate new phenomena beyond the observable phenomena of any individual object or subset of the objects

Expectations: the manner and procedures that the sender desires the receiver to use during execution

Explicit Communication**:** the purposeful exchange of a discrete message

Explicit Coordination**:** the explicit communication that is used to synchronize actions among team members

Fatigue**:** a physiological state of reduced mental or physical capacity, often resulting from sleep deprivation, circadian rhythm disruption, or long periods of high task demand

Feedback**:** information about the result of an action that is used to inform future action

Function**:** a process which accepts an input and generates an output

General Intelligence**:** a broad set of cognitive capabilities

Goal**:** a desired state or result

Goal Hierarchy**:** a method of organizing goals into broad high-level goals that are then subdivided into lower-level, more specific goals

Hard Dependence**:** when a single team member is unable to complete the goal without any aid from one or more teammates

Hard System**:** a system that has consistent, predictable behavior

Human–AI Teaming**:** the interdependent activities performed by members of a team, composed of both human and artificial agents, in pursuit of shared goals

Human–AI team**:** a system wherein one or more human agent(s) and one or more artificial intelligent agents work interdependently in a dynamic environment to pursue shared goals. Each teammate has defined roles and responsibilities where at least a portion of these responsibilities are shared, permitting each agent to contribute, based on their capabilities and capacities

Human Skill**:** a category within the intent estimation taxonomy that refers to the assumed knowledge of the human being observed

Implicit Communication**:** actions, behaviors, or expressions performed by an individual which conveys information to team members that permit them to gain insight into the state and intent of the individual

Implicit Coordination**:** the observable behaviors, other than explicit communication, that convey information between teammates

Information**:** situated data that enhances the receiver's awareness; in decision-making the human integrates information elements to support a decision

Informed**:** being kept apprised of task progress, decisions, and outcomes

Intelligent Agent**:** an agent which is capable of perceiving and responding to its environment (reactive), anticipating future environmental states and pursue goal-directed behavior (proactive), and effectively interacting with other agents to achieve its and their goals (social)

Intent**:** partial plans for future action where the formation of partial plans delays the need to commit to detailed plans

Intent Application**:** the third step in intent processing, in which the observer applies the intent estimate to aid decision-making and plan formation

Intent Detection**:** the first step in intent processing, in which the observer determines whether an action performed by another individual has purpose or is useful in intent estimation

Intent Estimation**:** the second step in intent processing, in which the observer applies information to estimate the goals or intent of another individual

***Intent Type*:** a category within the intent estimation taxonomy that defines the type of information which is estimated by the system

***Interactive Agent*:** an agent that is capable of interacting directly with other agents, ideally both human and other artificial agents

***Interdependence/Interdependent*:** a condition in which the parts of the system depend upon one another to achieve the system purpose or goal. That is, the system's purpose or goal cannot be achieved without interactions among the individual components of the system

***Interpersonal processes*:** processes teams use to manage interpersonal relationships

***Intuitive*:** a type of mental processing or decision-making that is performed unconsciously, which is believed to be fast and capable of integrating a large number of perceptual cues without involving working memory or conscious executive control

***Joint Activity*:** a situation in which multiple agents perform activities interdependently to accomplish a common goal

***Logical View*:** a conceptual view of the system in terms of the components and relationships, including interfaces and functionality

***Machine*:** a generic man-made physical object, consisting of multiple, mechanical parts, which operate together to perform a task

***Mental Model*:** an internal mental representation of elements of the external reality

***Model Validation*:** the process used to determine the degree to which a model is an accurate representation of the real world

***Model Verification*:** the process of determining if a model represents the conceptual description of the represented entity in the requirements documentation

***Object-Based Software*:** computer programming design methodology centered on data, or objects, rather than functions or logic; it consists of modeling the structural components such as classes, objects, methods, and attributes

***Observability*:** the ability to obtain information to understand the state of your teammate

***OPD Elements*:** information elements which must be observable, predictable, and directable among cognitive agents with a cognitive-intensive software system

***Open System*:** a system that is affected by external influences; items from the environment affect its performance as they pass across its boundary

***Option Display*:** a display in a decision support system that provides the user with options to select among when making a decision as well as the trades among important information elements that support decision selection

***Performance*:** a measure of how well a task is accomplished

***Plans*:** future-looking sequences of actions or tasks geared to achieving goals

***Pooled Interdependence*:** when the completion of a goal requiring two or more entities does not necessarily require communication as each entity performs separate functions that are not time dependent

***Predictability*:** the ability to understand what your teammate will do in the future

***Predictive Models*:** a representation of a real-world phenomenon that enables forecasting of future states, outcomes, or conditions

Preparation Phase**:** the initial phase of a team's activity where they engage primarily in transition processes in preparation for the action phase

Prescribed**:** the way in which an activity should be carried out, which is often taught via training and implemented via checklists and procedures

Proactive Agent**:** an agent that is capable of pursuing goal-directed behavior in an opportunistic fashion by anticipating a change within its environment and acting to change or avoid a future event

Production Team**:** a type of team that is highly integrated with their environment, such that the environment typically drives the pace of the team's activities, who work in an environment with limited variability, permitting the team to comprise individuals with similar capabilities reducing the need for a dynamic team composition as compared to action teams

Project Team**:** a type of team that is not highly integrated with their environment, such that the team controls the pace of their activities but typically comprises individuals with a diverse range of capabilities

Quality Goals**:** the desired degree (i.e., how-well) a state or result should occur, usually described in terms of performance, safety, social, or security qualities

Reactive Agent**:** an agent that is capable of perceiving (i.e., sense and interpret) its environment and responding to changes in a timely fashion

Reciprocal interdependence**:** when the completion of a goal requires the team members to each perform multiple interleaved tasks

Redirection Overhead**:** the coordination cost incurred due to the time team members spend following an out-of-date plan after a new plan is signaled but before all team members understand the change

Requests**:** negotiable directives which the sender expects the receiver to carry out, but the receiver may choose to reject or delay the implementation

Requirement**:** description of "what" the system will need to be capable of doing, including operating conditions and constraints

Resilience**:** the ability to rapidly recover from adverse circumstances

Responsibilities**:** an abstract item which must be accomplished to fulfill a role; responsibilities are intended to be functionally oriented, without any presumed temporal sequence or particular implementation in mind; responsibilities identify "who" is to conduct each activity

Responsible**:** having the obligation to ensure task completion

Risk**:** a set of conditions in which all action and all possible outcomes are known, but there is not a direct mapping from each action to each outcome

Role**:** the identifying behavior pattern expected in a particular social situation; the function or purpose of an entity in a system; a set of responsibilities necessary to fulfill a goal or a set of goals

Satisficing**:** a decision-making strategy which adopts the first potential option which is believed to have "acceptable" risk or to meet a minimum set of "acceptable" criteria

Scenario**:** a particular sequence of tasks which, when executed in a particular circumstance, fulfills a goal

Sequential Interdependence**:** when the completion of a goal requires one individual to perform an activity before another can perform their activity

***Shared Mental Models*:** a common understanding within a team of a representation or explanation of how something works, which enables improved communication or coordination

***Situation Awareness*:** a perception of information in the current situation, comprehension of the current situation, and projection of the future status

***Social Agent*:** an agent that is capable of interacting, when appropriate, with other agents to complete its own activities and, when able, to help other agents complete the activities for which they are responsible

***Sociotechnical systems*:** a theory of systems that integrates the social and technical aspects, considering not only the hardware and software components of a system but also human activities, relationships, and interactions

***Soft Dependence*:** when a team member can accomplish the goal on their own, but the goal can be achieved to greater effect, for example, more accurately, faster, or safer, when one or more teammates contribute toward completing the goal

***Soft System*:** a system that has variable behavior

***Specific/Narrow Intelligence*:** a cognitive ability that is highly proficient at a specific task(s) or in a limited domain(s)

***Status Display*:** a type of display used in a decision support system to inform a human with system status, providing the human information on the state of the world to permit them to determine when a decision or action might be necessary

***Strategy*:** a blueprint or idea used to accomplish a goal that is open for adaptation and change when needed. Often, strategies will include several plans where a plan consistent with the strategy is adopted on the basis of the environment or the behaviors of agents in the environment

***Structure*:** the elements of the system, including hardware, software, facilities, and people

***Suitability Evaluation*:** a category within the intent estimation taxonomy that refers to whether or not the system assumes that the human is capable of accurately forming intent or if the system assumes the human may make errors, requiring that their intent be corrected on the basis of a situated standard of intent

***Supporting*:** aiding in accomplishing task activities

***Suspicion*:** a doubt in the trustworthiness of the intentions of another; the tendency to characterize undesirable behaviors as malicious

***Symbiosis*:** the interaction between two different entities, which is mutually beneficial to each entity

***Synchronization Overhead*:** the coordination cost due to the time one team member spends waiting for another team member to complete a prerequisite task before beginning their task

***System*:** the holistic sum of interdependent parts which interact with one another to fulfill a specific purpose

***System Boundary*:** the division of the elements of a system that separates those items considered internal to the system from those considered external to the system

***Systems Engineering*:** an interdisciplinary field of engineering and engineering management that focuses on how to design, integrate, and manage complex systems over their life cycles

***Systems Markup Language (SysML)*:** a general-purpose graphical modeling language for specifying, analyzing, designing, and verifying complex systems that may include hardware, software, information, personnel, procedures, and facilities

***Task*:** an activity which is performed in real time during operation

***Task Analysis*:** the process of learning and understanding how a task is accomplished through interviews and/or observational analysis of the worker performing the task

***Team*:** two or more entities who train and work together interdependently, with diversified knowledge, skills, and abilities to accomplish work that is complex or critical enough that it requires greater knowledge, skill, ability, capacity, or redundancy than a single entity can provide within the available time

***Time Horizon*:** a category within the intent estimation taxonomy that defines the time scale of the intent estimate

***Transactive Memory Systems*:** group-level understanding or knowledge, enabling information to be shared and stored across team members

***Transition Process*:** periods of time when teams focus primarily on evaluating and planning activities to guide the accomplishment of a team goal or objective

***Trust*:** a belief that another agent will behave in a particular manner in a situation characterized by uncertainty and vulnerability

***Typical*:** situations containing the expected characteristics; the standards set of conditions we expect the system to operate in.

***Uncertain*:** a set of conditions in which, for a set of actions, the potential outcomes and the likelihood of each outcome are unknown and unknowable

***Unified Markup Language (UML)*:** a standardized modeling language enabling developers to specify, visualize, construct, and document artifacts of a software system

***Use case*:** the actions or events by which actors of a system interact with the system elements to produce outputs in order to achieve a goal

***Validation*:** the process used to determine the degree to which a system meets the needs of the user

***Verification*:** the process to determine if a constructed system meets the specifications or requirements

***Vigilance*:** the ability to maintain sustained attention when asked to detect temporary and infrequent signals, often in the presence of distractors, over prolonged periods of time

***Vigilance Decrement*:** the decline in performance due to an inability to maintain sustained attention

***Workload*:** the amount of effort that results when a human performs physical or cognitive tasks

Author Index

P

R

S

T

U

V

W

Y

Z

Subject Index

A

E

F

G

H

R

S

T

U

V

W

For Product Safety Concerns and Information please contact our EU representative GPSR@taylorandfrancis.com Taylor & Francis Verlag GmbH, Kaufingerstraße 24, 80331 München, Germany

Batch number: 10397790

Printed by Printforce, the Netherlands